ADVANCES IN IMAGING AND ELECTRON PHYSICS

VOLUME 108

MODERN MAP METHODS IN PARTICLE BEAM PHYSICS

EDITOR-IN-CHIEF

PETER W. HAWKES

*CEMES/Laboratoire d'Optique Electronique
du Centre National de la Recherche Scientifique
Toulouse, France*

ASSOCIATE EDITORS

BENJAMIN KAZAN

*Xerox Corporation
Palo Alto Research Center
Palo Alto, California*

TOM MULVEY

*Department of Electronic Engineering and Applied Physics
Aston University
Birmingham, United Kingdom*

Advances in
Imaging and Electron Physics

Modern Map Methods in Particle Beam Physics

MARTIN BERZ

Department of Physics
Michigan State University
East Lansing, Michigan

VOLUME 108

ACADEMIC PRESS

A Harcourt Science and Technology Company

San Diego San Francisco New York Boston
London Sydney Tokyo

This book is printed on acid-free paper.

Copyright © 1999 by Martin Berz

All rights reserved.
No part of this publication may be reproduced or transmitted in any form or by any means, electronic or mechanical, including photocopy, recording, or any information storage and retrieval system, without permission in writing from the publisher.

ACADEMIC PRESS
A Harcourt Science and Technology Company
525 B Street, Suite 1900, San Diego, CA 92101-4495, USA
http://www.academicpress.com

Academic Press
24–28 Oval Road, London NW1 7DX, UK
http://www.hbuk.co.uk/ap/

International Standard Serial Number: 1076-5670
International Standard Book Number: 0-12-014750-5

Printed in the United States of America
99 00 01 02 03 QW 9 8 7 6 5 4 3 2 1

CONTENTS

CHAPTER 1
Dynamics of Particles and Fields

1.1 Beams and Beam Physics 1
1.2 Differential Equations, Determinism, and Maps 2
    1.2.1 Existence and Uniqueness of Solutions 3
    1.2.2 Maps of Deterministic Differential Equations 6
1.3 Lagrangian Systems 7
    1.3.1 Existence and Uniqueness of Lagrangians 7
    1.3.2 Canonical Transformation of Lagrangians 8
    1.3.3 Connection to a Variational Principle 12
    1.3.4 Lagrangians for Particular Systems 14
1.4 Hamiltonian Systems 20
    1.4.1 Manipulations of the Independent Variable 23
    1.4.2 Existence and Uniqueness of Hamiltonians 26
    1.4.3 The Duality of Hamiltonians and Lagrangians 27
    1.4.4 Hamiltonians for Particular Systems 32
    1.4.5 Canonical Transformation of Hamiltonians 34
    1.4.6 Universal Existence of Generating Functions 44
    1.4.7 Flows of Hamiltonian Systems 53
    1.4.8 Generating Functions 54
    1.4.9 Time-Dependent Canonical Transformations 59
    1.4.10 The Hamilton–Jacobi Equation 60
1.5 Fields and Potentials 62
    1.5.1 Maxwell's Equations 62
    1.5.2 Scalar and Vector Potentials 65
    1.5.3 Boundary Value Problems 76

CHAPTER 2
Differential Algebraic Techniques

2.1 Function Spaces and Their Algebras 81
    2.1.1 Floating Point Numbers and Intervals 81
    2.1.2 Representations of Functions 82

	2.1.3 Algebras and Differential Algebras 84
2.2	Taylor Differential Algebras 85
	2.2.1 The Minimal Differential Algebra 85
	2.2.2 The Differential Algebra $_nD_v$ 91
	2.2.3 Generators, Bases, and Order 92
	2.2.4 The Tower of Ideals, Nilpotency, and Fixed Points 96
2.3	Advanced Methods 100
	2.3.1 Composition and Inversion 100
	2.3.2 Important Elementary Functions 102
	2.3.3 Power Series on $_nD_v$ 104
	2.3.4 ODE and PDE Solvers 108
	2.3.5 The Levi-Civita Field 111

Chapter 3
Fields

3.1	Analytic Field Representation 120
	3.1.1 Fields with Straight Reference Orbit 120
	3.1.2 Fields with Planar Reference Orbit 125
3.2	Practical Utilization of Field Information 126
	3.2.1 Multipole Measurements 127
	3.2.2 Midplane Field Measurements 128
	3.2.3 Electric Image Charge Methods 129
	3.2.4 Magnetic Image Charge Methods 132
	3.2.5 The Method of Wire Currents 139

Chapter 4
Maps: Properties

4.1	Manipulations 146
	4.1.1 Composition and Inversion 146
	4.1.2 Reversion 147
4.2	Symmetries 148
	4.2.1 Midplane Symmetry 148
	4.2.2 Rotational Symmetry 150
	4.2.3 Symplectic Symmetry 155
4.3	Representations 159
	4.3.1 Flow Factorizations 159
	4.3.2 Generating Functions 164

Chapter 5
Maps: Calculation

5.1	The Particle Optical Equations of Motion 168
	5.1.1 Curvilinear Coordinates 168

	5.1.2	The Lagrangian and Lagrange's Equations in Curvilinear Coordinates	174
	5.1.3	The Hamiltonian and Hamilton's Equations in Curvilinear Coordinates	179
	5.1.4	Arc Length as an Independent Variable for the Hamiltonian	185
	5.1.5	Curvilinear Coordinates for Planar Motion	188
5.2	Equations of Motion for Spin		190
5.3	Maps Determined by Algebraic Relations		195
	5.3.1	Lens Optics	195
	5.3.2	The Dipole	197
	5.3.3	Drifts and Kicks	201
5.4	Maps Determined by Differential Equations		201
	5.4.1	Differentiating ODE Solvers	201
	5.4.2	DA Solvers for Differential Equations	202
	5.4.3	Fast Perturbative Approximations	203

CHAPTER 6
Imaging Systems

6.1	Introduction	211
6.2	Aberrations and their Correction	214
6.3	Reconstructive Correction of Aberrations	217
	6.3.1 Trajectory Reconstruction	218
	6.3.2 Reconstruction in Energy Loss Mode	221
	6.3.3 Examples and Applications	223
6.4	Aberration Correction via Repetitive Symmetry	228
	6.4.1 Second-Order Achromats	230
	6.4.2 Map Representations	233
	6.4.3 Major Correction Theorems	238
	6.4.4 A Snake-Shaped Third-Order Achromat	240
	6.4.5 Repetitive Third- and Fifth Order Achromats	243

CHAPTER 7
Repetitive Systems

7.1	Linear Theory	250
	7.1.1 The Stability of the Linear Motion	250
	7.1.2 The Invariant Ellipse of Stable Symplectic Motion	260
	7.1.3 Transformations of Elliptical Phase Space	262
7.2	Parameter-Dependent Linear Theory	265
	7.2.1 The Closed Orbit	266
	7.2.2 Parameter Tune Shifts	267
	7.2.3 Chromaticity Correction	268
7.3	Normal Forms	270
	7.3.1 The DA Normal Form Algorithm	270
	7.3.2 Symplectic Systems	274
	7.3.3 Nonsymplectic Systems	277
	7.3.4 Amplitude Tune Shifts and Resonances	279

		7.3.5 Invariants and Stability Estimates	282
		7.3.6 Spin Normal Forms	288
7.4	Symplectic Tracking		292
		7.4.1 Generating Functions	293
		7.4.2 Prefactorization and Symplectic Extension	295
		7.4.3 Superposition of Local Generators	297
		7.4.4 Factorizations in Integrable Symplectic Maps	298
		7.4.5 Spin Tracking	299

PREFACE

The motion of charged particles in electromagnetic fields has been the subject of many contributions to this series—the first volume, published in 1948, contained chapters on the deflection of beams of charged particles and on particle accelerators by R.G.E. Hutter and M.S. Livingston respectively and later volumes have included chapters by many of the leading scientists in charged particle optics: H. Moss on cathode-ray tubes, P. Grivet on electron lenses, G. Liebmann on numerical methods for field plotting and ray tracing, for example, and, particularly relevant, A.J. Dragt and E. Forest on the Lie algebraic theory of charged-particle optics and electron microscopes. Supplement 7 by myself dealt with quadrupole optics and the three volumes of Supplement 13, edited by A. Septier, with applied charged-particle optics.

The present volume is entirely devoted to the single particle dynamics of beams and forms a modern, self-contained account of algebraic methods of analyzing the behavior of beams of charged particles in electric and magnetic fields. This approach is especially suitable for the description of nonlinear effects up to high orders in imaging systems and rings for arbitrary field arrangements. It is based on lectures given by the author at Michigan State University, including the Virtual University Program in Beam Physics, as well as at the U.S. Particle Accelerator School and in the CERN Academic Training Program. Much of the necessary physical and mathematical background material is included, with the result that it is reasonably self-contained. In an educational curriculum for beam physics, it is situated at an intermediate level, following an introduction to the general concepts of the subject such as [BMW99]; much of it could be mastered even without such previous knowledge. The level should be suitable for graduate students and much of the text is accessible to upper-division undergraduates.

The presentation is based on map theory for the description of motion and extensive use is made of differential algebraic methods. The tools and algorithms described here are available directly in the code COSY INFINITY. Many tools from the theory of dynamical systems and electrodynamics are included, notably a rigorous and modern account of Hamiltonian theory.

The opening chapter on the dynamics of particles and fields presents a wealth of material from classical mechanics, expressed in the terminology appropriate for beam physics and in particular, for the approach employed in subsequent chapters. Thus we meet the Lagrangians and Hamiltonians of particles in fields and the notion of maps. An entire chapter on the techniques of differential algebra follows. It is on these techniques that the book is founded and this extended, self-contained account is one of its principal attractions. The author is responsible for introducing and developing many of the ideas presented here. A short chapter on fields then defines the corresponding field and potential expansions and relates these and the numerical techniques to practical elements such as quadrupoles. Next come two long chapters on maps. The first of these presents the properties of the maps that relate the initial and final states of a system while the second and very important chapter explains how such maps are calculated in practice. The book ends with studies of two important classes of devices: imaging systems and repetitive systems such as storage rings and synchrotrons.

I am delighted that this extended account of an important contribution to particle physics appears in these Advances and I am sure that readers will be grateful for the author's efforts to render this complicated material easy to follow. I conclude with a list of forthcoming contributions to the series, three of which (by D. van Dyck, E. Kasper and M.A. O'Keefe) are also likely to occupy entire volumes.

Peter Hawkes

FORTHCOMING CONTRIBUTIONS

L. Alvarez Leon and J.-M. Morel (vol. 111)
Mathematical models for natural images

D. Antzoulatos
Use of the hypermatrix

W. Bacsa (vol. 110)
Interference scanning optical probe microscopy

N.D. Black, R. Millar, M. Kunt, F. Ziliani and M. Reid
Second generation image coding

N. Bonnet
Artificial intelligence and pattern recognition in microscope image processing

G. Borgefors
Distance transforms

A. van den Bos and A. Dekker
Resolution

O. Bostanjoglo (vol. 110)
High-speed electron microscopy

S. Boussakta and A.G.J. Holt (vol. 111)
Number-theoretic transforms and image processing

P.G. Casazza
Frames

J.A. Dayton
Microwave tubes in space

E.R. Dougherty and Y. Chen
Granulometric filters

J.M.H. Du Buf
Gabor filters and texture analysis

R.G. Forbes
Liquid metal ion sources

E. Förster and F.N. Chukhovsky
X-ray optics

A. Fox
The critical-voltage effect

M. Gabbouj
Stack filtering

M.J. Fransen (vol. 111)
The ZrO/W Schottky emitter

A. Gasteratos and I. Andreadis (vol. 110)
Soft morphology

W.C. Henneberger (vol. 112)
The Aharonov–Bohm effect

M.I. Herrera and L. Brú
The development of electron microscopy in Spain

K. Ishizuka
Contrast transfer and crystal images

C. Jeffries
Conservation laws in electromagnetics

M. Jourlin and J.-C. Pinoli
Logarithmic image processing

E. Kasper
Numerical methods in particle optics

A. Khursheed
Scanning electron microscope design

G. Kögel
Positron microscopy

K. Koike
Spin-polarized SEM

P. V. Kolev and M. Jamal Deen (vol. 109)
Development and applications of a new deep-level transient spectroscopy method and new averaging techniques

W. Krakow
Sideband imaging

A. van de Laak-Tijssen, E. Coets and T. Mulvey
Memoir of J.B. Le Poole

L.J. Latecki
Well-composed sets

J.-M. Lina, B. Goulard and P. Turcotte (vol. 109)
Complex wavelets

C. Mattiussi
The finite volume, finite element and finite difference methods

S. Mikoshiba and F. L. Curzon
Plasma displays

R.L. Morris
Electronic tools in parapsychology

J.G. Nagy
Restoration of images with space-variant blur

P.D. Nellist and S.J. Pennycook
Z contrast in the STEM and its applications

M.A. O'Keefe
Electron image simulation

G. Nemes
Phase-space treatment of photon beams

B. Olstad
Representation of image operators

M. Omote and S. Sakoda (vol. 110)
Aharonov–Bohm scattering

C. Passow
Geometric methods of treating energy transport phenomena

E. Petajan
HDTV

V. E. Ptitsin
Non-stationary thermal field emission

F.A. Ponce
Nitride semiconductors for high-brightness blue and green light emission

J.W. Rabalais
Scattering and recoil imaging and spectrometry

H. Rauch
The wave-particle dualism

D. Saldin
Electron holography

G.E. Sarty (vol. 111)
Reconstruction from non-Cartesian grids

G. Schmahl
X-ray microscopy

J.P.F. Sellschop
Accelerator mass spectroscopy

S. Shirai
CRT gun design methods

M. Shnaider and A.P. Paplinski (vol. 109)
Vector coding and wavelets

T. Soma
Focus-deflection systems and their applications

I. Talmon
Study of complex fluids by transmission electron microscopy

S. Tari (vol. 111)
Shape skeletons and greyscale images

J. Toulouse
New developments in ferroelectrics

T. Tsutsui and Z. Dechun
Organic electroluminescence, materials and devices

Y. Uchikawa
Electron gun optics

D. van Dyck
Very high resolution electron microscopy

J.S. Villarrubia
Mathematical morphology and scanned probe microscopy

L. Vincent
Morphology on graphs

N. White
Multi-photon microscopy

J.B. Wilburn (vol. 112)
Generalized ranked-order filters

C.D. Wright and E.W. Hill
Magnetic force microscopy

T. Yang (vol. 109)
Fuzzy cellular neural networks

T. Yang
Cellular Neural Networks: New Engines for Image Processing

ACKNOWLEDGMENTS

I am grateful to a number of individuals who have contributed in various ways to this text. Foremost I would like to thank the many students who participated in the various versions of these lectures and provided many useful suggestions. My thanks go to Kyoko Makino and Weishi Wan, who even actively contributed certain parts of this book. Specifically, I thank Kyoko for the derivation and documentation of the Hamiltonian in curvilinear coordinates (Section 5.1) and Weishi for being the main impetus in the development and documentation of the theory of nonlinear correction (Section 6.4). In the earlier stages of the preparation of the manuscript, we enjoyed various discussions with Georg Hoffstätter. To Khodr Shamseddine I am grateful for skillfully preparing many of the figures. For a careful reading of parts the manuscript, I would like to thank Kyoko Makino, Khodr Shamseddine, Bela Erdelyi, and Jens von Bergmann.

I am also thankful to Peter Hawkes, the editor of the series in which this book is appearing, for his ongoing encouragement and support, as well as the staff at Academic Press for the skilled and professional preparation of the final layout. Last but not least, I am grateful for financial support through grants from the U.S. Department of Energy and a fellowship from the Alfred P. Sloan foundation.

Martin Berz

ADVANCES IN IMAGING AND
ELECTRON PHYSICS

VOLUME 108

MODERN MAP METHODS IN
PARTICLE BEAM PHYSICS

Chapter 1

Dynamics of Particles and Fields

1.1 BEAMS AND BEAM PHYSICS

In a very general sense, the field of beam physics is concerned with the analysis of the dynamics of certain state vectors $\vec{z}$. These comprise the coordinates of interest, and their motion is described by a differential equation $d/dt\, \vec{z} = \vec{f}(\vec{z}, t)$. Usually it is necessary to analyze the manifold of solutions of the differential equation not only for one state vector $\vec{z}$ but also for an entire ensemble of state vectors. Different from other disciplines, in the field of beam physics the ensemble of state vectors is usually somewhat close together and also stays that way over the range of the dynamics. Such an ensemble of nearby state vectors is called a **beam**.

The study of the dynamics of beams has two important subcategories: the description of the motion to very high precision over short times and the analysis of the topology of the evolution of state space over long times. These subcategories correspond to the historical foundations of beam physics, which lie in the two seemingly disconnected fields of optics and celestial mechanics. In the case of optics, in order to make an image of an object or to spectroscopically analyze the distributions of colors, it is important to guide the individual rays in a light beam to very clearly specified final conditions. In the case of celestial mechanics, the precise calculation of orbits over short terms is also of importance, for example, for the prediction of eclipses or the assessment of orbits of newly discovered asteroids under consideration of the uncertainties of initial conditions. Furthermore, celestial mechanics is also concerned with the question of stability of a set of possible initial states over long periods of time, one of the first and fundamental concerns of the field of **nonlinear dynamics**; it may not be relevant where exactly the earth is located in a billion years as long as it is still approximately the same distance away from the sun. Currently, we recognize these fields as just two specific cases of the study of weakly nonlinear motion which are linked by the common concept of the so-called transfer map of the dynamics.

On the practical side, currently the study of beams is important because beams can provide and carry two fundamentally important scientific concepts, namely, **energy** and **information**. Energy has to be provided through acceleration, and the significance of this aspect is reflected in the name accelerator physics, which is often used almost synonymously with beam physics. Information is either generated by utilizing the beam's energy, in which case it is analyzed through spectroscopy, or it is transported at high rates and is thus relevant for the practical aspects of information science.

Applications of these two concepts are wide. The field of **high-energy physics** or particle physics utilizes both aspects of beams, and these aspects are so important that they are directly reflected in their names. First, common particles are brought to energies far higher than those found anywhere else on earth. Then this energy is utilized in collisions to produce particles that do not exist in our current natural environment, and information about such new particles is extracted.

In a similar way, **nuclear physics** uses the energy of beams to produce isotopes that do not exist in our natural environment and extracts information about their properties. It also uses beams to study the dynamics of the interaction of nuclei. Both particle physics and nuclear physics also re-create the state of our universe when it was much hotter, and beams are used to artificially generate the ambient temperature at these earlier times. Currently, two important questions relate to the understanding of the time periods close to the big bang as well as the understanding of nucleosynthesis, the generation of the currently existing different chemical elements.

In **chemistry** and **material science**, beams provide tools to study the details of the dynamics of chemical reactions and a variety of other questions. In many cases, these studies are performed using intense light beams, which are produced in conventional lasers, free-electron lasers, or synchrotron light sources.

Also, great progress continues to be made in the traditional roots of the field of beam physics, **optics**, and **celestial mechanics**. Modern glass lenses for cameras are better, more versatile, and used now more than ever before, and modern electron microscopes now achieve unprecedented resolutions in the Angstrom range. Celestial mechanics has made considerable progress in the understanding of the nonlinear dynamics of planets and the prospects for the long-term stability of our solar system; for example, while the orbit of earth does not seem to be in jeopardy at least for the medium term, there are other dynamical quantities of the solar system that behave chaotically.

Currently, the ability to transport information is being applied in the case of fiber optics, in which short light pulses provide very high transfer rates. Also, electron beams transport the information in the television tube, belonging to one of the most widespread consumer products that, for better or worse, has a significant impact on the values of our modern society.

1.2 Differential Equations, Determinism, and Maps

As discussed in the previous section, the study of beams is connected to the understanding of solutions of differential equations for an ensemble of related initial conditions. In the following sections, we provide the necessary framework to study this question.

1.2.1 Existence and Uniqueness of Solutions

Let us consider the **differential equation**

$$\frac{d}{dt}\vec{z} = \vec{f}(\vec{z}, t), \tag{1.1}$$

where $\vec{z}$ is a state vector that satisfies the **initial condition** $\vec{z}(t_1) = \vec{z}_1$. In practice the vector $\vec{z}$ can comprise positions, momenta or velocities, or any other quantity influencing the motion, such as a particle's spin, mass, or charge. We note that the ordinary differential equation (ODE) being of first order is immaterial because any **higher order differential equation**

$$\frac{d^n}{dt^n}\vec{z} = \vec{f}\left(\vec{z}, \frac{d}{dt}\vec{z}, \ldots, \frac{d^{n-1}}{dt^{n-1}}\vec{z}\right) \tag{1.2}$$

can be rewritten as a first-order differential equation; by introducing the $(n-1)$ new variables $\vec{z}_1 = d/dt\,\vec{z}$, $\vec{z}_2 = d^2/dt^2\,\vec{z}, \ldots, \vec{z}_{n-1} = d^{n-1}/dt^{n-1}\,\vec{z}$, we have the equivalent first-order system

$$\frac{d}{dt}\begin{pmatrix} \vec{z} \\ \vec{z}_1 \\ \vdots \\ \vec{z}_{n-1} \end{pmatrix} = \begin{pmatrix} \vec{z}_1 \\ \vec{z}_2 \\ \vdots \\ \vec{f}(\vec{z}, \vec{z}_1, \ldots, \vec{z}_{n-1}) \end{pmatrix}. \tag{1.3}$$

Furthermore, it is possible to rewrite the differential equation in terms of any **other independent variable** s that satisfies

$$\frac{ds}{dt} \neq 0 \tag{1.4}$$

and hence depends strictly monotonically on t. Indeed, in this case we have

$$\frac{d}{ds}\vec{z} = \frac{d}{dt}\vec{z} \cdot \frac{dt}{ds} = \vec{f}(\vec{z}, t(s)) \cdot \frac{dt}{ds}. \tag{1.5}$$

Similarly, the question of dependence on the independent variable is immaterial because an **explicitly t-dependent differential equation** can be rewritten as a t-independent differential equation by the introduction of an additional variable s with $d/dt\,s = 1$ and substitution of t by s in the equation:

$$\frac{d}{dt}\begin{pmatrix} \vec{z} \\ s \end{pmatrix} = \begin{pmatrix} \vec{f}(\vec{z}, s) \\ 1 \end{pmatrix}. \tag{1.6}$$

Frequently, one studies the evolution of a particular quantity that depends on the phase space variables along a solution of the ODE. For a given differentiable function $g(\vec{z}, t)$ from phase space into the real numbers (such a function is also often called an **observable**), it is possible to determine its derivative and hence an associated ODE describing its motion via

$$\frac{d}{dt}g(\vec{z}, t) = \vec{f} \cdot \vec{\nabla}g + \frac{\partial}{\partial t}g = L_{\vec{f}}(g); \qquad (1.7)$$

the differential operator $L_{\vec{f}}$ is usually called the **vector field** of the ODE. Other common names are **directional derivative** or **Lie derivative**. It plays an important role in the theory of dynamical systems and will be used repeatedly in the remainder of the text. Apparently, the differential operator can also be used to determine higher derivatives of g by repeated application of $L_{\vec{f}}$, and we have

$$\frac{d^n}{dt^n}g = L_{\vec{f}}^n g. \qquad (1.8)$$

An important question in the study of differential equations is whether any solutions exist and, if so, whether they are unique. The first part of the question is answered by the theorem of **Peano**; as long as the function $\vec{f}$ is **continuous** in a neighborhood of $(\vec{z}_0, t_0)$, the **existence of a solution is guaranteed** at least in a neighborhood.

The second question is more subtle than one might think at first glance. For example, it is ingrained in our thinking that classical mechanics is **deterministic**, which means that the value of a state at a given time t_0 uniquely specifies the value of the state at any later time t. However, this is not always the case; consider the Hamiltonian and its equations of motion:

$$H = \frac{p^2}{2} + V(q), \text{ where } V(q) = -|q|^{3/2} \qquad (1.9)$$

$$\dot{q} = p \text{ and } \dot{p} = \text{sign}(q)\frac{3}{2}|q|^{1/2}. \qquad (1.10)$$

The potential $V(q)$ is everywhere continuously differentiable and has an unstable equilibrium at the origin; it is schematically shown in Fig. 1.1. However, as will be shown, the stationary point at the origin gives rise to **nondeterministic dynamics**.

To this end, consider the solutions through the initial condition $(q_0, p_0) = (0, 0)$ at $t = 0$. We readily convince ourselves that the trajectory

$$(q, p) = (0, 0) \text{ for all } t \qquad (1.11)$$

is indeed a solution of the previous Hamiltonian equations. However, we observe that for any t_c, the trajectory

$$(q, p) = \begin{cases} (0, 0) & \text{for } t < t_c \\ \pm\left(\frac{1}{64}(t - t_c)^4, \frac{1}{16}(t - t_c)^3\right) & \text{for } t > t_c \end{cases} \qquad (1.12)$$

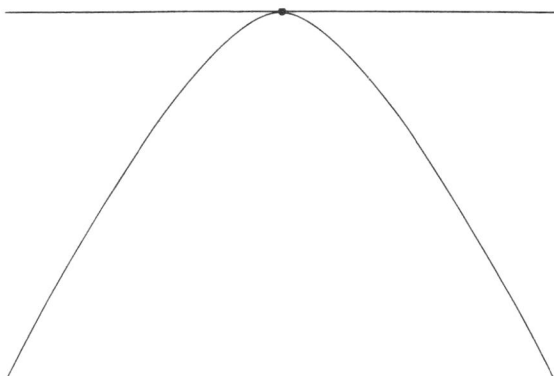

FIGURE 1.1. A nondeterministic potential.

is a solution of the differential equations! This means that the particle has a choice of staying on top of the equilibrium point or, at any time t_c it pleases, it can start to leave the equilibrium point, either toward the left or toward the right. Certainly, there is no determinism anymore.

In a deeper sense, the problem is actually more far-reaching than one may think. While the previous example may appear somewhat pathological, one can show that any potential $V(q)$ can be modified by adding a second term $aW(q)$ with $W(q)$ continuous and $|W(q)| < 1$ such that for any a, the potential $V(q)+aW(q)$ is nondeterministic. Choosing a smaller than the accuracy to which potentials can be measured, we conclude that determinism in classical mechanics is principally undecidable.

In a mathematical sense, the second question is answered by the **Picard-Lindelöf Theorem**, which asserts that if the right-hand side $\vec{f}$ of the differential equation (1.1) satisfies a **Lipschitz condition** with respect to $\vec{z}$, i.e., there is a k such that

$$|\vec{f}(\vec{z}_1, t) - \vec{f}(\vec{z}_2, t)| < k \cdot |\vec{z}_1 - \vec{z}_2| \tag{1.13}$$

for all $\vec{z}_1$, $\vec{z}_2$ and for all t, then there is a **unique solution** in a neighborhood of any $(\vec{z}_0, t_0)$.

Therefore, determinism of classical mechanics can be assured if one axiomatically assumes the Lipschitzness of the forces that can appear in classical mechanics. This axiom of determinism of mechanics, however, is more delicate than other axioms in physics because it cannot be checked experimentally in a direct way, as the previous example of the potential W shows.

1.2.2 Maps of Deterministic Differential Equations

Let us assume we are given a deterministic system

$$\frac{d}{dt}\vec{z} = \vec{f}(\vec{z}, t), \tag{1.14}$$

which means that for every initial condition $\vec{z}_1$ at t_1, there is a unique solution $\vec{z}(t)$ with $\vec{z}(t_1) = \vec{z}_1$. For a given time t_1, this allows to introduce a function $\mathcal{M}_{t_1,t_2}$ that for any initial condition $\vec{z}_1$ determines the value of the solution through $(\vec{z}_1, t_1)$ at the later time t_2; therefore we have

$$\vec{z}(t_2) = \mathcal{M}_{t_1,t_2}(\vec{z}_1). \tag{1.15}$$

The function $\mathcal{M}$ describes how individual state space points "flow" as time progresses, and in the theory of dynamical systems it is usually referred to as the **flow** of the differential equation. We also refer to it as the **propagator** (describing how state space points are propagated), the **transfer map** (describing how state space points are transferred), or simply the **map**. It is obvious that the transfer maps satisfy the property

$$\mathcal{M}_{t_1,t_3} = \mathcal{M}_{t_2,t_3} \circ \mathcal{M}_{t_1,t_2} \tag{1.16}$$

where "$\circ$" stands for the composition of functions. In particular, we have

$$\mathcal{I} = \mathcal{M}_{t_2,t_1} \circ \mathcal{M}_{t_1,t_2}, \tag{1.17}$$

and so the transfer maps are invertible.

An important case is the situation in which the ODE is autonomous, i.e., t independent. In this case, the transfer map depends only on the difference $\Delta t = t_2 - t_1$, and we have

$$\mathcal{M}_{\Delta t_1 + \Delta t_2} = \mathcal{M}_{\Delta t_2} \circ \mathcal{M}_{\Delta t_1}. \tag{1.18}$$

An important case of dynamical systems is the situation in which the motion is repetitive, which means that there is a Δt such that

$$\mathcal{M}_{t,t+\Delta t} = \mathcal{M}_{t+\Delta t, t+2\Delta t}. \tag{1.19}$$

In this case, the long-term behavior of the motion can be analyzed by merely studying the repeated action of $\mathcal{M}_{t,t+\Delta t}$. If the motion suits our needs for the discrete time steps Δt and occupies a suitable set S of state spaces there, then it is merely necessary to study once how this set S is transformed for intermediate times between t and $t + \Delta t$ and whether this satisfies our needs. This stroboscopic study of repetitive motion is an important example of the method of **Poincaré**

sections, of which there are several varieties including suitable changes of the independent variables but these always amount to the study of repetitive motion by discretization.

In the following sections, we embark on the study of one of the most important classes of dynamical systems, namely, those that can be expressed within the frameworks of the theories of Lagrange, Hamilton, and Jacobi.

1.3 LAGRANGIAN SYSTEMS

Let us assume we are given a dynamical system of n quantities $q_1, \ldots, q_n$ whose dynamics is described by a second-order differential equation of the form

$$\ddot{\vec{q}} = \vec{F}(t, \vec{q}, \dot{\vec{q}}). \tag{1.20}$$

If there exists a function L depending on $\vec{q}$ and $\dot{\vec{q}}$, i.e.,

$$L = L(\vec{q}, \dot{\vec{q}}, t), \tag{1.21}$$

such that the equations

$$\frac{d}{dt}\left(\frac{\partial L}{\partial \dot{q}_i}\right) - \frac{\partial L}{\partial q_i} = 0, \; i = 1, \ldots, n \tag{1.22}$$

are equivalent to the equations of motion (1.20) of the system, then L is called a **Lagrangian** for this system. In this case, equations (1.22) are known as Lagrange's equations of motion. There are usually many advantages if equations of motion can be rewritten in terms of a Lagrangian; the one that is perhaps most straightforwardly appreciated is that the motion is now described by a single scalar function L instead of the n components of $\vec{F}$. However, there are other advantages because Lagrangian systems have special properties, many of which are discussed in the following sections.

1.3.1 Existence and Uniqueness of Lagrangians

Before we proceed, we may ask ourselves whether a function $L(\vec{q}, \dot{\vec{q}}, t)$ always exists and, if so, whether it is unique. The latter part of the question apparently has to be negated because **Lagrangians are not unique**. Apparently with $L(\vec{q}, \dot{\vec{q}}, t)$ also $L(\vec{q}, \dot{\vec{q}}, t) + c$, where c is a constant, is also a Lagrangian for $\vec{F}$. Furthermore, a constant c can be added not only to L, but also more generally to any function $K(\vec{q}, \dot{\vec{q}}, t)$ that satisfies $d/dt(\partial K/\partial \dot{\vec{q}}) - \partial K/\partial \vec{q} = 0$ for all choices of $\vec{q}$ and $\dot{\vec{q}}$. Such functions do indeed exist; for example, the function $K(\vec{q}, \dot{\vec{q}}, t) = \vec{q} \cdot \dot{\vec{q}}$

apparently satisfies the requirement. Trying to generalize this example, one realizes that while the linearity with respect to $\dot{\vec{q}}$ is important, the dependence on $\vec{q}$ can be more general; also, a dependence on time is possible, if it is connected in the proper way to the position dependence. Altogether, if F is a three times differentiable function depending on position $\vec{q}$ and t, then

$$K(\vec{q}, \dot{\vec{q}}, t) = \frac{d}{dt} F(\vec{q}, t) = \sum_{j=1}^{n} \frac{\partial F}{\partial q_j} \cdot \dot{q}_j + \frac{\partial F}{\partial t} \quad (1.23)$$

satisfies $d/dt\,(\partial K/\partial \dot{\vec{q}}) - \partial K/\partial \vec{q} = 0$; indeed, studying the ith component of the condition yields

$$\frac{d}{dt} \frac{\partial}{\partial \dot{q}_i} \left(\sum_{j=1}^{n} \frac{\partial F}{\partial q_j} \cdot \dot{q}_j + \frac{\partial F}{\partial t} \right) - \frac{\partial}{\partial q_i} \left(\sum_{j=1}^{n} \frac{\partial F}{\partial q_j} \cdot \dot{q}_j + \frac{\partial F}{\partial t} \right)$$

$$= \frac{d}{dt} \left(\frac{\partial F}{\partial q_i} \right) - \sum_{j=1}^{n} \frac{\partial^2 F}{\partial q_i \partial q_j} \cdot \dot{q}_j - \frac{\partial^2 F}{\partial q_i \partial t}$$

$$= \sum_{j=1}^{n} \frac{\partial^2 F}{\partial q_j \partial q_i} \cdot \dot{q}_j + \frac{\partial^2 F}{\partial t \partial q_i} - \sum_{j=1}^{n} \frac{\partial^2 F}{\partial q_i \partial q_j} \cdot \dot{q}_j - \frac{\partial^2 F}{\partial q_i \partial t} = 0. \quad (1.24)$$

The question of **existence cannot be answered** in a general way, but fortunately for many systems of practical relevance to us, Lagrangians can be found. This usually requires some "guessing" for every special case of interest; many examples for this can be found in Section 1.3.4.

One of the important advantages of the Lagrangian formulation is that it is particularly easy to perform transformations to different sets of variables, as illustrated in the next section.

1.3.2 Canonical Transformation of Lagrangians

Given a system described by **coordinates** $(q_1, \ldots, q_n)$, its analysis can often be simplified by studying it in a set of **different coordinates**. One important case is motion that is subject to certain **constraints**; in this case, one can often perform a change of coordinates in such a way that a given constraint condition merely means that one of the new coordinates is to be kept constant. For example, if a particle is constrained to move along a circular loop, then expressing its motion in polar coordinates (r, ϕ) concentric with the loop instead of in Cartesian variables allows the simplification of the constraint condition to the simple statement that r is kept constant. This reduces the dimensionality of the problem and leads to a new system free of constraints, and thus this approach is frequently applied for the study of constrained systems. Currently, there is also an important field that studies such constrained systems or **differential algebraic equations** directly without removal of the constraints (Ascher and Petzold, 1998; Griepentrog and Roswitha, 1986; Brenan *et al.*, 1989; Matsuda, 1980).

Another important case is the situation in which certain **invariants** of the motion are known to exist; if these invariants can be introduced as new variables, it again follows that these variables are constant, and the dimensionality of the system can be reduced. More sophisticated changes of coordinates relevant to our goals will be discussed later.

It is worth noting that the choice of new variables does not have to be the same everywhere in variable space but may change from one region to another; this situation can be dealt with in a natural and organized way with the theory of **manifolds** in which the entire region of interest is covered by an atlas, which is a collection of local coordinate systems called charts that are patched together to cover the entire region of interest and to overlap smoothly. However, we do not need to discuss these issues in detail.

Let a Lagrangian system with coordinates $(q_1, \ldots, q_n)$ be given. Let $(Q_1, \ldots, Q_n)$ be another set of coordinates, and let $\mathcal{Q}$ be the transformation from the old to the new coordinates, which is allowed to be time dependent, such that

$$\vec{Q} = \mathcal{Q}(\vec{q}, t). \tag{1.25}$$

The transformation $\mathcal{Q}$ is demanded to be invertible, and hence we have

$$\vec{q} = \mathcal{Q}^{-1}(\vec{Q}, t); \tag{1.26}$$

furthermore, let both of the transformations be continuously differentiable. Altogether, the change of variables constitutes an isomorphism that is continuously differentiable in both directions, a so-called **diffeomorphism**.

In the following, we will often not distinguish between the symbol describing the transformation function $\mathcal{Q}$ and the symbol describing the new coordinates $\vec{Q}$ if it is clear from the context which one is meant. The previous transformation equations then appear as

$$\vec{Q} = \vec{Q}(\vec{q}, t) \text{ and } \vec{q} = \vec{q}(\vec{Q}, t), \tag{1.27}$$

where $\vec{Q}$ and $\vec{q}$ on the left side of the equations and inside the parentheses denote n-tuples, while in front of the parentheses they denote functions. While this latter convention admittedly has the danger of appearing logically paradoxical if not understood properly, it often simplifies the nomenclature to be discussed later.

If the coordinates transform according to Eq. (1.25), then the corresponding time derivatives $\dot{\vec{Q}}$ transform according to

$$\dot{\vec{Q}} = \text{Jac}(\mathcal{Q})(\vec{q}, t) \cdot \dot{\vec{q}} + \frac{\partial \mathcal{Q}}{\partial t}, \tag{1.28}$$

and we obtain the entire transformation $\mathcal{M}$ from $(\vec{q}, \dot{\vec{q}})$ to $(\vec{Q}, \dot{\vec{Q}})$ as

$$\begin{pmatrix} \vec{Q} \\ \dot{\vec{Q}} \end{pmatrix} = \begin{pmatrix} \mathcal{Q}(\vec{q}, t) \\ \text{Jac}(\mathcal{Q})(\vec{q}, t) \cdot \dot{\vec{q}} + \partial \mathcal{Q}/\partial t \end{pmatrix} = \mathcal{M}(\vec{q}, \dot{\vec{q}}, t). \quad (1.29)$$

It is a remarkable and most useful fact that in order to find a Lagrangian that describes the motion in the new variables $\vec{Q}$ and $\dot{\vec{Q}}$ from the Lagrangian describing the motion in the variables $\vec{q}$ and $\dot{\vec{q}}$, it is necessary only to insert the transformation rule (Eq. 1.26) as well as the resulting time derivative $\dot{\vec{q}} = \dot{\vec{q}}(\vec{Q}, \dot{\vec{Q}}, t)$ into the Lagrangian, and thus it depends on $\vec{Q}$ and $\dot{\vec{Q}}$. We say that the **Lagrangian is invariant under a coordinate transformation**. Therefore, the new Lagrangian will be given by

$$M(\vec{Q}, \dot{\vec{Q}}, t) = L(\vec{q}(\vec{Q}, t), \dot{\vec{q}}(\vec{Q}, \dot{\vec{Q}}, t), t), \quad (1.30)$$

or in terms of the transformation maps,

$$M = L(\mathcal{M}^{-1}) \text{ and } L = M(\mathcal{M}). \quad (1.31)$$

In order to prove this property, let us assume that $L = L(\vec{q}, \dot{\vec{q}}, t)$ is a Lagrangian describing the motion in the coordinates $\vec{q}$, i.e.,

$$\frac{d}{dt}\left(\frac{\partial L}{\partial \dot{q}_i}\right) - \frac{\partial L}{\partial q_i} = 0, \ i = 1, \ldots, n \quad (1.32)$$

yield the proper equations of motion of the system.

Differentiating Eqs. (1.25) and (1.26) with respect to time, as is necessary for the substitution into the Lagrangian, we obtain

$$\dot{\vec{Q}} = \dot{\vec{Q}}(\vec{q}, \dot{\vec{q}}, t)$$
$$\dot{\vec{q}} = \dot{\vec{q}}(\vec{Q}, \dot{\vec{Q}}, t). \quad (1.33)$$

We now perform the substitution of Eqs. (1.26) and (1.33) into (1.30) and let $M(\vec{Q}, \dot{\vec{Q}}, t)$ be the resulting function; that is,

$$M(\vec{Q}, \dot{\vec{Q}}, t) = L(\vec{q}(\vec{Q}, t), \dot{\vec{q}}(\vec{Q}, \dot{\vec{Q}}, t), t). \quad (1.34)$$

The goal is now to show that indeed

$$\frac{d}{dt}\left(\frac{\partial M}{\partial \dot{Q}_i}\right) - \frac{\partial M}{\partial Q_i} = 0, \ i = 1, \ldots, n. \quad (1.35)$$

LAGRANGIAN SYSTEMS

We fix a $j: 1 \leq j \leq n$. Then

$$\frac{\partial M}{\partial \dot{Q}_j} = \sum_{i=1}^{n} \frac{\partial L}{\partial \dot{q}_i} \frac{\partial \dot{q}_i}{\partial \dot{Q}_j} \qquad (1.36)$$

and

$$\frac{\partial M}{\partial Q_j} = \sum_{i=1}^{n} \frac{\partial L}{\partial q_i} \frac{\partial q_i}{\partial Q_j} + \sum_{i=1}^{n} \frac{\partial L}{\partial \dot{q}_i} \frac{\partial \dot{q}_i}{\partial Q_j}. \qquad (1.37)$$

Thus,

$$\frac{d}{dt} \frac{\partial M}{\partial \dot{Q}_j} = \frac{d}{dt} \left(\sum_{i=1}^{n} \frac{\partial L}{\partial \dot{q}_i} \frac{\partial \dot{q}_i}{\partial \dot{Q}_j} \right)$$

$$= \sum_{i=1}^{n} \left(\frac{d}{dt} \frac{\partial L}{\partial \dot{q}_i} \right) \frac{\partial \dot{q}_i}{\partial \dot{Q}_j} + \sum_{i=1}^{n} \frac{\partial L}{\partial \dot{q}_i} \left(\frac{d}{dt} \frac{\partial \dot{q}_i}{\partial \dot{Q}_j} \right).$$

Note that from

$$\dot{q}_i = \sum_{k=1}^{n} \frac{\partial q_i}{\partial Q_k} \dot{Q}_k + \frac{\partial q_i}{\partial t}, \qquad (1.38)$$

we obtain $\partial \dot{q}_i / \partial \dot{Q}_j = \partial q_i / \partial Q_j$. Hence,

$$\frac{d}{dt} \frac{\partial M}{\partial \dot{Q}_j} = \sum_{i=1}^{n} \left(\frac{d}{dt} \frac{\partial L}{\partial \dot{q}_i} \right) \frac{\partial q_i}{\partial Q_j} + \sum_{i=1}^{n} \frac{\partial L}{\partial \dot{q}_i} \frac{d}{dt} \left(\frac{\partial q_i}{\partial Q_j} \right). \qquad (1.39)$$

However, using Eq. (1.32), we have $d/dt(\partial L/\partial \dot{q}_i) = \partial L/\partial q_i$. Moreover,

$$\frac{d}{dt} \frac{\partial q_i}{\partial Q_j} = \sum_{k=1}^{n} \frac{\partial^2 q_i}{\partial Q_k \partial Q_j} \dot{Q}_k + \frac{\partial^2 q_i}{\partial Q_j \partial t}$$

$$= \frac{\partial}{\partial Q_j} \left(\sum_{k=1}^{n} \frac{\partial q_i}{\partial Q_k} \dot{Q}_k + \frac{\partial q_i}{\partial t} \right)$$

$$= \frac{\partial}{\partial Q_j} \left(\frac{d}{dt} q_i \right) = \frac{\partial \dot{q}_i}{\partial Q_j}.$$

Therefore,

$$\frac{d}{dt}\frac{\partial M}{\partial \dot{Q}_j} = \sum_{i=1}^{n} \frac{\partial L}{\partial q_i}\frac{\partial q_i}{\partial Q_j} + \sum_{i=1}^{n} \frac{\partial L}{\partial \dot{q}_i}\frac{\partial \dot{q}_i}{\partial Q_j}$$

$$= \frac{\partial M}{\partial Q_j},$$

and thus

$$\frac{d}{dt}\left(\frac{\partial M}{\partial \dot{Q}_j}\right) - \frac{\partial M}{\partial Q_j} = 0 \text{ for } j = 1,\ldots,n, \quad (1.40)$$

which we set out to prove. Therefore, in Lagrangian dynamics, the transformation from one set of coordinates to another again leads to Lagrangian dynamics, and the transformation of the Lagrangian can be accomplished in a particularly straightforward way by mere insertion of the transformation equations of the coordinates.

1.3.3 Connection to a Variational Principle

The particular form of the Lagrange equations (1.22) is the reason for an interesting cross-connection from the field of dynamical systems to the field of extremum problems treated by the so-called calculus of variations. As will be shown, if the problem is to find the path $\vec{q}(t)$ with fixed beginning $\vec{q}(t_1) = \vec{q}_1$ and end $\vec{q}(t_2) = \vec{q}_2$ such that the integral

$$I = \int_{t_1}^{t_2} L(\vec{q}(t), \dot{\vec{q}}(t), t)\, dt \quad (1.41)$$

assumes an extremum, then as shown by **Euler**, a necessary condition is that indeed the path $\vec{q}(t)$ satisfies

$$\frac{d}{dt}\left(\frac{\partial L}{\partial \dot{q}_i}\right) - \frac{\partial L}{\partial q_i} = 0,\ i = 1,\ldots,n, \quad (1.42)$$

which are just Lagrange's equations. Note that the converse is not true: The fact that a path $\vec{q}(t)$ satisfies Eq. (1.42) does not necessarily mean that the integral (1.41) assumes an extremum at $\vec{q}(t)$.

To show the necessity of Eq. (1.42) for the assumption of an extremum, let $\vec{q}(t)$ be a path for which the action $\int_{t_1}^{t_2} L(\vec{q}, \dot{\vec{q}}, t)\, dt$ is minimal. In particular, this implies that modifying $\vec{q}$ slightly in any direction increases the value of the integral. We now select n differentiable functions $\eta_j(t)$, $1 \leq j \leq n$, that satisfy

$$\eta_j(t_{1,2}) = 0 \text{ for } 1 \leq j \leq n. \quad (1.43)$$

Then, we define a family of varied curves via

$$\vec{q}_j(t, \alpha) = \vec{q}(t) + \alpha\, \eta_j(t) \cdot \vec{e}_j. \tag{1.44}$$

Thus, in $\vec{q}_j$, a variation only occurs in the direction of coordinate j; the amount of variation is controlled by the parameter α, and the value $\alpha = 0$ corresponds to the original minimizing path. We note that using Eqs. (1.43) and (1.44), we have $\vec{q}_j(t_{1,2}, \alpha) = \vec{q}_{1,2}$ for $1 \le j \le n$, and thus all the curves in the family go through the same starting and ending points, as demanded by the requirement of the minimization problem. Because $\vec{q}_j$ minimizes the action integral, when $\vec{q} + \alpha \eta_j \cdot \vec{e}_j$ is substituted into the integral, a minimum is assumed at $\alpha = 0$, and thus

$$0 = \frac{d}{d\alpha} \int_{t_1}^{t_2} L(\vec{q}_j(t,\alpha), \dot{\vec{q}}_j(t,\alpha), t)\, dt \bigg|_{\alpha=0} \quad \text{for all } 1 \le j \le n \tag{1.45}$$

Differentiating the integral with respect to α, we obtain the following according to the chain rule:

$$\frac{d}{d\alpha} \int_{t_1}^{t_2} L(\vec{q}_j(t,\alpha), \dot{\vec{q}}_j(t,\alpha), t)\, dt = \int_{t_1}^{t_2} \left(\frac{\partial L}{\partial q_j} \eta_j + \frac{\partial L}{\partial \dot{q}_j} \dot{\eta}_j \right) dt. \tag{1.46}$$

However, via integration by parts, the second term can be rewritten as

$$\int_{t_1}^{t_2} \frac{\partial L}{\partial \dot{q}_j} \dot{\eta}_j\, dt = \frac{\partial L}{\partial \dot{q}_j} \eta_j \bigg|_{t_1}^{t_2} - \int_{t_1}^{t_2} \frac{d}{dt}\left(\frac{\partial L}{\partial \dot{q}_j}\right) \eta_j\, dt. \tag{1.47}$$

Furthermore, since all the curves go through the same starting and ending points, we have

$$\frac{\partial L}{\partial \dot{q}_j} \eta_j \bigg|_{t_1}^{t_2} = 0. \tag{1.48}$$

Thus, we obtain

$$\frac{d}{d\alpha} \int_{t_1}^{t_2} L(\vec{q}_j(t,\alpha), \dot{\vec{q}}_j(t,\alpha), t)\, dt = \int_{t_1}^{t_2} \left(\frac{\partial L}{\partial q_j} - \frac{d}{dt}\frac{\partial L}{\partial \dot{q}_j} \right) \eta_j\, dt. \tag{1.49}$$

This condition must be true for every choice of η_j satisfying $\eta_j(t_{1,2}) = 0$, which can only be the case if indeed

$$\frac{\partial L}{\partial q_j} - \frac{d}{dt}\frac{\partial L}{\partial \dot{q}_j} = 0. \tag{1.50}$$

1.3.4 Lagrangians for Particular Systems

In this section, we discuss the motion of some important systems and try to assess the existence of a Lagrangian for each of them. In all cases, we study the motion in Cartesian variables, bearing in mind that the choice of different variables is straightforward according to the transformation properties of Lagrangians.

Newtonian Interacting Particles and the Principle of Least Action

We begin our study with the **nonrelativistic** motion of a **single particle** under Newtonian forces derivable from time-independent potentials, i.e., there is a $V(\vec{x})$ with $\vec{F} = -\vec{\nabla} V$. The Newtonian equation of motion of the particle is

$$\vec{F} = m \ddot{\vec{x}}. \tag{1.51}$$

We now try to "**guess**" a Lagrangian $L(\vec{x}, \dot{\vec{x}}, t)$. It seems natural to attempt to obtain $m\ddot{x}_k$ from $d/dt\,(\partial L/\partial \dot{x}_k)$ and F_k from $\partial L/\partial x_k$ for $k = 1, 2, 3$, and indeed the choice

$$L(\vec{x}, \dot{\vec{x}}, t) = \frac{1}{2} m\, \dot{\vec{x}}^2 - V(\vec{x}) \tag{1.52}$$

yields for $k = 1, 2, 3$ that

$$\frac{d}{dt} \frac{\partial L}{\partial \dot{x}_k} = m \ddot{x}_k \quad \text{and} \quad \frac{\partial L}{\partial x_k} = -\frac{\partial V}{\partial x_k} = F_k. \tag{1.53}$$

Thus, Lagrange's equations

$$\frac{d}{dt} \frac{\partial L}{\partial \dot{x}_k} - \frac{\partial L}{\partial x_k} = 0, \; k = 1, 2, 3 \tag{1.54}$$

yield the proper equations of motion.

Next, we study the nonrelativistic motion of a **system of N interacting particles** in which the external force $\vec{F}_i$ on the ith particle is derivable from a potential $V_i(\vec{x}_i)$ and the force on the ith particle due to the jth particle, $\vec{F}_{ji}$, is also derivable from a potential $V_{ji} = V_{ji}(|\vec{x}_i - \vec{x}_j|)$. That is,

$$\vec{F}_i = -\vec{\nabla} V_i \tag{1.55}$$

$$\vec{F}_{ji} = -\vec{\nabla}_i V_{ji} = \vec{\nabla}_j V_{ji} = \vec{\nabla}_j V_{ij} = -\vec{F}_{ij}, \tag{1.56}$$

where $\vec{\nabla}_k$ acts on the variable $\vec{x}_k$. Guided by the case of the single particle, we attempt

$$L = \sum_{i=1}^{N} \frac{1}{2} m_i\, \dot{\vec{x}}_i^2 - \sum_{i=1}^{N} V_i(\vec{x}_i) - \sum_{i<j} V_{ji}(|\vec{x}_i - \vec{x}_j|). \tag{1.57}$$

LAGRANGIAN SYSTEMS

Denoting the first term, often called the total kinetic energy, as T, and the second term, often called the total potential, as V, we have

$$L = T - V. \tag{1.58}$$

Let $x_{i,k}$ and $\dot{x}_{i,k}$ denote the kth components of the ith particle's position and velocity vectors, respectively. Then,

$$\frac{\partial L}{\partial x_{i,k}} = -\frac{\partial V_i}{\partial x_{i,k}} - \sum_{j, j \neq i} \frac{\partial V_{ji}}{\partial x_{i,k}}$$

$$= F_{i,k} + \sum_{j \neq i} F_{ji,k},$$

where $F_{i,k}$ and $F_{ji,k}$ denote the kth components of $\vec{F}_i$ and $\vec{F}_{ji}$, respectively. On the other hand,

$$\frac{\partial L}{\partial \dot{x}_{i,k}} = m_i \dot{x}_{i,k}. \tag{1.59}$$

Thus,

$$\frac{d}{dt}\frac{\partial L}{\partial \dot{x}_{i,k}} - \frac{\partial L}{\partial x_{i,k}} = 0 \tag{1.60}$$

is equivalent to

$$m_i \ddot{x}_{i,k} = F_{i,k} + \sum_{j \neq i} F_{ji,k}. \tag{1.61}$$

Therefore,

$$\frac{d}{dt}\frac{\partial L}{\partial \dot{x}_{i,k}} - \frac{\partial L}{\partial x_{i,k}} = 0 \text{ for } k = 1, 2, 3 \tag{1.62}$$

is equivalent to

$$m \ddot{\vec{x}}_i = \vec{F}_i + \sum_{j \neq i} \vec{F}_{ji}, \tag{1.63}$$

which is the proper equation of motion for the ith particle.

Considering the discussion in the previous section and the beauty and compactness of the integral relation (Eq. 1.41), one may be led to attempt to summarize Newtonian motion in the so-called **principle of least action** by Hamilton:

All mechanical motion initiating at $(t_1, \vec{q}_1)$ and terminating at $(t_2, \vec{q}_2)$ occurs such that it minimizes the so-called **action integral**

$$I = \int_{t_1}^{t_2} L\bigl(\vec{q}(t), \dot{\vec{q}}(t), t\bigr) \, dt. \tag{1.64}$$

In this case, the axiomatic foundation of mechanics is moved from the somewhat technical Newtonian view to Hamilton's least action principle with its mysterious beauty. However, many caveats are in order here. First, not all Newtonian systems can necessarily be written in Lagrangian form; as demonstrated above, obtaining a Lagrangian for a particular given system may or may not be possible. Second, as mentioned previously, Lagrange's equations alone do not guarantee an extremum, let alone a minimum, of the action integral as stated by Hamilton's principle. Finally, it generally appears advisable to select those statements that are to have axiomatic character in a physics theory in such a way that they can be checked as directly and as comprehensively as possible in the laboratory. In this respect, the Newton approach has an advantage because it requires only a check of forces and accelerations locally for any desired point in space. On the other hand, the test of the principle of least action requires the study of entire orbits during an extended time interval. Then, of all orbits that satisfy $\vec{q}(t_1) = \vec{q}_1$, a practically cumbersome preselection of only those orbits satisfying $\vec{q}(t_2) = \vec{q}_2$ is necessary. Finally, the verification that the experimentally appearing orbits really constitute those with the minimum action requires an often substantial and rather nontrivial computational effort.

Nonrelativistic Interacting Particles in General Variables

Now that the motion of nonrelativistic interacting particles in Cartesian variables can be expressed in terms of a Lagrangian, we can try to express it in terms of any coordinates $\vec{Q}$ for which $\vec{Q}(\vec{x})$ is a diffeomorphism, which is useful for a very wide variety of typical mechanics problems. All we need to do is to express $L = T - V$ in terms of the new variables, and thus we will have

$$L(\vec{Q}, \dot{\vec{Q}}) = T(\vec{Q}, \dot{\vec{Q}}) - V(\vec{Q}). \tag{1.65}$$

Without discussing specific examples, we make some general observations about the structure of the kinetic energy term $T(\vec{Q}, \dot{\vec{Q}})$ which will be very useful for later discussion. In Cartesian variables,

$$T(\dot{\vec{x}}) = \sum_{i=1}^{3N} \frac{m_i}{2} \dot{x}_i^2 = \frac{1}{2} \cdot \dot{\vec{x}}^t \cdot \begin{pmatrix} m_1 & & 0 \\ & \ddots & \\ 0 & & m_{3N} \end{pmatrix} \cdot \dot{\vec{x}}, \tag{1.66}$$

where the m_i's are now counted coordinatewise such that $m_1 = m_2 = m_3$ are all equal to the mass of the first particle, etc. Let the transformation between old and new variables be described by the map $\mathcal{M}$, i.e., $\vec{Q}(\vec{x}) = \mathcal{M}(\vec{x})$. Then in another set of variables, we have

$$\dot{\vec{Q}} = \text{Jac}(\mathcal{M}) \cdot \dot{\vec{x}} \,. \tag{1.67}$$

Since by requirement $\mathcal{M}$ is invertible, so is its Jacobian, and we obtain $\dot{\vec{x}} = \text{Jac}(\mathcal{M})^{-1} \cdot \dot{\vec{Q}}$, and so

$$T(\dot{\vec{Q}}) = \frac{1}{2} \cdot \dot{\vec{Q}}^t \cdot \hat{M}_Q \cdot \dot{\vec{Q}}, \tag{1.68}$$

where the new "mass matrix" $\hat{M}_Q$ has the form

$$\hat{M}_Q = \left(\text{Jac}(\mathcal{M})^{-1}\right)^t \cdot \begin{pmatrix} m_1 & & 0 \\ & \ddots & \\ 0 & & m_{3N} \end{pmatrix} \cdot \text{Jac}(\mathcal{M})^{-1}. \tag{1.69}$$

We observe that the matrix $\hat{M}_Q$ is again symmetric, and if none of the original masses vanish, it is also nonsingular. Furthermore, according to Eq. (1.68), in the new generalized coordinates the **kinetic energy is a quadratic form** in the generalized velocities $\dot{\vec{Q}}$.

Particles in Electromagnetic Fields

We study the nonrelativistic motion of a particle in an electromagnetic field. A discussion of the details of a backbone of electromagnetic theory as it is relevant for our purposes is presented below. The fields of mechanics and electrodynamics are coupled through the **Lorentz force**,

$$\vec{F} = e(\vec{E} + \dot{\vec{x}} \times \vec{B}), \tag{1.70}$$

which expresses how electromagnetic fields affect the motion. In this sense, it has axiomatic character and has been verified extensively and to great precision. From the theory of electromagnetic fields we need two additional pieces of knowledge, namely, that the electric and magnetic fields $\vec{E}$ and $\vec{B}$ can be derived from potentials $\phi(\vec{x}, t)$ and $\vec{A}(\vec{x}, t)$ as

$$\vec{E} = -\vec{\nabla}\phi - \frac{\partial \vec{A}}{\partial t}$$

and
$$\vec{B} = \vec{\nabla} \times \vec{A}.$$

The details of why this is indeed the case can be found below. Thus, in terms of the potentials ϕ and $\vec{A}$, the Lorentz force (Eq. 1.70) becomes

$$\vec{F} = e\left(-\vec{\nabla}\phi - \frac{\partial \vec{A}}{\partial t} + \dot{\vec{x}} \times (\vec{\nabla} \times \vec{A})\right). \tag{1.71}$$

Thus, the kth component of the force is

$$F_k = e\left(-\frac{\partial \phi}{\partial x_k} - \frac{\partial A_k}{\partial t} + \left(\dot{\vec{x}} \times \left(\vec{\nabla} \times \vec{A}\right)\right)_k\right). \tag{1.72}$$

However, using the common antisymmetric tensor ε_{ijk} and the Kronecker δ_{ij}, we see that

$$\left(\dot{\vec{x}} \times \left(\vec{\nabla} \times \vec{A}\right)\right)_k$$
$$= \sum_{i,j} \varepsilon_{ijk} \dot{x}_i \left(\vec{\nabla} \times \vec{A}\right)_j = \sum_{i,j} \varepsilon_{ijk} \dot{x}_i \sum_{l,m} \varepsilon_{lmj} \frac{\partial A_m}{\partial x_l}$$
$$= \sum_{i,j,l,m} \varepsilon_{kij} \varepsilon_{lmj} \dot{x}_i \frac{\partial A_m}{\partial x_l} = \sum_{i,j,l,m} \left(\varepsilon_{kij}\right)^2 \left(\delta_{kl}\delta_{im} - \delta_{km}\delta_{il}\right) \dot{x}_i \frac{\partial A_m}{\partial x_l}$$
$$= \sum_{i,j} \left(\varepsilon_{kij}\right)^2 \left(\dot{x}_i \frac{\partial A_i}{\partial x_k} - \dot{x}_i \frac{\partial A_k}{\partial x_i}\right) = \sum_i \left(\dot{x}_i \frac{\partial A_i}{\partial x_k} - \dot{x}_i \frac{\partial A_k}{\partial x_i}\right)$$
$$= \frac{\partial}{\partial x_k}(\dot{\vec{x}} \cdot \vec{A}) - (\dot{\vec{x}} \cdot \vec{\nabla}) A_k. \tag{1.73}$$

On the other hand, the total time derivative of A_k is

$$\frac{dA_k}{dt} = \frac{\partial A_k}{\partial t} + (\dot{\vec{x}} \cdot \vec{\nabla}) A_k. \tag{1.74}$$

Thus, Eq. (1.73) can be rewritten as

$$\left(\dot{\vec{x}} \times \left(\vec{\nabla} \times \vec{A}\right)\right)_k = \frac{\partial}{\partial x_k}(\dot{\vec{x}} \cdot \vec{A}) - \frac{dA_k}{dt} + \frac{\partial A_k}{\partial t}. \tag{1.75}$$

Substituting Eq. (1.75) in Eq. (1.72), we obtain for the kth component of the force

$$F_k = e\left(-\frac{\partial}{\partial x_k}(\phi - \dot{\vec{x}} \cdot \vec{A}) - \frac{dA_k}{dt}\right). \tag{1.76}$$

While this initally appears to be more complicated, it is now actually easier to guess a Lagrangian; in fact, the partial $\partial/\partial x_k$ suggests that $\phi - \dot{\vec{x}} \cdot \vec{A}$ is the term responsible for the force. However, because of the velocity dependence, there is also a contribution from $d/dt(\partial/\partial \dot{x}_k)$; this fortunately produces only the required term dA_k/dt in Eq. (1.76). The terms $m\ddot{\vec{x}}$ can be produced as before, and altogether we arrive at

$$L(\vec{x}, \dot{\vec{x}}, t) = \frac{1}{2}m\dot{\vec{x}}^2 - e\phi(\vec{x}, t) + e\vec{A}(\vec{x}, t) \cdot \dot{\vec{x}}. \quad (1.77)$$

Indeed, $d/dt(\partial L/\partial \dot{x}_k) - \partial L/\partial x_k = 0$ for all $k = 1, 2, 3$ is equivalent to $F_k = m\ddot{x}_k$ for all $k = 1, 2, 3$, and hence Lagrange's equations yield the right Lorentz force law.

It is worthwhile to see what happens if we consider the motion of an ensemble of nonrelativistic interacting particles in an electromagnetic field, where the interaction forces $\vec{F}_{ji}$, $i \neq j$, are derivable from potentials $V_{ji} = V_{ji}(|\vec{x}_i - \vec{x}_j|)$. From the previous examples, we are led to try

$$L = \sum_{i=1}^{N} \frac{1}{2}m_i \dot{\vec{x}}_i^2 - \sum_{i=1}^{N} e_i \phi_i(\vec{x}_i, t) + \sum_{i=1}^{N} e_i \vec{A}_i(\vec{x}_i, t) \cdot \dot{\vec{x}}_i - \frac{1}{2}\sum_{i \neq j=1}^{N} V_{ji}. \quad (1.78)$$

Indeed, in this case $d/dt(\partial L/\partial \dot{x}_{i,k}) - \partial L/\partial x_{i,k} = 0$ is equivalent to $m_i \ddot{x}_{i,k} = F_{i,k} + \sum_{j \neq i} F_{ji,k}$, and therefore

$$\frac{d}{dt}\frac{\partial L}{\partial \dot{x}_{i,k}} - \frac{\partial L}{\partial x_{i,k}} = 0 \text{ for all } k = 1, 2, 3 \quad (1.79)$$

is equivalent to

$$m_i \ddot{\vec{x}}_i = \vec{F}_i + \sum_{j \neq i} \vec{F}_{ji}, \quad (1.80)$$

which again yield the right equations of motion for the ith particle.

Now we proceed to **relativistic motion**. In this case, we restrict our discussion to the motion of a single particle. The situation for an ensemble is much more subtle for a variety of reasons. First, the interaction potentials would have to include retardation effects. Second, particles moving relativistically also produce complicated magnetic fields; therefore the interaction is not merely derivable from scalar potentials. In fact, the matter is so complicated that it is not fully understood, and there are even seemingly **paradoxical situations** in which particles interacting relativistically keep accelerating, gaining energy beyond bounds (Parrott, 1987; Rohrlich, 1990).

As a first step, let us consider the relativistic motion of a particle under forces derivable from potentials only. The equation of motion is given by

$$\vec{F} = \frac{d}{dt}\left(\frac{m\dot{\vec{x}}}{\sqrt{1-\dot{\vec{x}}^2/c^2}}\right). \quad (1.81)$$

We try to find a Lagrangian $L(\vec{x}, \dot{\vec{x}})$ such that $\partial L/\partial \dot{x}_k$ yields $m\dot{x}_k/\sqrt{1-\dot{\vec{x}}^2/c^2}$ and $\partial L/\partial x_k$ gives the kth component of the force, F_k, for $k = 1, 2, 3$. Let

$$L(\vec{x}, \dot{\vec{x}}, t) = -mc^2\sqrt{1-\dot{\vec{x}}^2/c^2} - V(\vec{x}, t). \quad (1.82)$$

Differentiating L with respect to x_k, $k = 1, 2, 3$, we obtain

$$\frac{\partial L}{\partial x_k} = -\frac{\partial V}{\partial x_k} = F_k. \quad (1.83)$$

Differentiating L with respect to $\dot{x}_k$, $k = 1, 2, 3$, we obtain

$$\frac{\partial L}{\partial \dot{x}_k} = -mc^2\left(\frac{1}{2}\right)\left(1-\frac{\dot{\vec{x}}^2}{c^2}\right)^{-1/2}\left(\frac{-2\dot{x}_k}{c^2}\right)$$

$$= \frac{m\dot{x}_k}{\sqrt{1-\dot{\vec{x}}^2/c^2}} \quad (1.84)$$

Thus, Lagrange's equations yield the proper equation of motion.

Next we study the relativistic motion of a particle in a **full electromagnetic field**. Based on previous experience, we expect that we merely have to combine the terms that lead to the Lorentz force with those that lead to the relativistic version of the Newton acceleration term, and hence we are led to try

$$L = -mc^2\sqrt{1-\dot{\vec{x}}^2/c^2} + e\dot{\vec{x}}\cdot\vec{A} - e\phi, \quad (1.85)$$

where ϕ is a scalar potential for the electric field and $\vec{A}$ is a vector potential for the magnetic field. Since the last term does not contribute to $d/dt(\partial L/\partial \dot{x}_k)$, the verification that the Lagrangian is indeed the proper one follows like in the previous examples.

1.4 HAMILTONIAN SYSTEMS

Besides the case of differential equations that can be expressed in terms of a Lagrangian function that was discussed in the previous section, another important class of differential equations includes those that can be expressed in terms of a so-called **Hamiltonian** function. We begin our study with the introduction of the $2n \times 2n$ matrix $\hat{J}$ that has the form

$$\hat{J} = \begin{pmatrix} \hat{0} & \hat{I} \\ -\hat{I} & \hat{0} \end{pmatrix}, \tag{1.86}$$

where $\hat{I}$ is the appropriate unity matrix. Before proceeding, we establish some **elementary properties** of the matrix $\hat{J}$. We apparently have

$$\hat{J}^t = -\hat{J}, \quad \hat{J} \cdot \hat{J}^t = \hat{I}, \text{ and } \hat{J}^{-1} = \hat{J}^t = -\hat{J}. \tag{1.87}$$

From the last condition one immediately sees that $\det(\hat{J}) = \pm 1$, but we also have $\det(\hat{J}) = +1$, which follows from simple rules for determinants. We first interchange row 1 with row $(n+1)$, row 2 with row $(n+2)$, etc. Each of these yields a factor of -1, for a total of $(-1)^n$. We then multiply the first n rows by (-1), which produces another factor of $(-1)^n$, and we have now arrived at the unity matrix; therefore,

$$\det \begin{pmatrix} \hat{0} & \hat{I} \\ -\hat{I} & \hat{0} \end{pmatrix} = (-1)^n \cdot \det \begin{pmatrix} -\hat{I} & \hat{0} \\ \hat{0} & \hat{I} \end{pmatrix}$$

$$= (-1)^n \cdot (-1)^n \cdot \det \begin{pmatrix} \hat{I} & \hat{0} \\ \hat{0} & \hat{I} \end{pmatrix} = 1 \tag{1.88}$$

We are now ready to define when we call a system **Hamiltonian**, or **canonical**. Let us consider a dynamical system described by $2n$ first-order differential equations. For the sake of notation, let us denote the first n variables by the column vector $\vec{q}$ and the second n variables by $\vec{p}$, so the dynamical system has the form

$$\frac{d}{dt}\begin{pmatrix} \vec{q} \\ \vec{p} \end{pmatrix} = \vec{F}\begin{pmatrix} \vec{q} \\ \vec{p} \end{pmatrix}. \tag{1.89}$$

The fact that we are apparently restricting ourselves to first-order differential equations is no reason for concern: As seen in Eq. (1.3), any higher order system can be transformed into a first-order system. For example, any set of n second-order differential equations in $\vec{q}$ can be transformed into a set of $2n$ first-order differential equations by the mere choice of $\vec{p} = \dot{\vec{q}}$. Similarly, in principle an odd-dimensional system can be made even dimensional by the introduction of a new variable that does not affect the equations of motion and that itself can satisfy an arbitrary differential equation.

We now call the system (Eq. 1.89) of $2n$ first-order differential equations **Hamiltonian, or canonical** if there exists a function H of $\vec{q}$ and $\vec{p}$ such that the function $\vec{F}$ is connected to the gradient $(\partial H/\partial \vec{q}, \partial H/\partial \vec{p})$ of the function H via

$$\vec{F}\begin{pmatrix} \vec{q} \\ \vec{p} \end{pmatrix} = \hat{J} \cdot (\partial H/\partial \vec{q}, \partial H/\partial \vec{p})^t. \tag{1.90}$$

In this case, we call the function H a **Hamiltonian function** or a **Hamiltonian** of the motion. The equations of motion then of course assume the form

$$\frac{d}{dt}\begin{pmatrix} \vec{q} \\ \vec{p} \end{pmatrix} = \hat{J} \cdot \begin{pmatrix} \partial H/\partial \vec{q} \\ \partial H/\partial \vec{p} \end{pmatrix}. \tag{1.91}$$

Written out in components, this reads

$$\frac{d}{dt}\vec{q} = \frac{\partial H}{\partial \vec{p}} \tag{1.92}$$

and

$$\frac{d}{dt}\vec{p} = -\frac{\partial H}{\partial \vec{q}}. \tag{1.93}$$

In the case that the motion is Hamiltonian, the variables $(\vec{q}, \vec{p})$ are referred to as **canonical variables**, and the $2n$ dimensional space consisting of the ordered pairs $\vec{z} = (\vec{q}, \vec{p})$ is called **phase space**. Frequently, we also say that the $\vec{p}$ are the **conjugate momenta** belonging to $\vec{q}$, and the $\vec{q}$ are the **canonical positions** belonging to $\vec{p}$.

Note that in the case of Hamiltonian motion, different from the Lagrangian case, the second variable, $\vec{p}$, is not directly connected to $\vec{q}$ by merely being the time derivative of $\vec{q}$, but rather plays a quite independent role. Therefore, there is often no need to distinguish the two sets of variables, and frequently they are summarized in the single phase space vector $\vec{z} = (\vec{q}, \vec{p})$. While this makes some of the notation more compact, we often maintain the convention of $(\vec{q}, \vec{p})$ for the practical reason that in this case it is clear whether vectors are row or column vectors, the distinction of which is important in many of the later derivations.

We may also ask ourselves whether the function $H(\vec{q}, \vec{p}, t)$ has any special significance; while similar to L and mostly a tool for computation, it has an interesting property. Let $(\vec{q}(t), \vec{p}(t))$ be a solution curve of the Hamilton differential equations; then the function $H(\vec{q}(t), \vec{p}(t), t)$ evolving with the solution curve satisfies

$$\begin{aligned}\frac{dH}{dt} &= \frac{\partial H}{\partial \vec{q}} \cdot \frac{d\vec{q}}{dt} + \frac{\partial H}{\partial \vec{p}} \cdot \frac{d\vec{p}}{dt} + \frac{\partial H}{\partial t} \\ &= \frac{\partial H}{\partial \vec{q}} \cdot \frac{\partial H}{\partial \vec{p}} - \frac{\partial H}{\partial \vec{p}} \cdot \frac{\partial H}{\partial \vec{q}} + \frac{\partial H}{\partial t} = \frac{\partial H}{\partial t}.\end{aligned} \tag{1.94}$$

Therefore, in the special case that the system is **autonomous**, the Hamiltonian is a preserved quantity.

1.4.1 Manipulations of the Independent Variable

As will be shown in this section, Hamiltonian systems allow various convenient manipulations connected to the independent variable. First, we discuss a method that allows **the independent variable** to be **changed**. As shown in Eq. (1.5), in principle this is possible with any ODE; remarkably, however, if we exchange t with a suitable position variable, then the **Hamiltonian structure** of the motion can be **preserved**. To see this, let us assume we are given a Hamiltonian $H = H(\vec{q}, \vec{p}, t)$ and suppose that in the region of phase space of interest one of the coordinates, e.g., q_1, is strictly monotonic in time and satisfies

$$\frac{dq_1}{dt} \neq 0 \tag{1.95}$$

everywhere. Then each value of time corresponds to a unique value of q_1, and thus it is possible to use q_1 instead of t to parametrize the motion.

We begin by introducing a new quantity called p_t, which is the negative of the Hamiltonian, i.e.,

$$p_t = -H. \tag{1.96}$$

We then observe that

$$\frac{\partial p_t}{\partial p_1} = -\frac{\partial H}{\partial p_1} = -\dot{q}_1 \neq 0, \tag{1.97}$$

which according to the implicit function theorem, entails that for every value of the coordinates $q_1, q_2, \ldots, q_n, p_2, \ldots, p_n, t$, it is possible to invert the dependence of p_t on p_1 and to obtain p_1 as a function of p_t. In this process, p_t becomes a variable; in fact, the canonical momentum p_t belongs to the new position variable t. Also, the motion with respect to the new independent variable q_1 will be shown to be described by the Hamiltonian

$$K(t, q_2, \ldots, q_n, p_t, p_2, \ldots, p_n, q_1) = -p_1(t, q_2, \ldots, q_n, p_t, p_2, \ldots, p_n, q_1). \tag{1.98}$$

We observe that by virtue of Eq. (1.97), we have

$$\frac{\partial K}{\partial p_t} = -\frac{\partial p_1}{\partial p_t} = \frac{1}{\dot{q}_1} = \frac{dt}{dq_1}. \tag{1.99}$$

Because K is obtained by inverting the relationship between p_1 and p_t, it follows that if we reinsert $p_t(q_1, \ldots, q_n, p_1, \ldots, p_n, t) = -H(q_1, \ldots, q_n, p_1, \ldots, p_n, t)$ into K, all the dependencies on $t, q_2, \ldots, q_n, p_t, p_2, \ldots, p_n, q_1$ disappear; so if we define

$$I = K(t, q_2, \ldots, q_n, p_t(q_1, \ldots, q_n, p_t, \ldots, p_n, t), p_2, \ldots, p_n, q_1), \tag{1.100}$$

then in fact

$$I = -p_1. \tag{1.101}$$

In particular, this entails that

$$0 = \frac{\partial I}{\partial q_k} = \frac{\partial K}{\partial q_k} + \frac{\partial K}{\partial p_t} \cdot \frac{\partial p_t}{\partial q_k} \quad \text{for } k = 2, \ldots, n \tag{1.102}$$

$$0 = \frac{\partial I}{\partial p_k} = \frac{\partial K}{\partial p_k} + \frac{\partial K}{\partial p_t} \cdot \frac{\partial p_t}{\partial p_k} \quad \text{for } k = 2, \ldots, n \tag{1.103}$$

From Eq. (1.102), it follows under utilization of Eq. (1.99) that

$$\frac{\partial K}{\partial q_k} = -\frac{\partial K}{\partial p_t} \cdot \frac{\partial p_t}{\partial q_k} = \frac{1}{\dot{q}_1} \cdot \frac{\partial H}{\partial q_k} = -\frac{dt}{dq_1} \cdot \frac{dp_k}{dt}$$

$$= -\frac{dp_k}{dq_1} \quad \text{for } k = 2, \ldots, n, \tag{1.104}$$

and similarly

$$\frac{\partial K}{\partial p_k} = -\frac{\partial K}{\partial p_t} \cdot \frac{\partial p_t}{\partial p_k} = \frac{1}{\dot{q}_1} \cdot \frac{\partial H}{\partial p_k} = \frac{dt}{dq_1} \cdot \frac{dq_k}{dt}$$

$$= \frac{dq_k}{dq_1} \quad \text{for } k = 2, \ldots, n. \tag{1.105}$$

Since we also have $0 = \partial I/\partial t = \partial K/\partial p_t \cdot \partial p_t/\partial t + \partial K/\partial t$, it follows using Eq. (1.94) that we have

$$\frac{\partial K}{\partial t} = -\frac{\partial K}{\partial p_t} \cdot \frac{\partial p_t}{\partial t} = \frac{dt}{dq_1} \cdot \frac{dH}{dt} = \frac{dH}{dq_1}. \tag{1.106}$$

Together with Eq. (1.99), Eqs. (1.104), (1.105), and (1.106) represent a set of Hamiltonian equations for the Hamiltonian $K(t, q_2, \ldots, q_n, p_t, p_2, \ldots, p_n, q_1)$, in which now q_1 plays the role of the independent variable. Thus, the interchange of t and q_1 is complete.

Next, we observe that it is possible to transform any **nonautonomous** Hamiltonian system to an **autonomous** system by introducing new variables, similar to the case of ordinary differential equations in Eq. (1.6). In fact, let $H(\vec{q}, \vec{p}, t)$ be the $2n$-dimensional time-dependent or nonautonomous Hamiltonian of a dynamical system. For ordinary differential equations of motion, we remind ourselves that all that was needed in Eq. (1.6) to make the system autonomous was to introduce a new variable s that satisfies the new differential equation $d/dt\, s = 1$ and the initial condition $s(t_0) = t_0$, which has the solution $s = t$ and hence allows replacement of all occurrences of t in the original differential equations by s.

In the Hamiltonian picture it is not immediately clear that something similar is possible because one has to assert that the resulting differential equations can somehow be represented by a new Hamiltonian. It is also necessary to introduce a matching canonical momentum p_s to the new variable s, and hence the dimensionality increases by 2. However, now consider the autonomous Hamiltonian of $2(n+1)$ variables

$$\hat{H}(\vec{q}, s, \vec{p}, p_s) = H(\vec{q}, \vec{p}, s) + p_s. \tag{1.107}$$

The resulting Hamiltonian equations are

$$\frac{d}{dt} q_i = \frac{\partial}{\partial p_i} H = \frac{\partial}{\partial p_i} \hat{H} \quad \text{for } i = 1, \ldots, n \tag{1.108}$$

$$\frac{d}{dt} p_i = -\frac{\partial}{\partial q_i} H = -\frac{\partial}{\partial q_i} \hat{H} \quad \text{for } i = 1, \ldots, n \tag{1.109}$$

$$\frac{d}{dt} s = 1 = \frac{\partial}{\partial p_s} \hat{H} \tag{1.110}$$

$$\frac{d}{dt} p_s = -\frac{d}{dt} H = -\frac{\partial}{\partial t} H = -\frac{\partial}{\partial s} \hat{H} \tag{1.111}$$

The third equation ensures that indeed the solution for the variable s is $s = t$, as needed, and hence the replacement of t by s in the first two equations leads to the same equations as previously described. Therefore, the autonomous Hamiltonian (Eq. 1.107) of $2(n+1)$ variables indeed describes the same motion as the original nonautonomous Hamiltonian of $2n$ variables H. The fourth equation is not primarily relevant since the dynamics of p_s do not affect the other equations. It is somewhat illuminating, however, that in the new Hamiltonian the **conjugate of the time** can be chosen as the old Hamiltonian.

The transformation to autonomous Hamiltonian systems is performed frequently in practice, and the $2(n+1)$ dimensional space $(\vec{q}, s, \vec{p}, p_s)$ is referred to as the **extended phase space**.

In conclusion, we show that despite the unusual intermixed form characteristic of Hamiltonian dynamics, there is a rather straightforward and intuitive **geometric interpretation** of Hamiltonian motion. To this end, let H be the Hamiltonian of an

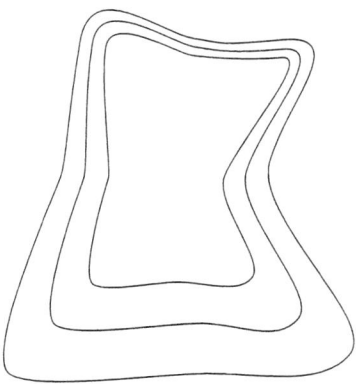

FIGURE 1.2. Contour surfaces of the Hamiltonian.

autonomous system, either originally or after transition to extended phase space, with the equations of motion

$$\frac{d}{dt}\begin{pmatrix} \vec{q} \\ \vec{p} \end{pmatrix} = \hat{J} \cdot \begin{pmatrix} \partial H/\partial \vec{q} \\ \partial H/\partial \vec{p} \end{pmatrix}. \quad (1.112)$$

Then, we apparently have that

$$(\partial H/\partial \vec{q}, \partial H/\partial \vec{p}) \cdot \frac{d}{dt}\begin{pmatrix} \vec{q} \\ \vec{p} \end{pmatrix} \quad (1.113)$$

$$= (\partial H/\partial \vec{q}, \partial H/\partial \vec{p}) \cdot \hat{J} \cdot \begin{pmatrix} \partial H/\partial \vec{q} \\ \partial H/\partial \vec{p} \end{pmatrix} = 0. \quad (1.114)$$

Thus, the direction of the motion is **perpendicular to the gradient** of H and is constrained to the contour surfaces of H. Furthermore, the **velocity** of motion is proportional to the **magnitude** $|(\partial H/\partial \vec{q}, \partial H/\partial \vec{p})|$ of the gradient. Since the distance between contour surfaces is antiproportional to the gradient, this is reminiscent to the motion of an incompressible liquid within a structure of guidance surfaces: The surfaces can never be crossed, and the closer they lie together, the faster the liquid flows. Figure 1.2 illustrates this behavior, which is often useful for an intuitive understanding of the dynamics of a given system.

1.4.2 Existence and Uniqueness of Hamiltonians

Before we proceed, we first try to answer the question of existence and uniqueness of a Hamiltonian for a given motion. Apparently, the Hamiltonian is **not unique** because with H, also $H + c$, where c is any constant, it yields the same equations of motion; aside from this, any other candidate to be a Hamiltonian must have

HAMILTONIAN SYSTEMS

the same gradient as H, and this entails that it differs from H by not more than a constant. Therefore, the question of uniqueness can be answered much more clearly than it could in the Lagrangian case.

The same is true for the question of **existence**, at least in a strict sense: In order for Eq. (1.90) to be satisfied, since $\hat{J}^{-1} = -\hat{J}$, it must hold that

$$\begin{pmatrix} \partial H/\partial \vec{q} \\ \partial H/\partial \vec{p} \end{pmatrix} = \vec{G}\begin{pmatrix} \vec{q} \\ \vec{p} \end{pmatrix} = -\hat{J} \cdot \vec{F}\begin{pmatrix} \vec{q} \\ \vec{p} \end{pmatrix}. \tag{1.115}$$

This, however, is a problem from **potential theory**, which is addressed in detail later. In Eq. (1.300) it is shown that the condition $\partial G_i/\partial x_j = \partial G_j/\partial x_i$ for all i, j is necessary and sufficient for this potential to exist. Once these $n \cdot (n-1)/2$ conditions are satisfied, the existence of a Hamiltonian is assured, and it can even be constructed explicitly by integration, much in the same way as the potential is obtained from the electric field.

In a broader sense, one may wonder if a given system is not Hamiltonian whether it is perhaps possible to make a diffeomorphic transformation of variables such that the motion in the new variables is Hamiltonian. This is indeed frequently possible, but it is difficult to classify those systems that are Hamiltonian up to coordinate transformations or to establish a general mechanism to find a suitable transformation. However, a very important special case of such transformations will be discussed in the next section.

1.4.3 The Duality of Hamiltonians and Lagrangians

As we mentioned in Section 1.4.2, the question of the existence of a Hamiltonian to a given motion is somewhat restrictive, and in general it is not clear whether there may be certain changes of variables that will lead to a Hamiltonian system. However, for the very important case of Lagrangian systems, there is a transformation that usually allows the construction of a Hamiltonian. Conversely, the transformation can also be used to construct a Lagrangian to a given Hamiltonian. This transformation is named after Legendre, and we will develop some of its elementary theory in the following subsection.

Legendre Transformations

Let f be a real function of n variables $\vec{x}$ that is twice differentiable. Let $\vec{G} = \vec{\nabla} f$ be the gradient of the function f, which is a function from R^n to R^n. Let $\vec{G}$ be a diffeomorphism, i.e., invertible and differentiable everywhere. Then, we define the **Legendre transformation** $\mathcal{L}(f)$ of the function f as

$$\mathcal{L}(f) = \vec{G}^{-1} \cdot \vec{I} - f(\vec{G}^{-1}). \tag{1.116}$$

Here, $\vec{I}$ is the identity function, and the dot "·" denotes the scalar product of vectors.

Legendre transformations are a useful tool for many applications in mathematics not directly connected to our needs, for example, for the solution of differential equations. Of all the important properties of Legendre transformations, we restrict ourselves to the observation that if the Legendre transformation of f exists, then the Legendre transformation of $\mathcal{L}(f)$ exists and satisfies

$$\mathcal{L}(\mathcal{L}(f)) = f. \qquad (1.117)$$

Therefore, the **Legendre transformation is idempotent** or self-inverse. To show this, we first observe that all partials of $\mathcal{L}(f)$ must exist since $\vec{G}$ was required to be a diffeomorphism and f is differentiable. Using the product and chain rules, we then obtain from Eq. (1.116) that

$$\vec{\nabla}\mathcal{L}(f) = \mathrm{Jac}(\vec{G}^{-1}) \cdot \vec{I} + \vec{G}^{-1} - \vec{G}(\vec{G}^{-1}) \cdot \mathrm{Jac}(\vec{G}^{-1})$$
$$= \vec{G}^{-1} \qquad (1.118)$$

and so

$$\mathcal{L}(\mathcal{L}(f)) = (\vec{G}^{-1})^{-1} \cdot \vec{I} - \mathcal{L}(f)((\vec{G}^{-1})^{-1})$$
$$= \vec{G} \cdot \vec{I} - \mathcal{L}(f)(\vec{G})$$
$$= \vec{G} \cdot \vec{I} - \vec{I} \cdot \vec{G} + f = f, \qquad (1.119)$$

as required.

Legendre Transformation of the Lagrangian

We now apply the Legendre transformation to the Lagrangian of the motion in a particular way. We will transform all $\dot{\vec{q}}$ variables but leave the original $\vec{q}$ variables unchanged. Let us assume that the system can be described by a Lagrangian $L(\vec{q}, \dot{\vec{q}}, t)$, and the n second-order differential equations are obtained from Lagrange's equations as

$$\frac{d}{dt}\left(\frac{\partial L}{\partial \dot{q}_i}\right) - \frac{\partial L}{\partial q_i} = 0, \ i = 1, \ldots, n. \qquad (1.120)$$

In order to perform the Legendre transformation with respect to the variables $\dot{\vec{q}}$, we first perform the gradient of L with respect to $\dot{\vec{q}}$; we call the resulting function $\vec{p} = (p_1, p_2, \ldots, p_n)$ and thus have

$$p_i = \frac{\partial L(\vec{q}, \dot{\vec{q}}, t)}{\partial \dot{q}_i}. \qquad (1.121)$$

HAMILTONIAN SYSTEMS

The p_i will play the role of the canonical momenta belonging to q_i. The Legendre transformation exists if the function $\vec{p}(\dot{\vec{q}})$ is a diffeomorphism, and in particular if it is invertible. In fact, it will be possible to obtain a Hamiltonian for the system only if this is the case. There are important situations where this is not so, for example, for the case of the relativistically covariant Lagrangian.

If $\vec{p}(\dot{\vec{q}})$ is a diffeomorphism, then the Legendre transformation can be constructed, and we define the function $H(\vec{q}, \vec{p})$ to be the Legendre transformation of L. Therefore, we have

$$H(\vec{q}, \vec{p}) = \mathcal{L}(L)(\vec{q}, \vec{p}) = \dot{\vec{q}}(\vec{p}) \cdot \vec{p} - L(\vec{q}, \dot{\vec{q}}(\vec{p}), t) \qquad (1.122)$$

As we shall now demonstrate, the newly defined function H actually does play the role of a Hamiltonian of the motion. Differentiating H with respect to an as yet unspecified arbitrary variable a, we obtain by the chain rule

$$\frac{dH}{da} = \sum_{i=1}^{n} \frac{\partial H}{\partial q_i} \frac{dq_i}{da} + \sum_{i=1}^{n} \frac{\partial H}{\partial p_i} \frac{dp_i}{da} + \frac{\partial H}{\partial t} \frac{dt}{da}. \qquad (1.123)$$

On the other hand, from the definition of H (Eq. 1.122) via the Legendre transformation, we obtain

$$\begin{aligned}
\frac{dH}{da} &= \sum_{i=1}^{n} \dot{q}_i \frac{dp_i}{da} + \sum_{i=1}^{n} p_i \frac{d\dot{q}_i}{da} - \sum_{i=1}^{n} \frac{\partial L}{\partial q_i} \frac{dq_i}{da} - \sum_{i=1}^{n} \frac{\partial L}{\partial \dot{q}_i} \frac{d\dot{q}_i}{da} - \frac{\partial L}{\partial t} \frac{dt}{da} \\
&= \sum_{i=1}^{n} \dot{q}_i \frac{dp_i}{da} - \sum_{i=1}^{n} \frac{\partial L}{\partial q_i} \frac{dq_i}{da} - \frac{\partial L}{\partial t} \frac{dt}{da} \\
&= \sum_{i=1}^{n} \dot{q}_i \frac{dp_i}{da} - \sum_{i=1}^{n} \dot{p}_i \frac{dq_i}{da} - \frac{\partial L}{\partial t} \frac{dt}{da},
\end{aligned} \qquad (1.124)$$

where the definition of the conjugate momenta (Eq. 1.121) as well as the Lagrange equations (1.120) were used. Choosing $a = q_i, p_i, t, i = 1, \ldots, n$ successively and comparing Eqs. (1.123) and (1.124), we obtain:

$$\dot{q}_i = \frac{\partial H}{\partial p_i} \qquad (1.125)$$

$$-\dot{p}_i = \frac{\partial H}{\partial q_i} \qquad (1.126)$$

$$-\frac{\partial L}{\partial t} = \frac{\partial H}{\partial t}. \qquad (1.127)$$

The first two of these equations are indeed Hamilton's equations, showing that the function H defined in Eq. (1.122) is a Hamiltonian for the motion in the variables

($\vec{q}, \vec{p}$). The third equation represents a noteworthy connection between the time dependences of Hamiltonians and Lagrangians.

In some cases of the transformations from Lagrangians to Hamiltonians, it may be of interest to perform a Legendre transformation of only a select group of velocities and to leave the other velocities in their original form. For the sake of notational convenience, let us assume that the coordinates that are to be transformed are the first k velocities. The transformation is possible if the function in k variables that maps $(\dot{q}_1, \ldots, \dot{q}_k)$ into

$$p_i = \frac{\partial L}{\partial \dot{q}_i} \text{ for } i = 1, \ldots, k \tag{1.128}$$

is a diffeomorphism. If this is the case, the Legendre transformation would lead to a function

$$R(q_1, \ldots, q_n, p_1, \ldots, p_k, \dot{q}_{k+1}, \ldots, \dot{q}_n)$$
$$= \sum_{i=1}^{k} \dot{q}_i(\vec{p}) \cdot p_i - L(q_1, \ldots, q_n, \dot{q}_1(\vec{p}), \ldots, \dot{q}_k(\vec{p}), \dot{q}_{k+1}, \ldots, \dot{q}_n)$$
$$\tag{1.129}$$

that depends on the positions as well as the first k momenta and the last $(n-k)$ velocities. This function is often called the **Routhian** of the motion, and it satisfies Hamiltonian equations for the first k coordinates and Lagrange equations for the second $(n-k)$ coordinates

$$\dot{q}_i = \frac{\partial R}{\partial p_i}, \quad \dot{p}_i = -\frac{\partial R}{\partial q_i} \text{ for } i = 1, \ldots, k$$
$$0 = \frac{d}{dt} \frac{\partial R}{\partial \dot{q}_i} - \frac{\partial R}{\partial q_i} \text{ for } i = k+1, \ldots, n. \tag{1.130}$$

Legendre Transformation of the Hamiltonian

Since the Legendre transformation is self-inverse, it is interesting to study what will happen if it is applied to the Hamiltonian. As is to be expected, if the Hamiltonian was generated by a Legendre transformation from a Lagrangian, then this Lagrangian can be recovered. Moreover, even for Hamiltonians for which no Lagrangian was known per se, a **Lagrangian can be generated** if the transformation can be executed, which we will demonstrate.

As the first step, we calculate the derivatives of the Hamiltonian with respect to the momenta and call them $\dot{q}_i$. We get

$$\dot{q}_i = \frac{\partial H}{\partial p_i} \tag{1.131}$$

HAMILTONIAN SYSTEMS

(which, in the light of Hamilton's equations, seems somewhat tautological). If the transformation between the $\dot{q}_i$ and p_i is a diffeomorphism, the Legendre transformation can be executed; we call the resulting function $L(\vec{q}, \dot{\vec{q}}, t)$ and have

$$L(\vec{q}, \dot{\vec{q}}, t) = \dot{\vec{q}} \cdot \vec{p}(\vec{q}, \dot{\vec{q}}) - H(\vec{q}, \vec{p}(\vec{q}, \dot{\vec{q}}), t). \tag{1.132}$$

In case the Hamiltonian was originally obtained from a Lagrangian, clearly the transformation between the $\dot{q}_i$ and p_i is a diffeomorphism, and Eq. (1.132) leads to the old Lagrangian, since it is the same as Eq. (1.122) solved for $L(\vec{q}, \dot{\vec{q}}, t)$. On the other hand, if H was not originally obtained from a Lagrangian, we want to show that indeed the newly created function $L(\vec{q}, \dot{\vec{q}}, t)$ satisfies Lagrange's equations. To this end, we compute

$$\frac{\partial L}{\partial q_i} = \sum_{j=1}^{n} \dot{q}_j \frac{\partial p_j}{\partial q_i} - \frac{\partial H}{\partial q_i} - \sum_{j=1}^{n} \frac{\partial H}{\partial p_j} \frac{\partial p_j}{\partial q_i} = -\frac{\partial H}{\partial q_i} \tag{1.133}$$

$$\frac{d}{dt} \frac{\partial L}{\partial \dot{q}_i} = \frac{d}{dt} p_i, \tag{1.134}$$

and hence Lagrange's equations are indeed satisfied.

This observation is often important in practice since it offers one mechanism to **change variables** in a Hamiltonian while preserving the Hamiltonian structure. To this end, one first performs a transformation to a Lagrangian, then the desired and straightforward canonical Lagrangian transformation, and finally the transformation back to the new Hamiltonian. An additional interesting aspect is that the Lagrangian L of a system is not unique, but for any function $F(\vec{q}, t)$, according to Eq. (1.23), the function $\hat{L}$ with

$$\hat{L} = L + \frac{d}{dt} F(\vec{q}, t) = L + \sum_{j=1}^{n} \frac{\partial F}{\partial q_j} \cdot \dot{q}_j + \frac{\partial F}{\partial t} \tag{1.135}$$

also yields the same equations of motion. Such a change of Lagrangian can be used to influence the new variables because it affects the momenta via

$$\hat{p}_i = \frac{\partial \hat{L}}{\partial \dot{q}_i} = \frac{\partial L}{\partial \dot{q}_i} + \frac{\partial F}{\partial q_i} = p_i + \frac{\partial F}{\partial q_i}. \tag{1.136}$$

We conclude by noting that instead of performing a Legendre transformation of the momentum variable $\vec{p}$ to a velocity variable $\dot{\vec{q}}$, because of the symmetry of the Hamiltonian equations it is also possible to compute a Legendre transformation of the position variable $\vec{q}$ to a variable $\dot{\vec{p}}$. In this case, one obtains a system with Lagrangian $L(\vec{p}, \dot{\vec{p}}, t)$.

1.4.4 Hamiltonians for Particular Systems

In this section, the method of Legendre transforms is applied to find Hamiltonians for several Lagrangian systems discussed in Section 1.3.4.

Nonrelativistic Interacting Particles in General Variables

We attempt to determine a Hamiltonian for the nonrelativistic motion of a **system of N interacting particles**. We begin by investigating only Cartesian coordinates. According to Eq. (1.57), the Lagrangian for the motion has the form

$$L = \sum_{i=1}^{N} \frac{1}{2} m_i \, \dot{\vec{x}}_i^{\,2} - \sum_{i=1}^{N} V_i(\vec{x}_i) - \sum_{i<j} V_{ji}(|\vec{x}_i - \vec{x}_j|). \tag{1.137}$$

According to Eq. (1.121), the canonical momenta are obtained via $\vec{p}_i = \partial L / \partial \dot{\vec{x}}_i$; here, we merely have

$$\vec{p}_i = m_i \, \dot{\vec{x}}_i, \tag{1.138}$$

which is the same as the conventional momentum. The relations between the canonical momenta and the velocities can obviously be inverted and they represent a diffeomorphism. For the inverse, we have $\dot{\vec{x}}_i = \vec{p}_i / m_i$; according to Eq. (1.122), the Hamiltonian can be obtained via

$$\begin{aligned} H(\vec{x}_i, \vec{p}_i) &= \sum_{i=1}^{N} \vec{p}_i \cdot \dot{\vec{x}}_i \,(\vec{x}_i, \vec{p}_i) - L(\vec{x}_i, \dot{\vec{x}}_i \,(\vec{x}_i, \vec{p}_i)) \\ &= \sum_{i=1}^{N} \frac{\vec{p}_i^{\,2}}{m_i} - \sum_{i=1}^{N} \frac{1}{2} \frac{\vec{p}_i^{\,2}}{m_i} + \sum_{i=1}^{N} V_i(\vec{x}_i) + \sum_{i<j} V_{ji}(|\vec{x}_i - \vec{x}_j|) \\ &= \sum_{i=1}^{N} \frac{\vec{p}_i^{\,2}}{2m_i} + \sum_{i=1}^{N} V_i(\vec{x}_i) + \sum_{i<j} V_{ji}(|\vec{x}_i - \vec{x}_j|). \end{aligned} \tag{1.139}$$

In this case, the entire process of obtaining the Hamiltonian was next to trivial, mostly because the relationship between velocities and canonical momenta is exceedingly simple and can easily be inverted. We see that the first term is the kinetic energy T, and the second term is the potential energy V, so that we have

$$H = T + V. \tag{1.140}$$

Let us now study the case of $3N$ generalized variables $\vec{Q}$; the Lagrangian still has the form

$$L = T(\dot{\vec{Q}}) - V(\vec{Q}), \tag{1.141}$$

where $T(\dot{\vec{Q}})$ now has the form

$$T(\dot{\vec{Q}}) = \frac{1}{2} \cdot \dot{\vec{Q}}^t \cdot \hat{M}_Q \cdot \dot{\vec{Q}}, \tag{1.142}$$

where according to Eq. (1.69), $\hat{M}_Q$ is a nonsingular symmetric matrix. We then have

$$\vec{P} = \frac{\partial L}{\partial \dot{\vec{Q}}} = \frac{\partial T}{\partial \dot{\vec{Q}}} = \hat{M}_Q \cdot \dot{\vec{Q}}, \tag{1.143}$$

and since $\hat{M}_Q$ is nonsingular, this relationship can be inverted; therefore, the Legendre transformation can be performed, and we obtain

$$\begin{aligned}
H &= \sum_{i=1}^{3N} P_i \cdot \dot{Q}_i - L \\
&= \vec{P}^t \cdot \hat{M}_Q^{-1} \cdot \vec{P} - \frac{1}{2} \cdot \vec{P}^t \cdot \left(\hat{M}_Q^{-1}\right)^t \cdot \hat{M}_Q \cdot \hat{M}_Q^{-1} \cdot \vec{P} + V \\
&= \frac{1}{2} \cdot \vec{P}^t \cdot \hat{M}_Q^{-1} \cdot \vec{P} + V \\
&= T + V;
\end{aligned} \tag{1.144}$$

Again, the Hamiltonian is given by $T + V$ for this special case.

Particles in Electromagnetic Fields

We now consider the relativistic motion of a **single particle in an electromagnetic field**. Recall that the following suitable Lagrangian for this problem was obtained in Eq. (1.85) and found to be

$$L(\vec{x}, \dot{\vec{x}}, t) = -mc^2\sqrt{1 - \dot{\vec{x}}^2/c^2} + e\dot{\vec{x}} \cdot \vec{A}(\vec{x}, t) - e\phi(\vec{x}, t). \tag{1.145}$$

Now we proceed to find the Hamiltonian for the motion of the particle. Let $\vec{p}$ denote the canonical momentum of the particle; then, for $k = 1, 2, 3$, we have $p_k = \partial L/\partial \dot{x}_k = \gamma m\dot{x}_k + eA_k$ with $\gamma = 1/\sqrt{1 - \dot{\vec{x}}^2/c^2}$, or in vector form,

$$\vec{p} = \gamma m \dot{\vec{x}} + e\vec{A}. \tag{1.146}$$

In this case, the canonical momentum is different from the conventional momentum. However, utilizing the Legendre transformation, it is possible to proceed in a rather automated way. We first study whether it is possible to solve Eq. (1.146) for $\dot{\vec{x}}$. Apparently, $(\vec{p} - e\vec{A}) = \gamma m\,\dot{\vec{x}}$, and thus $(\vec{p} - e\vec{A})^2 = m^2\,\dot{\vec{x}}^2/(1 - \dot{\vec{x}}^2/c^2)$. After some simple arithmetic, we obtain

$$\dot{\vec{x}}^2 = \frac{(\vec{p} - e\vec{A})^2 c^2}{(\vec{p} - e\vec{A})^2 + m^2 c^2}, \tag{1.147}$$

and noting that according to Eq. (1.146), $\dot{\vec{x}}$ is always parallel to $(\vec{p} - e\vec{A})$, we can even conclude that

$$\dot{\vec{x}} = \frac{c\,(\vec{p} - e\vec{A})}{\sqrt{(\vec{p} - e\vec{A})^2 + m^2 c^2}}. \tag{1.148}$$

Indeed, the transformation (Eq. 1.146) could be solved for $\dot{\vec{x}}$ and it represents a diffeomorphism, which ensures the existence of the Hamiltonian. Proceeding further, we see that

$$\sqrt{1 - \dot{\vec{x}}^2/c^2} = \frac{mc}{\sqrt{(\vec{p} - e\vec{A})^2 + m^2 c^2}}.$$

Substituting these last two results into the Lagrangian (Eq. 1.145), we obtain the Hamiltonian:

$$\begin{aligned}
H &= \vec{p} \cdot \dot{\vec{x}} - L(\vec{x}, \dot{\vec{x}}\,(\vec{x}, \vec{p})) \\
&= \frac{\vec{p} \cdot c(\vec{p} - e\vec{A})}{\sqrt{(\vec{p} - e\vec{A})^2 + m^2 c^2}} + \frac{m^2 c^3}{\sqrt{(\vec{p} - e\vec{A})^2 + m^2 c^2}} - \frac{c(e\vec{p} - e^2\vec{A}) \cdot \vec{A}}{\sqrt{(\vec{p} - e\vec{A})^2 + m^2 c^2}} + e\phi \\
&= c \cdot \frac{\vec{p}^2 - 2e\vec{p}\vec{A} + m^2 c^2 + e^2\vec{A}^2}{\sqrt{(\vec{p} - e\vec{A})^2 + m^2 c^2}} + e\phi \\
&= c \cdot \sqrt{(\vec{p} - e\vec{A})^2 + m^2 c^2} + e\phi \tag{1.149}
\end{aligned}$$

1.4.5 Canonical Transformation of Hamiltonians

For similar reasons as those discussed in Section 1.3.2 for Lagrangians, for Hamiltonians it is important to study how the Hamiltonian structure can be maintained under a **change of coordinates**. For example, if it can be determined by a change of variable that the Hamiltonian does not explicitly depend on q_i, then $\dot{p}_i = 0$, and the conjugate momentum corresponding to q_i is a constant.

In the case of the Lagrangian, it turned out that regardless of what new coordinates $\vec{Q}$ are used to describe the system, in order to obtain the Lagrangian for the new variables $\vec{Q}$ and $\dot{\vec{Q}}$, it is necessary only to insert the functional dependence of $\vec{q}$ on $\vec{Q}$ as well as the functional dependence of $\dot{\vec{q}}$ on $\vec{Q}$ and $\dot{\vec{Q}}$ resulting from differentiation into the old Lagrangian.

The question now arises whether there is a similarly simple method for the case of Hamiltonians. In the case of the Lagrangian, when changing from $\vec{q}$ to $\vec{Q}$, the velocity-like coordinates $\dot{\vec{q}}$ and $\dot{\vec{Q}}$ are directly linked to the respective position coordinates $\vec{q}$ and $\vec{Q}$ by virtue of the fact that they are the time derivatives of the positions. In the Hamiltonian case, the situation is more complicated: If we want to transform from $\vec{q}$ to $\vec{Q}$, it is not automatically determined how the canonical momentum $\vec{P}$ belonging to the new coordinate $\vec{Q}$ is to be chosen, but it seems clear that only very specific choices would preserve the Hamiltonian structure of the motion. Furthermore, the fact that $\vec{p}$ is not directly tied to $\vec{q}$ and, as shown in the last section, the roles of $\vec{q}$ and $\vec{p}$ are rather interchangeable, suggests that it may be possible to change the momentum $\vec{p}$ to our liking to a more suitable momentum $\vec{P}$ and, if we do so, to choose a "matching" $\vec{Q}$.

In order to establish the transformation properties of Hamiltonians, it is in general **necessary to transform positions and momenta simultaneously**, i.e., to study $2n$ dimensional diffeomorphisms

$$(\vec{Q}, \vec{P}) = \mathcal{M}(\vec{q}, \vec{p}). \tag{1.150}$$

We now study what happens if it is our desire to change from the old variables $(\vec{q}, \vec{p})$ to the new variables $(\vec{Q}, \vec{P})$. It is our goal to obtain the Hamiltonian for the variables $(\vec{Q}, \vec{P})$ in the same way as occurred automatically in the case of the Lagrangian in Eq. (1.34), i.e., by mere insertion of the transformation rules, so that

$$K(\vec{Q}, \vec{P}) = H(\vec{q}(\vec{Q}, \vec{P}), \vec{p}(\vec{Q}, \vec{P})). \tag{1.151}$$

In terms of the map $\mathcal{M}$, we want

$$K = H(\mathcal{M}^{-1}) \text{ and } H = K(\mathcal{M}). \tag{1.152}$$

If a transformation $\mathcal{M}$ satisfies this condition for every Hamiltonian H, we call the transformation **canonical,** indicating its preservation of the canonical structure.

The set of canonical transformations of phase space form a **group** under composition. Obviously, the identity transformation is canonical. Furthermore, since each transformation is a diffeomorphism, it is invertible, and the inverse is a canonical transformation since it merely returns to the original Hamiltonian system. Finally, associativity is satisfied because composition is associative.

Elementary Canonical Transformations

There are a few straightforward examples for canonical transformations called elementary canonical transformations. First, it is apparently possible to change the labeling of the variables; hence, for a given k and l, the transformation

$$\begin{pmatrix} Q_k \\ P_k \end{pmatrix} = \begin{pmatrix} q_l \\ p_l \end{pmatrix}, \quad \begin{pmatrix} Q_l \\ P_l \end{pmatrix} = \begin{pmatrix} q_k \\ p_k \end{pmatrix}$$

and

$$\begin{pmatrix} Q_i \\ P_i \end{pmatrix} = \begin{pmatrix} q_i \\ p_i \end{pmatrix} \quad \text{for } i \neq k, l \tag{1.153}$$

is canonical. There are apparently $n(n-1)/2$ such transformations.

A more interesting transformation is the following one: for a given k, set

$$\begin{pmatrix} Q_k \\ P_k \end{pmatrix} = \begin{pmatrix} p_k \\ -q_k \end{pmatrix}, \quad \text{and} \quad \begin{pmatrix} Q_i \\ P_i \end{pmatrix} = \begin{pmatrix} q_i \\ p_i \end{pmatrix} \quad \text{for } i \neq k. \tag{1.154}$$

To verify that this is in fact canonical, consider an arbitrary Hamiltonian $H(\vec{q}, \vec{p})$ with Hamiltonian equations

$$\begin{pmatrix} \dot{q}_i \\ \dot{p}_i \end{pmatrix} = \begin{pmatrix} \partial H / \partial p_i \\ -\partial H / \partial q_i \end{pmatrix}. \tag{1.155}$$

In light of (1.153), we may assume that $k = 1$. The new Hamiltonian $\hat{H}$ then is given by

$$\hat{H}(Q_1, Q_2, \ldots, Q_n, P_1, P_2, \ldots, P_n)$$
$$= H(p_1, q_2, \ldots, q_n, -q_1, p_2, \ldots, p_n), \tag{1.156}$$

and we have

$$\frac{\partial \hat{H}}{\partial P_1} = -\frac{\partial H}{\partial q_1} = \dot{p}_1 = \dot{Q}_1$$

$$\frac{\partial \hat{H}}{\partial Q_1} = \frac{\partial H}{\partial p_1} = \dot{q}_1 = -\dot{P}_1, \tag{1.157}$$

which are the proper Hamilton equations. For the other coordinates nothing changes, and altogether the transformation is canonical. While essentially trivial, the transformation is of interest in that it stresses that positions and momenta play essentially identical roles in the Hamiltonian; this is different from the case of the Lagrangian, in which the $\dot{\vec{q}}$ variable is always obtained as a quantity derived from $\vec{q}$.

Canonical Transformation and Symplecticity

We now derive a fundamental condition that canonical transformations must satisfy—the so-called symplectic condition or condition of symplecticity. Let us assume we are given a diffeomorphism of phase space of the form

$$(\vec{Q}, \vec{P}) = \mathcal{M}(\vec{q}, \vec{p}). \tag{1.158}$$

Let the Jacobian of $\mathcal{M}$ be given by

$$\text{Jac}(\mathcal{M}) = \begin{pmatrix} \partial \vec{Q}/\partial \vec{q} & \partial \vec{Q}/\partial \vec{p} \\ \partial \vec{P}/\partial \vec{q} & \partial \vec{P}/\partial \vec{p} \end{pmatrix} \tag{1.159}$$

We want to study the motion in the new variables. Apparently, we have $\dot{Q}_i = \sum_{k=1}^{n} (\partial Q_i/\partial q_k \cdot \dot{q}_k + \partial Q_i/\partial p_k \cdot \dot{p}_k)$ and $\dot{P}_i = \sum_{k=1}^{n} (\partial P_i/\partial q_k \cdot \dot{q}_k + \partial P_i/\partial p_k \cdot \dot{p}_k)$; using the Jacobian matrix of $\mathcal{M}$, we have

$$\frac{d}{dt} \begin{pmatrix} \vec{Q} \\ \vec{P} \end{pmatrix} = \text{Jac}(\mathcal{M}) \cdot \frac{d}{dt} \begin{pmatrix} \vec{q} \\ \vec{p} \end{pmatrix} = \text{Jac}(\mathcal{M}) \cdot \hat{J} \cdot \begin{pmatrix} \partial H/\partial \vec{q} \\ \partial H/\partial \vec{p} \end{pmatrix}. \tag{1.160}$$

Furthermore, because $H = K(\mathcal{M})$, we have $\partial H/\partial q_i = \sum_{k=1}^{n} \partial K/\partial Q_k \cdot \partial Q_k/\partial q_i + \partial K/\partial P_k \cdot \partial P_k/\partial q_i$ and $\partial H/\partial p_i = \sum_{k=1}^{n} \partial K/\partial Q_k \cdot \partial Q_k/\partial p_i + \partial K/\partial P_k \cdot \partial P_k/\partial p_i$. In matrix form this can be written as

$$\begin{pmatrix} \partial H/\partial \vec{q} \\ \partial H/\partial \vec{p} \end{pmatrix} = \text{Jac}(\mathcal{M})^t \cdot \begin{pmatrix} \partial K/\partial \vec{Q} \\ \partial K/\partial \vec{P} \end{pmatrix}, \tag{1.161}$$

and combined with the previous equation we have

$$\frac{d}{dt} \begin{pmatrix} \vec{Q} \\ \vec{P} \end{pmatrix} = \left(\text{Jac}(\mathcal{M}) \cdot \hat{J} \cdot \text{Jac}(\mathcal{M})^t \right) \cdot \begin{pmatrix} \partial K/\partial \vec{Q} \\ \partial K/\partial \vec{P} \end{pmatrix}. \tag{1.162}$$

This equation describes how the motion in the new variables can be expressed in terms of the gradient of the Hamiltonian expressed in the new variables. If we want to maintain the Hamiltonian structure, it is necessary to have

$$\frac{d}{dt} \begin{pmatrix} \vec{Q} \\ \vec{P} \end{pmatrix} = \hat{J} \cdot \begin{pmatrix} \partial K/\partial \vec{Q} \\ \partial K/\partial \vec{P} \end{pmatrix}. \tag{1.163}$$

Apparently, this requirement can be met if the transformation map $\mathcal{M}$ satisfies

$$\left(\text{Jac}(\mathcal{M}) \cdot \hat{J} \cdot \text{Jac}(\mathcal{M})^t \right) = \hat{J}, \tag{1.164}$$

which is commonly known as the **symplectic condition** or **condition of symplecticity**. A map $\mathcal{M}$ is called **symplectic** if it satisfies the symplectic condition.

Thus, any symplectic map produces a transformation that yields new coordinates in which the motion is again Hamiltonian.

We have seen that for the transformation from the old to the new coordinates, for a given Hamiltonian, symplecticity is sufficient to preserve the Hamiltonian structure of the motion. Therefore, **symplectic maps are canonical.** However, there are situations in which symplecticity is not necessary to provide a transformation resulting again in a Hamiltonian form; if we consider a two-dimensional Hamiltonian $H(q_1, q_2, p_1, p_2)$ that in fact does depend on only q_1 and p_1, then while it is important to transform q_1 and p_1 properly, the transformation of q_2 and p_2 is insignificant and hence there are many nonsymplectic choices preserving the Hamiltonian form.

On the other hand, if the demand is that the transformation $\mathcal{M} : (\vec{q}, \vec{p}) \to (\vec{Q}, \vec{P})$ transform every Hamiltonian system with a given Hamiltonian H into a new Hamiltonian system with Hamiltonian $K = H(\mathcal{M}^{-1})$, then it is also necessary that $\mathcal{M}$ be symplectic. Therefore, **canonical transformations are symplectic.** Indeed, considering the $2n$ Hamiltonians $K_i = Q_i$ for $i = 1, \ldots, n$ and $K_{n+i} = P_i$ for $i = 1, \ldots, n$, shows that for each of the $2n$ columns of the respective matrices in Eq. (1.162), we must have $(\text{Jac}(\mathcal{M}) \cdot \vec{J} \cdot \text{Jac}(\mathcal{M})^t) = \hat{J}$.

The symplectic condition can be cast in a **variety of forms**. Denoting $\hat{M} = \text{Jac}(\mathcal{M})$, we arrived at the form

$$\hat{M} \cdot \hat{J} \cdot \hat{M}^t = \hat{J}, \qquad (1.165)$$

which is also probably the most common form. Multiplying with $\hat{J}^{-1} = -\hat{J}$ from the right, we see that

$$\hat{M}^{-1} = -\hat{J} \cdot \hat{M}^t \cdot \hat{J}, \qquad (1.166)$$

which is a convenient formula for the inverse of the Jacobian. Multiplying with $\hat{J}$ from the right or left, respectively, we obtain

$$\hat{J} \cdot \hat{M}^t = \hat{M}^{-1} \cdot \hat{J} \text{ and } \hat{M}^t \cdot \hat{J} = \hat{J} \cdot \hat{M}^{-1}. \qquad (1.167)$$

Transposing these equations yields

$$\hat{M} \cdot \hat{J} = \hat{J} \cdot (\hat{M}^{-1})^t \text{ and } \hat{J} \cdot \hat{M} = (\hat{M}^{-1})^t \cdot \hat{J}, \qquad (1.168)$$

while multiplying the second equation in Eq. (1.167) with $\hat{M}$ from the right, we have

$$\hat{M}^t \cdot \hat{J} \cdot \hat{M} = \hat{J}. \qquad (1.169)$$

There is actually a high degree of symmetry in these equations, which also greatly simplifies memorization: All equations stay valid if $\hat{J}$ or $\hat{M}$ are replaced by their transpose.

Properties of Symplectic Maps

It is worthwhile to study some of the properties of symplectic maps in detail. First, we observe that symplectic diffeomorphisms form a **group** under composition. The identity map is clearly symplectic. Furthermore, for a given symplectic diffeomorphism $\mathcal{M}$, we have by virtue of Eq. (1.166) that

$$\mathrm{Jac}(\mathcal{M}^{-1}) = \mathrm{Jac}(\mathcal{M})^{-1} = -\hat{J} \cdot \mathrm{Jac}(\mathcal{M}^t) \cdot \hat{J}, \qquad (1.170)$$

and we obtain

$$\begin{aligned}
& \mathrm{Jac}(\mathcal{M}^{-1}) \cdot \hat{J} \cdot \mathrm{Jac}(\mathcal{M}^{-1})^t \\
&= (-\hat{J} \cdot \mathrm{Jac}(\mathcal{M})^t \cdot \hat{J}) \cdot \hat{J} \cdot (-\hat{J} \cdot \mathrm{Jac}(\mathcal{M})^t \cdot \hat{J})^t \\
&= -\hat{J} \cdot \mathrm{Jac}(\mathcal{M})^t \cdot \hat{J} \cdot \mathrm{Jac}(\mathcal{M}) \cdot \hat{J} \\
&= -\hat{J} \cdot \hat{J} \cdot \hat{J} = \hat{J},
\end{aligned} \qquad (1.171)$$

which shows that the inverse of $\mathcal{M}$ is symplectic. Finally, let $\mathcal{M}_1$ and $\mathcal{M}_2$ be symplectic maps; then we have $\mathrm{Jac}(\mathcal{M}_1 \circ \mathcal{M}_2) = \mathrm{Jac}(\mathcal{M}_1) \cdot \mathrm{Jac}(\mathcal{M}_2)$, and so

$$\begin{aligned}
& \mathrm{Jac}(\mathcal{M}_1 \circ \mathcal{M}_2) \cdot \hat{J} \cdot \mathrm{Jac}(\mathcal{M}_1 \circ \mathcal{M}_2)^t \\
&= \mathrm{Jac}(\mathcal{M}_1) \cdot \mathrm{Jac}(\mathcal{M}_2) \cdot \hat{J} \cdot \mathrm{Jac}(\mathcal{M}_2)^t \cdot \mathrm{Jac}(\mathcal{M}_1)^t \\
&= \mathrm{Jac}(\mathcal{M}_1) \cdot \hat{J} \cdot \mathrm{Jac}(\mathcal{M}_1)^t = \hat{J},
\end{aligned} \qquad (1.172)$$

where the symplecticity of $\mathcal{M}_1$ and $\mathcal{M}_2$ has been used.

A particularly important class of symplectic maps are the linear symplectic transformations, which form a subgroup of the symplectic maps. In this case, the Jacobian of the map is constant everywhere and just equals the matrix of the linear transformation, and the composition of maps corresponds to multiplication of these matrices. This **group** of $2n \times 2n$ **symplectic matrices** is usually denoted by $Sp(2n)$.

In a similar way as the orthogonal transformations preserve the scalar product of vectors, the symplectic matrices in $Sp(2n)$ preserve the **antisymmetric scalar product** of the column vectors $\vec{F}$ and $\vec{G}$, defined by

$$\langle \vec{F}, \vec{G} \rangle = \vec{F}^t \cdot \hat{J} \cdot \vec{G}. \qquad (1.173)$$

Indeed, for any matrix $\hat{A} \in Sp(2n)$, we have

$$\langle \hat{A} \cdot \vec{F}, \hat{A} \cdot \vec{G} \rangle = \left(\hat{A} \cdot \vec{F} \right)^t \cdot \hat{J} \cdot \left(\hat{A} \cdot \vec{G} \right) \qquad (1.174)$$

$$= \vec{F}^t \cdot \left(\hat{A}^t \cdot \hat{J} \cdot \hat{A} \right) \cdot \vec{G} = \vec{F}^t \cdot \hat{J} \cdot \vec{G} = \langle \vec{F}, \vec{G} \rangle. \qquad (1.175)$$

Below, we will recognize that the antisymmetric scalar product is just a special case of the Poisson bracket for the case of linear functions.

It is also interesting to study the **determinant** of symplectic matrices. Because $\det(\hat{J}) = 1$, we have from the symplectic condition that $1 = \det(\hat{J}) = \det(\text{Jac}(\mathcal{M}) \cdot \hat{J} \cdot \text{Jac}(\mathcal{M})^t) = \det(\text{Jac}(\mathcal{M}))^2$, and so

$$\det(\text{Jac}(\mathcal{M})) = \pm 1. \tag{1.176}$$

However, we can show that the determinant is always $+1$. This is connected to a quite remarkable and useful property of the antisymmetric scalar product.

Let us consider $2n$ arbitrary vectors in phase space $\vec{z}^{(j)}$, $j = 1, \ldots, 2n$. Let us arrange these $2n$ vectors into the columns of a matrix $\hat{Z}$, so that

$$\hat{Z} = \begin{pmatrix} z_1^{(1)} & & z_1^{(2n)} \\ & \ddots & \\ z_{2n}^{(1)} & & z_{2n}^{(2n)} \end{pmatrix}. \tag{1.177}$$

Let π denote a permutation of $\{1, \ldots, 2n\}$ and $\sigma(\pi)$ its signature. We now study the expression

$$C = \sum_{\pi} \sigma(\pi) \cdot \langle \vec{z}^{\pi(1)}, \vec{z}^{\pi(2)} \rangle \cdot \ldots \cdot \langle \vec{z}^{\pi(2n-1)}, \vec{z}^{\pi(2n)} \rangle. \tag{1.178}$$

Performing all products explicitly using the matrix elements $\hat{J}_{i,j}$ of the matrix $\hat{J}$, we obtain

$$C = \sum_{i_1, \ldots, i_{2n}} \hat{J}_{i_1, i_2} \cdot \hat{J}_{i_3, i_4} \cdot \ldots \cdot \hat{J}_{i_{2n-1}, i_{2n}} \cdot \sum_{\pi} \sigma(\pi) \cdot z_{i_1}^{\pi(1)} z_{i_2}^{\pi(2)} \cdots z_{i_{2n}}^{\pi(2n)} \tag{1.179}$$

Observe that the last term is the determinant of the matrix obtained from $\hat{Z}$ by rearranging the rows in the order $i_1, i_2, \ldots, i_{2n}$. Because this determinant vanishes if any two rows are equal, only those choices $i_1, i_2, \ldots, i_{2n}$ contribute where the numbers are pairwise distinct, and hence a permutation $\tilde{\pi}$ of $\{1, \ldots, 2n\}$. In this case, the value of the last product equals $\sigma(\tilde{\pi}) \cdot \det(\hat{Z})$, and we have

$$C = \left(\sum_{\tilde{\pi}} \hat{J}_{\tilde{\pi}(1), \tilde{\pi}(2)} \cdot \hat{J}_{\tilde{\pi}(3), \tilde{\pi}(4)} \cdot \ldots \cdot \hat{J}_{\tilde{\pi}(2n-1), \tilde{\pi}(2n)} \cdot \sigma(\tilde{\pi}) \right) \cdot \det(\hat{Z})$$

$$= K_n \cdot \det(\hat{Z}) \tag{1.180}$$

Let us now study the sum K_n, which is independent of the vectors $\vec{z}^{(j)}$. To obtain a contribution to the sum, all factors $\hat{J}_{\nu, \mu}$ must be nonzero, which requires $\mu = \nu \pm n$. One such contribution is

$$((i_1, i_2), (i_3, i_4), \ldots, (i_{2n-1, 2n})) = ((1, n+1), (2, n+2), \ldots, (n, 2n)), \tag{1.181}$$

which has a signature of $(-1)^{n \cdot (n-1)/2}$. It corresponds to the case in which for all n factors $\mu = \nu + n$ and hence the product of the J matrix elements is 1. The other contributions are obtained in the following way. First, it is possible to exchange entire subpairs, which does not affect the signature and still yields a product of $\hat{J}$ matrix elements of 1; there apparently are $n!$ ways for such an exchange. Secondly, it is possible to flip two entries within a subpair $(j, j+n)$, each of which changes both the signature and the sign of the product of the $\hat{J}$ matrix elements, thus still yielding the contribution 1; there are apparently 2^n ways for such a flip. Altogether, we have $2^n \cdot n!$ permutations that yield contributions, all of which have the same magnitude $(-1)^{n \cdot (n-1)/2}$ as the original one, and we thus obtain

$$K_n = (-1)^{n \cdot (n-1)/2} \cdot 2^n \cdot n! \qquad (1.182)$$

Therefore, the antisymmetric scalar product allows for the determination of the value of a determinant merely by subsequent scalar multiplications of the columns of the matrix.

While interesting in its own right, this fact can be used as a key to our argument since the antisymmetric scalar product is invariant under a symplectic transformation. Let $\hat{M}$ be such a transformation. For $\hat{Z}$ we choose the special case of the unit matrix, and argue as follows:

$$\det(\hat{M}) = \det(\hat{M} \cdot \hat{Z}) = \det\left(\hat{M} \cdot \vec{z}_1, \ldots, \hat{M} \cdot \vec{z}_{2n}\right)$$

$$= \sum_\pi \frac{\sigma(\pi)}{K_n} \cdot \langle \hat{M} \cdot \vec{z}^{\pi(1)}, \hat{M} \cdot \vec{z}^{\pi(2)} \rangle \cdots \langle \hat{M} \cdot \vec{z}^{\pi(2n-1)}, \hat{M} \cdot \vec{z}^{\pi(2n)} \rangle$$

$$= \sum_\pi \frac{\sigma(\pi)}{K_n} \cdot \langle \vec{z}^{\pi(1)}, \vec{z}^{\pi(2)} \rangle \cdots \langle \vec{z}^{\pi(2n-1)}, \vec{z}^{\pi(2n)} \rangle$$

$$= \det(\hat{Z}) = 1. \qquad (1.183)$$

A very important consequence of the fact that the determinant of the Jacobian of a symplectic map is unity is that this means that the map **preserves volume** in phase space. We study a measurable set S_1 of phase space, and let $S_2 = \mathcal{M}(S_1)$ be the set that is obtained by sending S_1 through the symplectic map $\mathcal{M}$. Then the volumes of S_1 and S_2, denoted by V_1 and V_2, are equal. In fact, according to the substitution rule of multidimensional integration, we have

$$V_2 = \int_{S_2} d^n \vec{Q} \, d^n \vec{P}$$

$$= \int_{S_1} \det(\text{Jac}(\mathcal{M})) \, d^n \vec{q} \, d^n \vec{p}$$

$$= \int_{S_1} d^n \vec{q} \, d^n \vec{p} = V_1. \qquad (1.184)$$

Poisson Brackets and Symplecticity

Let f and g be observables, i.e., differentiable functions from phase space into R. We define the **Poisson bracket** of f and g as

$$[f, g] = (\partial f/\partial \vec{q}, \partial f/\partial \vec{p}) \cdot \hat{J} \cdot (\partial g/\partial \vec{q}, \partial g/\partial \vec{p})^t$$
$$= \sum_{i=1}^{n} \left(\frac{\partial f}{\partial q_i} \frac{\partial g}{\partial p_i} - \frac{\partial f}{\partial p_i} \frac{\partial g}{\partial q_i} \right). \tag{1.185}$$

For example, consider the special choices for the functions f and g of the forms q_i and p_j. We obtain that for all $i, j = 1, \ldots, n$,

$$[q_i, q_j] = 0, \quad [q_i, p_j] = \delta_{ij}, \quad [p_i, p_j] = 0. \tag{1.186}$$

Among other uses, the Poisson brackets allow a convenient representation of the **vector field** (Eq. 1.7) of a Hamiltonian system. Let $g(\vec{z}, t)$ be a function of phase space; according to Eqs. (1.91) and (1.185) we obtain for the vector field $L_{\vec{f}}$ that

$$L_{\vec{f}} g = \dot{g} = \sum_{i=1}^{n} \left(\frac{\partial g}{\partial q_i} \dot{q}_i + \frac{\partial g}{\partial p_i} \dot{p}_i \right) + \frac{\partial g}{\partial t}$$
$$= \sum_{i=1}^{n} \left(\frac{\partial g}{\partial q_i} \frac{\partial H}{\partial p_i} - \frac{\partial g}{\partial p_i} \frac{\partial H}{\partial q_i} \right) + \frac{\partial g}{\partial t}$$
$$= [g, H] + \frac{\partial g}{\partial t}. \tag{1.187}$$

We observe in passing that replacing g by q_i and p_i, $i = 1, \ldots, n$, we recover Hamilton's equations of motion, namely $\dot{q}_i = \partial H/\partial p_i$ and $\dot{p}_i = -\partial H/\partial q_i$.

Introducing $: h :$ as a "Poisson bracket waiting to happen," i.e., as an operator on the space of functions $g(\vec{q}, \vec{p}, t)$, that acts as $: h : g = [h, g]$, the vector field can also be written as

$$L_{\vec{f}} = - : H : + \partial/\partial t. \tag{1.188}$$

Above, we recognized that the antisymmetric scalar product is a special case of the Poisson bracket when the functions f and g are linear. In this case, we can write $f(\vec{z}) = \vec{F}^t \cdot \vec{z}$ and $g(\vec{z}) = \vec{G}^t \cdot \vec{z}$, and

$$[f, g] = \vec{F}^t \cdot \hat{J} \cdot \vec{G}. \tag{1.189}$$

Similar to the way in which the symplectic matrices preserve the antisymmetric product, **symplectic maps preserve Poisson brackets.** By this we mean that if

$\mathcal{M}$ is a symplectic map, then for any observables, i.e., differentiable functions of phase space f and g, we have

$$[f \circ \mathcal{M}, g \circ \mathcal{M}] = [f, g] \circ \mathcal{M}. \tag{1.190}$$

This fact follows quite readily from the chain rule and the symplecticity of $\mathcal{M}$:

$$\begin{aligned}
[f \circ \mathcal{M}, g \circ \mathcal{M}] &= \vec{\nabla}(f \circ \mathcal{M}) \cdot \hat{J} \cdot (\vec{\nabla}(g \circ \mathcal{M}))^t \\
&= ((\vec{\nabla} f) \circ \mathcal{M}) \cdot \mathrm{Jac}(\mathcal{M}) \cdot \hat{J} \cdot \mathrm{Jac}(\mathcal{M})^t \cdot ((\vec{\nabla} g) \circ \mathcal{M})^t \\
&= ((\vec{\nabla} f) \circ \mathcal{M}) \cdot \hat{J} \cdot ((\vec{\nabla} g) \circ \mathcal{M})^t = [f, g] \circ \mathcal{M}. \tag{1.191}
\end{aligned}$$

In particular, this entails that if we take for f and g the components q_i and p_i and write the symplectic map as $\mathcal{M} = (\mathcal{Q}, \mathcal{P})$, then we see from Eq. (1.186) that

$$[\mathcal{Q}_i, \mathcal{Q}_j] = 0, \quad [\mathcal{Q}_i, \mathcal{P}_j] = \delta_{ij}, \quad [\mathcal{P}_i, \mathcal{P}_j] = 0 \tag{1.192}$$

for all $i, j = 1, \ldots, n$. This is the same structure as Eq. (1.186), and thus we speak of the preservation of the so-called elementary Poisson brackets. Conversely, we show now that **preservation of all elementary Poisson brackets implies symplecticity.**

Let $\mathcal{M} = (\mathcal{Q}, \mathcal{P})$ be an arbitrary diffeomorphism on phase space; then we apparently have

$$\begin{aligned}
[f \circ \mathcal{M}, g \circ \mathcal{M}] &= \vec{\nabla}(f \circ \mathcal{M}) \cdot \hat{J} \cdot (\vec{\nabla}(g \circ \mathcal{M}))^t \\
&= ((\vec{\nabla} f) \circ \mathcal{M}) \cdot \mathrm{Jac}(\mathcal{M}) \cdot \hat{J} \cdot \mathrm{Jac}(\mathcal{M})^t \\
&\quad \cdot ((\vec{\nabla} g) \circ \mathcal{M})^t \tag{1.193}
\end{aligned}$$

Let $\mathcal{M}$ preserve the elementary Poisson brackets. Now we consider the $2n \cdot 2n$ cases of choosing q_i and p_j for both f and g. We observe that $q_i \circ \mathcal{M} = \mathcal{Q}_i$ and $p_i \circ \mathcal{M} = \mathcal{P}_i$. Furthermore, $\vec{\nabla} q_i$ and $\vec{\nabla} p_i$ are unit vectors with a 1 in coordinate i and $i + n$, respectively, and 0 everywhere else. Each of the $2n \cdot 2n$ choices for f and g, hence, project out a different matrix element of $\mathrm{Jac}(\mathcal{M}) \cdot \hat{J} \cdot \mathrm{Jac}(\mathcal{M})^t$ on the right of the equation, while on the left we have the respective elementary Poisson brackets. Since these are preserved by assumption, we have

$$\mathrm{Jac}(\mathcal{M}) \cdot \hat{J} \cdot \mathrm{Jac}(\mathcal{M})^t = \hat{J} \tag{1.194}$$

and the map $\mathcal{M}$ is symplectic.

To finalize our discussion of the Poisson bracket, we easily see from Eq. (1.185) that for all observables f and g, we have

$$[f, g] = -[g, f] \tag{1.195}$$

and

$$[f+g, h] = [f, h] + [g, h] \text{ and } [c \cdot f, g] = c \cdot [f, g] \tag{1.196}$$

Thus [,] is an antisymmetric bilinear form. Moreover, [,] satisfies the **Jacobi identity**

$$[f, [g, h]] + [g, [h, f]] + [h, [f, g]] = 0 \text{ for all } f, g, h, \tag{1.197}$$

which follows from a straightforward but somewhat involved computation. Altogether, the space of functions on phase space with multiplication $f \odot g = [f, g]$, forms a **Lie Algebra**.

1.4.6 Universal Existence of Generating Functions

After having studied some of the properties of canonical transformations, in particular their symplecticity and the preservation of the Poisson bracket, we now address a **representation** of the canonical transformation via a so-called **generating function**. Later, we will also show that any generating function yields a canonical transformation without having to impose further conditions; thus, generating functions provide a convenient and exhaustive representation of canonical transformations. We begin by restating the symplectic condition in one of its equivalent forms (Eq. 1.167):

$$\hat{J} \cdot \hat{M}^t = \hat{M}^{-1} \cdot \hat{J} \tag{1.198}$$

Writing out the equations reads

$$\begin{pmatrix} 0 & \hat{I} \\ -\hat{I} & 0 \end{pmatrix} \cdot \begin{pmatrix} (\partial \vec{Q}/\partial \vec{q})^t & (\partial \vec{P}/\partial \vec{q})^t \\ (\partial \vec{Q}/\partial \vec{p})^t & (\partial \vec{P}/\partial \vec{p})^t \end{pmatrix}$$
$$= \begin{pmatrix} (\partial \vec{q}/\partial \vec{Q}) & (\partial \vec{q}/\partial \vec{P}) \\ (\partial \vec{p}/\partial \vec{Q}) & (\partial \vec{p}/\partial \vec{P}) \end{pmatrix} \cdot \begin{pmatrix} 0 & \hat{I} \\ -\hat{I} & 0 \end{pmatrix} \tag{1.199}$$

and so

$$\begin{pmatrix} (\partial \vec{Q}/\partial \vec{p})^t & (\partial \vec{P}/\partial \vec{p})^t \\ -(\partial \vec{Q}/\partial \vec{q})^t & -(\partial \vec{P}/\partial \vec{q})^t \end{pmatrix} = \begin{pmatrix} -(\partial \vec{q}/\partial \vec{P}) & (\partial \vec{q}/\partial \vec{Q}) \\ -(\partial \vec{p}/\partial \vec{P}) & (\partial \vec{p}/\partial \vec{Q}) \end{pmatrix}. \tag{1.200}$$

Writing the conditions expressed by the submatrices explicitly yields

$$\frac{\partial Q_j}{\partial p_i} = -\frac{\partial q_i}{\partial P_j}, \quad \frac{\partial P_j}{\partial p_i} = \frac{\partial q_i}{\partial Q_j}$$
$$\frac{\partial Q_j}{\partial q_i} = \frac{\partial p_i}{\partial P_j}, \quad \frac{\partial P_j}{\partial q_i} = -\frac{\partial p_i}{\partial Q_j}. \tag{1.201}$$

At first glance, these equations may appear to be a strange mixture of partials of the coordinate transformation diffeomorphism and its inverse. First, a closer inspection shows that they are all related; the upper right condition follows from the upper left if we subject the map under consideration to the elementary canonical transformation $(\vec{Q}, \vec{P}) \to (\vec{P}, -\vec{Q})$, and the lower row follows from the upper one by the transformation of the initial conditions via $(\vec{q}, \vec{p}) \to (\vec{p}, -\vec{q})$.

Because of this fact we may restrict our attention to only one of them, the lower right condition. This relationship can be viewed as a condition on a map $\mathcal{G}$ in which the final and initial momenta $\vec{P}$ and $\vec{p}$ are expressed in terms of the final and initial positions $\vec{Q}$ and $\vec{q}$ via

$$\begin{pmatrix} \vec{P} \\ \vec{p} \end{pmatrix} = \begin{pmatrix} \mathcal{G}_{\vec{P}}(\vec{Q}, \vec{q}) \\ \mathcal{G}_{\vec{p}}(\vec{Q}, \vec{q}) \end{pmatrix} = \begin{pmatrix} \mathcal{G}_{\vec{P}} \\ \mathcal{G}_{\vec{p}} \end{pmatrix} \begin{pmatrix} \vec{Q} \\ \vec{q} \end{pmatrix}, \tag{1.202}$$

if such a map $\mathcal{G}$ exists. In this case, the conditions are reminiscent of the condition for the existence of a potential (Eq. 1.300). Indeed, let us assume that

$$\frac{\partial P_j}{\partial q_i} = -\frac{\partial p_i}{\partial Q_j}, \tag{1.203}$$

and

$$\frac{\partial P_j}{\partial Q_i} = \frac{\partial P_i}{\partial Q_j} \tag{1.204}$$

and

$$\frac{\partial p_j}{\partial q_i} = \frac{\partial p_i}{\partial q_j}. \tag{1.205}$$

Setting $q_i^* = -q_i$, Eqs. (1.203), (1.204), and (1.205) apparently represent integrability conditions (1.300) asserting the existence of a function $F^*(\vec{Q}, \vec{q}^*)$ such that $\mathcal{G}_{P,i} = \partial F^*/\partial Q_i$ and $\mathcal{G}_{p,i} = \partial F^*/\partial q_i^*$. Introducing the function of the original variables $F(\vec{Q}, \vec{q}) = F^*(\vec{Q}, -\vec{q}^*)$, we obtain

$$\mathcal{G}_{P,i} = \frac{\partial F}{\partial Q_i} \text{ and } \mathcal{G}_{p,i} = -\frac{\partial F}{\partial q_i}. \tag{1.206}$$

In passing we note that the minus sign appearing in the right equation is immaterial, as the choice of $F(\vec{Q}, \vec{q}) = -F^*(\vec{Q}, -\vec{q}^*)$ would lead to a minus sign in the left equation; clearly, however, one of the equations has to have a minus sign.

This representation of the canonical transformation by a single function is reminiscent of the fact that every canonical dynamical system can be represented by a single Hamiltonian function H. However, here the representation is somewhat indirect in the sense that old and new coordinates are mixed.

Of course the map $\mathcal{G}$ mentioned previously **need not exist a priori**; for example, if the canonical transformation under consideration is simply the identity map, it is not possible to find such a $\mathcal{G}$. However, in the following we establish that $\mathcal{G}$ and **a generating function always exists**, as long as the underlying map is first subjected to a suitable combination of elementary canonical transformations.

Let $\mathcal{M}$ be a canonical transformation, let $\mathcal{Q}$ and $\mathcal{P}$ denote the position and momentum parts of $\mathcal{M}$, and let $\mathcal{I}_q$ and $\mathcal{I}_p$ denote the position and momentum parts of the identity map. Then, we have

$$\begin{pmatrix} \vec{Q} \\ \vec{P} \end{pmatrix} = \begin{pmatrix} \mathcal{Q}(\vec{q}, \vec{p}) \\ \mathcal{P}(\vec{q}, \vec{p}) \end{pmatrix} = \begin{pmatrix} \mathcal{Q} \\ \mathcal{P} \end{pmatrix} \begin{pmatrix} \vec{q} \\ \vec{p} \end{pmatrix} \qquad (1.207)$$

and hence

$$\begin{pmatrix} \vec{Q} \\ \vec{q} \end{pmatrix} = \begin{pmatrix} \mathcal{Q} \\ \mathcal{I}_q \end{pmatrix} \begin{pmatrix} \vec{q} \\ \vec{p} \end{pmatrix}. \qquad (1.208)$$

If the map $(\mathcal{Q}, \mathcal{I}_q)^t$ can be inverted, we have

$$\begin{pmatrix} \vec{q} \\ \vec{p} \end{pmatrix} = \begin{pmatrix} \mathcal{Q} \\ \mathcal{I}_q \end{pmatrix}^{-1} \begin{pmatrix} \vec{Q} \\ \vec{q} \end{pmatrix}, \qquad (1.209)$$

from which we then obtain the desired relationship

$$\begin{pmatrix} \vec{P} \\ \vec{p} \end{pmatrix} = \begin{pmatrix} \mathcal{P} \\ \mathcal{I}_p \end{pmatrix} \circ \begin{pmatrix} \mathcal{Q} \\ \mathcal{I}_q \end{pmatrix}^{-1} \begin{pmatrix} \vec{Q} \\ \vec{q} \end{pmatrix} \qquad (1.210)$$

which expresses the old and new momenta in terms of the old and new positions.

First, we have to establish the **invertibility condition** for the map $(\mathcal{Q}, \mathcal{I}_q)^t$. According to the implicit function theorem, this is possible in even a neighborhood of a point as long as the Jacobian does not vanish at the point. Since the Jacobian of $\mathcal{I}_q$ contains the identity matrix in the first n columns and the zero matrix in the second n columns, we conclude that

$$\det\left[\mathrm{Jac}\begin{pmatrix} \mathcal{Q} \\ \mathcal{I}_q \end{pmatrix}\right] = (-1)^n \cdot \begin{vmatrix} \partial \mathcal{Q}_1/\partial p_1 & \cdots & \partial \mathcal{Q}_1/\partial p_n \\ \vdots & & \vdots \\ \partial \mathcal{Q}_n/\partial p_1 & \cdots & \partial \mathcal{Q}_n/\partial p_n \end{vmatrix}. \qquad (1.211)$$

The determinant on the right is denoted by D. In general, it cannot be ensured that $D \neq 0$ for a given canonical transformation. In fact, as already mentioned, in the case of the identity transformation the final $\vec{Q}$ does not depend on the initial $\vec{p}$, and the determinant will certainly be zero. However, as we now show, for any given point $\vec{z}$ in phase space it is possible to subject the coordinates to a sequence of elementary canonical transformations discussed previously, consisting only of rearrangements of the pairs (q_i, p_i) of the coordinates and of exchanges of (q_i, p_i) with $(p_i, -q_i)$, such that the **determinant is nonzero** at $\vec{z}$.

We prove this statement by **induction**. As the first step, we observe that since $[\mathcal{Q}_1, \mathcal{P}_1] = 1$ because of symplecticity of $\mathcal{M}$, not all partial derivatives of $\mathcal{Q}_1$ with respect to q_i, p_i can vanish. We then **rearrange** the variables $(\vec{q}, \vec{p})$ and possibly perform an **exchange** of q_1 and p_1, such that $\partial \mathcal{Q}_1 / \partial p_1 \neq 0$.

We now assume the statement is true for $m < n$; i.e., the subdeterminant D_m defined by

$$D_m = \begin{vmatrix} \partial\mathcal{Q}_1/\partial p_1 & \cdots & \partial\mathcal{Q}_1/\partial p_m \\ \vdots & & \vdots \\ \partial\mathcal{Q}_m/\partial p_1 & \cdots & \partial\mathcal{Q}_m/\partial p_m \end{vmatrix} \tag{1.212}$$

satisfies $D_m \neq 0$ for all $m < n$. We then show that, after suitable elementary canonical transformations, we will also have $D_{m+1} \neq 0$. To this end, we consider all $2n - m$ determinants

$$D_{m+1}^{(p)} = \begin{vmatrix} \partial\mathcal{Q}_1/\partial p_1 & \cdots & \partial\mathcal{Q}_1/\partial p_m & \partial\mathcal{Q}_1/\partial p_k \\ \vdots & & \vdots & \vdots \\ \partial\mathcal{Q}_m/\partial p_1 & \cdots & \partial\mathcal{Q}_m/\partial p_m & \partial\mathcal{Q}_m/\partial p_k \\ \partial\mathcal{Q}_{m+1}/\partial p_1 & \cdots & \partial\mathcal{Q}_{m+1}/\partial p_m & \partial\mathcal{Q}_{m+1}/\partial p_k \end{vmatrix}, \tag{1.213}$$

$$D_{m+1}^{(q)} = \begin{vmatrix} \partial\mathcal{Q}_1/\partial p_1 & \cdots & \partial\mathcal{Q}_1/\partial p_m & \partial\mathcal{Q}_1/\partial q_k \\ \vdots & & \vdots & \vdots \\ \partial\mathcal{Q}_m/\partial p_1 & \cdots & \partial\mathcal{Q}_m/\partial p_m & \partial\mathcal{Q}_m/\partial q_k \\ \partial\mathcal{Q}_{m+1}/\partial p_1 & \cdots & \partial\mathcal{Q}_{m+1}/\partial p_m & \partial\mathcal{Q}_{m+1}/\partial q_k \end{vmatrix}, \tag{1.214}$$

where the index k runs from $m + 1$ to n. Now the goal is to show that at least one of these is nonzero. If this is the case, then by performing a reordering among the q_k and p_k and possibly an exchange of q_k and p_k, indeed $D_{m+1}^{(p)}$ is nonzero for $k = m + 1$.

To find this matrix with nonvanishing determinant, we consider the $(m+1) \times 2n$ matrix

$$\hat{A} = \begin{pmatrix} \frac{\partial\mathcal{Q}_1}{\partial p_1} & \cdots & \frac{\partial\mathcal{Q}_1}{\partial p_m} & \frac{\partial\mathcal{Q}_1}{\partial p_{m+1}} & \cdots & \frac{\partial\mathcal{Q}_1}{\partial p_n} & \frac{\partial\mathcal{Q}_1}{\partial q_1} & \cdots & \frac{\partial\mathcal{Q}_1}{\partial q_n} \\ \vdots & & \vdots & \vdots & & \vdots & \vdots & & \vdots \\ \frac{\partial\mathcal{Q}_m}{\partial p_1} & \cdots & \frac{\partial\mathcal{Q}_m}{\partial p_m} & \frac{\partial\mathcal{Q}_m}{\partial p_{m+1}} & \cdots & \frac{\partial\mathcal{Q}_m}{\partial p_n} & \frac{\partial\mathcal{Q}_m}{\partial q_1} & \cdots & \frac{\partial\mathcal{Q}_m}{\partial q_n} \\ \frac{\partial\mathcal{Q}_{m+1}}{\partial p_1} & \cdots & \frac{\partial\mathcal{Q}_{m+1}}{\partial p_m} & \frac{\partial\mathcal{Q}_{m+1}}{\partial p_{m+1}} & \cdots & \frac{\partial\mathcal{Q}_{m+1}}{\partial p_n} & \frac{\partial\mathcal{Q}_{m+1}}{\partial q_1} & \cdots & \frac{\partial\mathcal{Q}_{m+1}}{\partial q_n} \end{pmatrix}. \tag{1.215}$$

The rank of the matrix $\hat{A}$ must then be $m + 1$ since it consists of the upper $m + 1$ rows of the Jacobian of the canonical transformation, which is known to have

48 DYNAMICS OF PARTICLES AND FIELDS

determinant one. Now consider the $(m+1) \times (2n-m)$ matrix $\hat{A}^*$ obtained by striking from $\hat{A}$ the m columns containing derivatives with respect to q_k for $k = 1, \ldots, m$. So $\hat{A}^*$ has the form

$$\hat{A}^* = \begin{pmatrix} \frac{\partial \mathcal{Q}_1}{\partial p_1} & \cdots & \frac{\partial \mathcal{Q}_1}{\partial p_m} & \frac{\partial \mathcal{Q}_1}{\partial p_{m+1}} & \cdots & \frac{\partial \mathcal{Q}_1}{\partial p_n} & \frac{\partial \mathcal{Q}_1}{\partial q_{m+1}} & \cdots & \frac{\partial \mathcal{Q}_1}{\partial q_n} \\ \vdots & & \vdots & \vdots & & \vdots & \vdots & & \vdots \\ \frac{\partial \mathcal{Q}_m}{\partial p_1} & \cdots & \frac{\partial \mathcal{Q}_m}{\partial p_m} & \frac{\partial \mathcal{Q}_m}{\partial p_{m+1}} & \cdots & \frac{\partial \mathcal{Q}_m}{\partial p_n} & \frac{\partial \mathcal{Q}_m}{\partial q_{m+1}1} & \cdots & \frac{\partial \mathcal{Q}_m}{\partial q_n} \\ \frac{\partial \mathcal{Q}_{m+1}}{\partial p_1} & \cdots & \frac{\partial \mathcal{Q}_{m+1}}{\partial p_m} & \frac{\partial \mathcal{Q}_{m+1}}{\partial p_{m+1}} & \cdots & \frac{\partial \mathcal{Q}_{m+1}}{\partial p_n} & \frac{\partial \mathcal{Q}_{m+1}}{\partial q_{m+1}1} & \cdots & \frac{\partial \mathcal{Q}_{m+1}}{\partial q_n} \end{pmatrix}.$$

(1.216)

We now show that even the matrix $\hat{A}^*$ has rank $m+1$. We proceed indirectly: Assume that the rank of $\hat{A}^*$ is not $m+1$. Because of $D_m \neq 0$, it must have rank m, and furthermore its last row can be expressed as a linear combination of the first m rows. This means that there are coefficients α_i such that

$$\frac{\partial \mathcal{Q}_{m+1}}{\partial p_j} = \sum_{i=1}^m \alpha_i \cdot \frac{\partial \mathcal{Q}_i}{\partial p_j} \quad \text{for } j = 1, \ldots, n$$

$$\frac{\partial \mathcal{Q}_{m+1}}{\partial q_j} = \sum_{i=1}^m \alpha_i \cdot \frac{\partial \mathcal{Q}_i}{\partial q_j} \quad \text{for } j = m+1, \ldots, n \qquad (1.217)$$

Because of the symplecticity of $\mathcal{M} = (\mathcal{Q}, \mathcal{P})$, Poisson brackets between any two parts of $\mathcal{Q}$ vanish. This entails

$$0 = [\mathcal{Q}_{m+1}, \mathcal{Q}_k] = \sum_{j=1}^n \frac{\partial \mathcal{Q}_{m+1}}{\partial q_j} \frac{\partial \mathcal{Q}_k}{\partial p_j} - \sum_{j=1}^n \frac{\partial \mathcal{Q}_{m+1}}{\partial p_j} \frac{\partial \mathcal{Q}_k}{\partial q_j}$$

$$= \sum_{j=1}^m \frac{\partial \mathcal{Q}_{m+1}}{\partial q_j} \frac{\partial \mathcal{Q}_k}{\partial p_j} + \sum_{j=m+1}^n \sum_{i=1}^m \alpha_i \cdot \frac{\partial \mathcal{Q}_i}{\partial q_j} \frac{\partial \mathcal{Q}_k}{\partial p_j} - \sum_{j=1}^n \sum_{i=1}^m \alpha_i \cdot \frac{\partial \mathcal{Q}_i}{\partial p_j} \frac{\partial \mathcal{Q}_k}{\partial q_j}$$

$$= \sum_{j=1}^m \frac{\partial \mathcal{Q}_{m+1}}{\partial q_j} \frac{\partial \mathcal{Q}_k}{\partial p_j} + \sum_{i=1}^m \alpha_i \cdot [\mathcal{Q}_i, \mathcal{Q}_k] - \sum_{j=1}^m \sum_{i=1}^m \alpha_i \cdot \frac{\partial \mathcal{Q}_i}{\partial q_j} \frac{\partial \mathcal{Q}_k}{\partial p_j}$$

$$= \sum_{j=1}^m \left(\frac{\partial \mathcal{Q}_{m+1}}{\partial q_j} - \sum_{i=1}^m \alpha_i \cdot \frac{\partial \mathcal{Q}_i}{\partial q_j} \right) \frac{\partial \mathcal{Q}_k}{\partial p_j}.$$

Because $D_m \neq 0$, this even implies that

$$\frac{\partial \mathcal{Q}_{m+1}}{\partial q_j} = \sum_{i=1}^m \alpha_i \cdot \frac{\partial \mathcal{Q}_i}{\partial q_j} \quad \text{for } j = 1, \ldots, m. \qquad (1.218)$$

HAMILTONIAN SYSTEMS

Together with Eq. (1.217), this entails that even in the matrix $\hat{A}$, the last row can be expressed in terms of a linear combination of the upper m rows, which in turn means that $\hat{A}$ has rank m. But this is a contradiction, showing that the assumption of $\hat{A}^*$ not having rank $m+1$ is false. Thus, $\hat{A}^*$ has rank $m+1$.

Since the first m columns of $\hat{A}^*$ are linearly independent because $D_m \neq 0$, it is possible to select one of the $2n - m$ rightmost columns that cannot be expressed as a linear combination of the first m columns. We now perform a reordering and possibly a q, p exchange such that this column actually appears as the $(m+1)$st column. Then indeed, the determinant

$$D_{m+1} = \begin{vmatrix} \frac{\partial Q_1}{\partial p_1} & \cdots & \frac{\partial Q_1}{\partial p_m} & \frac{\partial Q_1}{\partial p_{m+1}} \\ \vdots & & \vdots & \vdots \\ \frac{\partial Q_m}{\partial p_1} & \cdots & \frac{\partial Q_m}{\partial p_m} & \frac{\partial Q_m}{\partial p_{m+1}} \\ \frac{\partial Q_{m+1}}{\partial p_1} & \cdots & \frac{\partial Q_{m+1}}{\partial p_m} & \frac{\partial Q_{m+1}}{\partial p_{m+1}} \end{vmatrix} \quad (1.219)$$

is nonzero. This completes the induction, and we conclude that

$$D = \begin{vmatrix} \frac{\partial Q_1}{\partial p_1} & \cdots & \frac{\partial Q_1}{\partial p_n} \\ \vdots & & \vdots \\ \frac{\partial Q_n}{\partial p_1} & \cdots & \frac{\partial Q_n}{\partial p_n} \end{vmatrix} \neq 0, \quad (1.220)$$

asserting that it is possible to determine $\mathcal{G}$ by inversion.

To show that it is possible to represent the $\mathcal{G}_{P,i}$ and the $\mathcal{G}_{p,i}$ via the generating function F as in Eq. (1.206), it is sufficient to show that the $\mathcal{G}_{P,i}$ and the $\mathcal{G}_{p,i}$ satisfy the integrability conditions (1.203), (1.204) and (1.205). To show these conditions, we proceed as follows. From the definition of $\mathcal{M} = (\mathcal{Q}, \mathcal{P})$ and $\mathcal{G}$, we have

$$Q_i = Q_i(\vec{q}, \mathcal{G}_{\vec{p}}(\vec{Q}, \vec{q})). \quad (1.221)$$

Recognizing that both sides are now expressions in the variables $(\vec{Q}, \vec{q})$ and the left side is independent of q_k, we obtain by partial differentiation with respect to q_k the relation

$$0 = \frac{\partial Q_i}{\partial q_k} + \sum_{j=1}^{n} \frac{\partial Q_i}{\partial p_j} \cdot \frac{\partial \mathcal{G}_{p_j}}{\partial q_k} \quad (1.222)$$

for all $i, k = 1, \ldots, n$. Multiplying this relation with $\partial Q_m / \partial p_k$ and summing over k, we have

$$-\sum_{k=1}^{n} \frac{\partial Q_m}{\partial p_k} \cdot \frac{\partial Q_i}{\partial q_k} = \sum_{j,k=1}^{n} \frac{\partial Q_m}{\partial p_k} \cdot \frac{\partial Q_i}{\partial p_j} \cdot \frac{\partial \mathcal{G}_{p_j}}{\partial q_k} \quad (1.223)$$

This relation is modified by interchanging m and i on both sides, and k and j on the right-hand side. We obtain

$$-\sum_{k=1}^{n} \frac{\partial \mathcal{Q}_m}{\partial q_k} \cdot \frac{\partial \mathcal{Q}_i}{\partial p_k} = \sum_{j,k=1}^{n} \frac{\partial \mathcal{Q}_m}{\partial p_k} \cdot \frac{\partial \mathcal{Q}_i}{\partial p_j} \cdot \frac{\partial \mathcal{G}_{p_k}}{\partial q_j} \quad (1.224)$$

for all $i, m = 1, \ldots, n$. We now subtract the two previous equations from each other while observing that because of the symplecticity of $\mathcal{M}$,

$$0 = [\mathcal{Q}_i, \mathcal{Q}_m] = \sum_{k=1}^{n} \frac{\partial \mathcal{Q}_i}{\partial q_k} \frac{\partial \mathcal{Q}_m}{\partial p_k} - \frac{\partial \mathcal{Q}_i}{\partial p_k} \frac{\partial \mathcal{Q}_m}{\partial q_k}$$

and obtain

$$0 = \sum_{j,k=1}^{n} \frac{\partial \mathcal{Q}_m}{\partial p_k} \frac{\partial \mathcal{Q}_i}{\partial p_j} \left(\frac{\partial \mathcal{G}_{p_j}}{\partial q_k} - \frac{\partial \mathcal{G}_{p_k}}{\partial q_j} \right)$$

for all $i, m = 1, \ldots, n$. We now abbreviate

$$\lambda_{ik} = \sum_{j=1}^{n} \frac{\partial \mathcal{Q}_i}{\partial p_j} \left(\frac{\partial \mathcal{G}_{p_j}}{\partial q_k} - \frac{\partial \mathcal{G}_{p_k}}{\partial q_j} \right) \quad (1.225)$$

and write the condition as

$$0 = \sum_{k=1}^{n} \frac{\partial \mathcal{Q}_m}{\partial p_k} \cdot \lambda_{ik}, \quad (1.226)$$

which again holds for all $i, m = 1, \ldots, n$.

Now consider this relationship for any fixed i. Since the functional determinant D in Eq. (1.220) is nonzero, this actually implies that we must have $\lambda_{ik} = 0$ for all $k = 1, \ldots, n$, and hence we have

$$\lambda_{ik} = 0 \text{ for all } i, k = 1, \ldots, n. \quad (1.227)$$

Now we consider the definition of $\lambda_{i,k}$ in Eq. (1.225) and again use the fact that $D \neq 0$, which shows that we must have

$$\frac{\partial \mathcal{G}_{p_j}}{\partial q_k} = \frac{\partial \mathcal{G}_{p_k}}{\partial q_j} \text{ for all } j, k = 1, \ldots, n, \quad (1.228)$$

and hence we have shown Eq. (1.205).

HAMILTONIAN SYSTEMS 51

Now consider the relationship $Q_i = Q_i(\vec{q}, \vec{p})$, which by using $\vec{p} = \mathcal{G}_{\vec{p}}(\vec{Q}, \vec{q})$ can be written as $Q_i = Q_i(\vec{q}, \mathcal{G}_{\vec{p}}(\vec{Q}, \vec{q}))$. Partial differentiation with respect to Q_k shows that for all $i, k = 1, \ldots, n$, we have

$$\delta_{ik} = \sum_{j=1}^{n} \frac{\partial Q_i}{\partial p_j} \frac{\partial \mathcal{G}_{p_j}}{\partial Q_k}. \tag{1.229}$$

We also have $P_k = \mathcal{G}_{P_k}(\vec{Q}, \vec{q})$, which by using $\vec{Q} = Q(\vec{q}, \vec{p})$ can be written as $P_k = \mathcal{G}_{P_k}(Q(\vec{q}, \vec{p}), \vec{q})$. We use this fact to obtain

$$-\delta_{ik} = [\mathcal{P}_k, Q_i] = [\mathcal{G}_{P_k}(Q(\vec{q}, \vec{p}), \vec{q}), \; Q_i(\vec{q}, \vec{p})]$$

$$= \sum_{j=1}^{n} \left\{ \left(\frac{\partial \mathcal{G}_{P_k}}{\partial q_j} + \sum_{l=1}^{n} \frac{\partial \mathcal{G}_{P_k}}{\partial Q_l} \frac{\partial Q_l}{\partial q_j} \right) \frac{\partial Q_i}{\partial p_j} - \sum_{l=1}^{n} \frac{\partial \mathcal{G}_{P_k}}{\partial Q_l} \frac{\partial Q_l}{\partial p_j} \frac{\partial Q_i}{\partial q_j} \right\}$$

$$= \sum_{j=1}^{n} \frac{\partial \mathcal{G}_{P_k}}{\partial q_j} \frac{\partial Q_i}{\partial p_j} + \sum_{l=1}^{n} \frac{\partial \mathcal{G}_{P_k}}{\partial Q_l} \sum_{j=1}^{n} \left(\frac{\partial Q_l}{\partial q_j} \frac{\partial Q_i}{\partial p_j} - \frac{\partial Q_l}{\partial p_j} \frac{\partial Q_i}{\partial q_j} \right)$$

$$= \sum_{j=1}^{n} \frac{\partial \mathcal{G}_{P_k}}{\partial q_j} \frac{\partial Q_i}{\partial p_j} \tag{1.230}$$

for all $i, k = 1, \ldots, n$, where in the last step, use has been made of $[Q_l, Q_i] = 0$. Combining this result and Eq. (1.229) yields

$$0 = \sum_{j=1}^{n} \frac{\partial Q_i}{\partial p_j} \left(\frac{\partial \mathcal{G}_{p_j}}{\partial Q_k} + \frac{\partial \mathcal{G}_{P_k}}{\partial q_j} \right) \tag{1.231}$$

for all $i, k = 1, \ldots, n$. Again, because the functional determinant D does not vanish, this can only be true if

$$\frac{\partial \mathcal{G}_{p_j}}{\partial Q_k} = -\frac{\partial \mathcal{G}_{P_k}}{\partial q_j} \text{ for all } j, k = 1, \ldots, n, \tag{1.232}$$

which is Eq. (1.203).

In a similar manner, we now write $\mathcal{P}_j = \mathcal{G}_{P_j}(Q(\vec{q}, \vec{p}), \vec{q})$ and obtain

$$0 = [\mathcal{P}_i, \mathcal{P}_j] = [\mathcal{P}_i, \mathcal{G}_{P_j}(Q(\vec{q}, \vec{p}), \vec{q})]$$

$$= \sum_{k=1}^{n} \left\{ \frac{\partial \mathcal{P}_i}{\partial q_k} \left(\sum_l \frac{\partial \mathcal{G}_{P_j}}{\partial Q_l} \frac{\partial Q_l}{\partial p_k} \right) - \frac{\partial \mathcal{P}_i}{\partial p_k} \left(\frac{\partial \mathcal{G}_{P_j}}{\partial q_k} + \sum_{l=1}^{n} \frac{\partial \mathcal{G}_{P_j}}{\partial Q_l} \frac{\partial Q_l}{\partial q_k} \right) \right\}$$

$$= -\sum_{k=1}^{n} \frac{\partial \mathcal{P}_i}{\partial p_k} \frac{\partial \mathcal{G}_{P_j}}{\partial q_k} + \sum_{l=1}^{n} \frac{\partial \mathcal{G}_{P_j}}{\partial \mathcal{Q}_l} \sum_{k=1}^{n} \left(\frac{\partial \mathcal{P}_i}{\partial q_k} \frac{\partial \mathcal{Q}_l}{\partial p_k} - \frac{\partial \mathcal{P}_i}{\partial p_k} \frac{\partial \mathcal{Q}_l}{\partial q_k} \right)$$

$$= -\sum_{k=1}^{n} \frac{\partial \mathcal{P}_i}{\partial p_k} \frac{\partial \mathcal{G}_{P_j}}{\partial q_k} - \sum_{l=1}^{n} \frac{\partial \mathcal{G}_{P_j}}{\partial \mathcal{Q}_l} \cdot \delta_{il}$$

for all $i, j = 1, \ldots, n$, where we use $[\mathcal{P}_i, \mathcal{Q}_k] = -\delta_{ik}$. Using Eq. (1.232), this can be written as

$$\frac{\partial \mathcal{G}_{P_j}}{\partial \mathcal{Q}_i} = \sum_{k=1}^{n} \frac{\partial \mathcal{P}_i}{\partial p_k} \frac{\partial \mathcal{G}_{p_k}}{\partial \mathcal{Q}_j}. \tag{1.233}$$

On the other hand, from $\mathcal{G}_{P_i}(\vec{Q}, \vec{q}) = \mathcal{P}_i(\mathcal{G}_{\vec{p}}(\vec{Q}, \vec{q}), \vec{q})$ we obtain after partial differentiation with respect to Q_j that

$$\frac{\partial \mathcal{G}_{P_i}}{\partial \mathcal{Q}_j} = \sum_{k=1}^{n} \frac{\partial \mathcal{P}_i}{\partial p_k} \frac{\partial \mathcal{G}_{p_k}}{\partial \mathcal{Q}_j} \tag{1.234}$$

Comparing these two equations now shows that

$$\frac{\partial \mathcal{G}_{P_i}}{\partial \mathcal{Q}_j} = \frac{\partial \mathcal{G}_{P_j}}{\partial \mathcal{Q}_i} \text{ for all } i, j = 1, \ldots, n \tag{1.235}$$

The equations (1.228), (1.232), and (1.235) now assert the existence of a function $F(\vec{q}, \vec{Q})$ that satisfies the condition

$$\mathcal{G}_{P,i} = \frac{\partial F}{\partial Q_i} \text{ and } \mathcal{G}_{p,i} = -\frac{\partial F}{\partial q_i}. \tag{1.236}$$

Given the functions $\mathcal{G}_{P,i}$ and $\mathcal{G}_{p,i}$, the function F can be calculated by mere integration by virtue of Eq. (1.304). Furthermore, the nonvanishing of the functional determinant D describing the dependence of $\vec{Q}$ on $\vec{p}$ also implies that the determinant of the dependence of $\vec{p}$ on $\vec{Q}$ is nonzero. Therefore, according to the implicit function theorem, the conditions

$$p_i = -\frac{\partial F}{\partial q_i}, \quad P_i = \frac{\partial F}{\partial Q_i} \tag{1.237}$$

can be solved for the coordinate $\vec{Q}$ in a neighborhood, and hence the canonical transformation $(\vec{Q}, \vec{P})$ can be determined.

1.4.7 Flows of Hamiltonian Systems

In the previous sections, we established various necessary and sufficient conditions for a map to be canonical, and we established a standard representation for given canonical transformations. In this and the following section, we address the question of how to obtain canonical transformations.

An important way of generating canonical transformations is via the flow $\mathcal{M}(\vec{r}_0, t)$ of a Hamiltonian system, which describes how initial coordinates $\vec{r}_0$ and t_0 are transformed to later times t via

$$\vec{r}(t) = \mathcal{M}(\vec{r}_0, t). \tag{1.238}$$

As we now show, if H is three times continuously differentiable, the **flow is symplectic** for any choice of t_0 and t, and hence satisfies

$$\mathrm{Jac}(\mathcal{M}) \cdot \hat{J} \cdot \mathrm{Jac}(\mathcal{M})^t = \hat{J} \tag{1.239}$$

To this end, let $\hat{M} = \mathrm{Jac}(\mathcal{M})$. Set

$$\hat{P}(\vec{r}_0, t) = \hat{M}(\vec{r}_0, t) \cdot \hat{J} \cdot \hat{M}^t(\vec{r}_0, t). \tag{1.240}$$

Then, $\hat{P}(\vec{r}_0, t_0) = \hat{J}$. Consider the equations of motion

$$\frac{d}{dt}\vec{r} = \vec{f}(\vec{r}, t), \tag{1.241}$$

which imply

$$\frac{d}{dt}\mathcal{M}(\vec{r}_0, t) = \vec{f}(\mathcal{M}(\vec{r}_0, t), t). \tag{1.242}$$

Now we have

$$\frac{d}{dt}\hat{M}(\vec{r}_0, t) = \frac{d}{dt}\mathrm{Jac}(\mathcal{M}(\vec{r}_0, t)) = \mathrm{Jac}\left(\frac{d}{dt}\mathcal{M}(\vec{r}_0, t)\right) = \mathrm{Jac}(\vec{f}(\mathcal{M}(\vec{r}_0, t), t))$$

$$= \mathrm{Jac}(\vec{f})(\mathcal{M}(\vec{r}_0, t), t) \cdot \mathrm{Jac}(\mathcal{M})(\vec{r}_0, t)$$

$$= \mathrm{Jac}(\vec{f})(\mathcal{M}(\vec{r}_0, t), t) \cdot \hat{M}(\vec{r}_0, t), \tag{1.243}$$

where in the second transformation we have used

$$\frac{d}{dt}\frac{\partial r_i}{\partial r_{0j}} = \sum_{k=1}^{n} \frac{\partial^2 r_i}{\partial r_{0k}\partial r_{0j}}\dot{r}_{0k} + \frac{\partial^2 r_i}{\partial r_{0j}\partial t}$$

$$= \frac{\partial}{\partial r_{0j}}\left(\sum_{k=1}^{n}\frac{\partial r_i}{\partial r_{0k}}\dot{r}_{0k} + \frac{\partial r_i}{\partial t}\right) = \frac{\partial}{\partial r_{0j}}\left(\frac{d}{dt}r_i\right) = \frac{\partial \dot{r}_i}{\partial r_{0j}}.$$

Because of this, it follows that

$$\begin{aligned}\frac{d}{dt}\hat{P}(\vec{r}_0,t) &= \frac{d}{dt}(\hat{\mathcal{M}}(\vec{r}_0,t)\cdot\hat{J}\cdot\hat{\mathcal{M}}^t(\vec{r}_0,t))\\ &= \mathrm{Jac}(\vec{f})(\mathcal{M}(\vec{r}_0,t),t)\cdot\hat{\mathcal{M}}(\vec{r}_0,t)\cdot J\cdot\hat{\mathcal{M}}^t(\vec{r}_0,t)\\ &\quad+\hat{\mathcal{M}}(\vec{r}_0,t)\cdot\hat{J}\cdot\hat{\mathcal{M}}^t(\vec{r}_0,t)\cdot\mathrm{Jac}(\vec{f})^t(\mathcal{M}(\vec{r}_0,t),t)\\ &= \mathrm{Jac}(\vec{f})(\mathcal{M}(\vec{r}_0,t),t)\cdot\hat{P}(\vec{r}_0,t)\\ &\quad+\hat{P}(\vec{r}_0,t)\cdot\mathrm{Jac}(\vec{f})^t(\mathcal{M}(\vec{r}_0,t),t)\end{aligned}\qquad(1.244)$$

Now we utilize that the underlying differential equation is Hamiltonian. We then show that $\mathrm{Jac}(\vec{f})\cdot\hat{J}$ is symmetric. In fact,

$$\vec{f} = \begin{pmatrix}\partial H/\partial\vec{p}\\ -\partial H/\partial\vec{q}\end{pmatrix}\qquad(1.245)$$

and so

$$\begin{aligned}\mathrm{Jac}(\vec{f})\cdot\hat{J} &= \begin{pmatrix}\partial^2 H/\partial\vec{q}\,\partial\vec{p} & \partial^2 H/\partial\vec{p}\,\partial\vec{p}\\ -\partial^2 H/\partial\vec{q}\,\partial\vec{q} & -\partial^2 H/\partial\vec{p}\,\partial\vec{q}\end{pmatrix}\cdot\hat{J}\\ &= \begin{pmatrix}-\partial^2 H/\partial\vec{p}\,\partial\vec{p} & \partial^2 H/\partial\vec{q}\,\partial\vec{p}\\ \partial^2 H/\partial\vec{p}\,\partial\vec{q} & -\partial^2 H/\partial\vec{q}\,\partial\vec{q}\end{pmatrix}\end{aligned}\qquad(1.246)$$

showing symmetry as claimed. Because $\hat{J}^t = -\hat{J}$, we also have that $\mathrm{Jac}(\vec{f})\cdot\hat{J} + \hat{J}\cdot\mathrm{Jac}(\vec{f})^t = 0$. This entails that one solution of the differential equation for $\hat{P}(\vec{r}_0,t)$ that satisfies the necessary initial condition $\hat{P}(\vec{r}_0,t_0) = \hat{J}$ is

$$\hat{P}(\vec{r}_0,t) = \hat{J} \text{ for all } \vec{r}_0, t.\qquad(1.247)$$

However, because of the uniqueness theorem for ordinary differential equations, here applied to the system consisting of the $4n^2$ matrix elements of $\hat{P}$, this is also the only solution, and we have proved what we wanted to show.

Thus, the flows of Hamiltonian systems are symplectic. Since symplectic transformations preserve the volume of phase space according to Eq. (1.184), so does the flow of a Hamiltonian system, which is known as **Liouville's theorem**.

1.4.8 Generating Functions

In Section 1.4.6, it was shown that every canonical transformation and every symplectic map can be represented by a generating function. We now address the converse question, namely, whether **any mixed-variable function** that can be solved for the final variables and that is twice continuously differentiable **provides a canonical transformation**.

Without loss of generality we assume the function has the form $F(\vec{Q}, \vec{q})$; if it is represented in terms of other mixed variables, it can be brought into this form via elementary canonical transformations. We now assume that

$$p_i = -\frac{\partial F}{\partial q_i}, \quad P_i = \frac{\partial F}{\partial Q_i} \quad (1.248)$$

can be solved for $\vec{Q}$ in terms of $(\vec{q}, \vec{p})$ to provide a diffeomorphism

$$\begin{pmatrix} \vec{Q} \\ \vec{P} \end{pmatrix} = \begin{pmatrix} \mathcal{Q}(\vec{q}, \vec{p}) \\ \mathcal{P}(\vec{q}, \vec{p}) \end{pmatrix} = \mathcal{M}\begin{pmatrix} \vec{q} \\ \vec{p} \end{pmatrix} \quad (1.249)$$

that generates a coordinate transformation. Combining this with Eqs. (1.248), we obtain the relationships

$$p_i + F_{q_i}(\mathcal{Q}(\vec{q}, \vec{p}), \vec{q}) = 0 \quad (1.250)$$

$$\mathcal{P}_i(\vec{q}, \vec{p}) - F_{Q_i}(\mathcal{Q}(\vec{q}, \vec{p}), \vec{q}) = 0. \quad (1.251)$$

We note that according to the implicit function theorem, invertibility of the dependence of $\vec{p}$ on $\vec{Q}$ in $\vec{p} = -\partial F/\partial \vec{q}\,(\vec{q}, \vec{Q})$ entails that for any value of $\vec{q}$, the $n \times n$ Jacobian of the dependence of $\vec{p}$ on $\vec{Q}$ is regular, i.e.,

$$\begin{vmatrix} \partial^2 F/\partial q_1 \partial Q_1 & \cdots & \partial^2 F/\partial q_1 \partial Q_n \\ \vdots & & \vdots \\ \partial^2 F/\partial q_n \partial Q_1 & \cdots & \partial^2 F/\partial q_n \partial Q_n \end{vmatrix} \neq 0. \quad (1.252)$$

Moreover, the nonvanishing of the determinant and the fact that the determinant is continuous asserts the existence of an inverse even in a neighborhood.

To show that the transformation is canonical, we derive relationships between various derivatives of F and $(\mathcal{Q}, \mathcal{P})$, and use these to show that the elementary Poisson bracket relations

$$[\mathcal{Q}_i, \mathcal{P}_j] = \delta_{ij}, \quad [\mathcal{P}_i, \mathcal{P}_j] = 0, \quad [\mathcal{Q}_i, \mathcal{Q}_j] = 0 \quad (1.253)$$

are satisfied, which according to Eq. (1.192) asserts that $\mathcal{M}$ is indeed canonical.

We begin by partially differentiating Eq. (1.250) with respect to q_j and p_l, respectively, and obtain

$$F_{q_i q_j} + \sum_{k=1}^{n} F_{q_i Q_k} \frac{\partial Q_k}{\partial q_j} = 0 \quad (1.254)$$

$$\delta_{il} + \sum_{k=1}^{n} F_{q_i Q_k} \frac{\partial Q_k}{\partial p_l} = 0. \tag{1.255}$$

Writing $F_{q_i q_j} = \sum_{l=1}^{n} F_{q_l q_j} \cdot \delta_{il}$ in the first equation and inserting for δ_{il} the value from the second equation, we obtain

$$\sum_{k=1}^{n} F_{q_i Q_k} \cdot \left(\frac{\partial Q_k}{\partial q_j} - \sum_{l=1}^{n} F_{q_l q_j} \frac{\partial Q_k}{\partial p_l} \right) = 0.$$

Because this equation is satisfied for all $i = 1, \ldots, n$ and because of Eq. (1.252), it follows that

$$\frac{\partial Q_k}{\partial q_j} - \sum_{l=1}^{n} F_{q_l q_j} \frac{\partial Q_k}{\partial p_l} = 0 \text{ for all } j, k = 1, \ldots, n. \tag{1.256}$$

Next, we multiply Eq. (1.255) by $F_{q_l Q_j}$ and sum over l to obtain

$$0 = \sum_{l=1}^{n} F_{q_l Q_j} \delta_{il} + \sum_{k,l=1}^{n} F_{q_l Q_j} F_{q_i Q_k} \frac{\partial Q_k}{\partial p_l}$$

$$= \sum_{k=1}^{n} F_{q_i Q_k} \delta_{jk} + \sum_{k,l=1}^{n} F_{q_l Q_j} F_{q_i Q_k} \frac{\partial Q_k}{\partial p_l}$$

$$= \sum_{k=1}^{n} F_{q_i Q_k} \left(\delta_{jk} + \sum_{l=1}^{n} F_{q_l Q_j} \frac{\partial Q_k}{\partial p_l} \right). \tag{1.257}$$

Observing that the relation is satisfied for all $i = 1, \ldots, n$, because of (1.252) we have

$$\delta_{jk} + \sum_{l=1}^{n} F_{q_l Q_j} \frac{\partial Q_k}{\partial p_l} = 0 \text{ for all } j, k = 1, \ldots, n. \tag{1.258}$$

To conclude our derivation of conditions between the derivatives of F and $(\mathcal{Q}, \mathcal{P})$, we utilize Eq. (1.251) and differentiate with respect to p_l to obtain

$$\frac{\partial \mathcal{P}_i}{\partial p_l} - \sum_{k=1}^{n} F_{Q_i Q_k} \frac{\partial Q_k}{\partial p_l} = 0.$$

HAMILTONIAN SYSTEMS

Multiplying with $F_{q_l Q_j}$, summing over l, and considering Eq. (1.258), we obtain

$$0 = \sum_{l=1}^{n} F_{q_l Q_j} \left\{ \frac{\partial \mathcal{P}_i}{\partial p_l} - \sum_{k=1}^{n} F_{Q_i Q_k} \frac{\partial Q_k}{\partial p_l} \right\}$$

$$= \sum_{l=1}^{n} F_{q_l Q_j} \frac{\partial \mathcal{P}_i}{\partial p_l} - \sum_{k=1}^{n} F_{Q_i Q_k} \left\{ \sum_{l=1}^{n} F_{q_l Q_j} \frac{\partial Q_k}{\partial p_l} \right\}$$

$$= \sum_{l=1}^{n} F_{q_l Q_j} \frac{\partial \mathcal{P}_i}{\partial p_l} + F_{Q_i Q_j} \quad \text{for all } i, j = 1, \ldots, n. \quad (1.259)$$

Now we begin the computation of the elementary Poisson brackets. We take the Poisson bracket between $\mathcal{Q}_i$ and both sides of Eq. (1.250), and utilizing Eq. (1.256), we obtain

$$0 = [\mathcal{Q}_i, p_k + F_{q_k}(Q(\vec{q}, \vec{p}), \vec{q})]$$

$$= \frac{\partial \mathcal{Q}_i}{\partial q_k} + \sum_{l=1}^{n} \left\{ \frac{\partial \mathcal{Q}_i}{\partial q_l} \cdot \sum_{j=1}^{n} F_{q_k Q_j} \cdot \frac{\partial \mathcal{Q}_j}{\partial p_l} - \frac{\partial \mathcal{Q}_i}{\partial p_l} \left(\sum_{j=1}^{n} F_{q_k Q_j} \frac{\partial \mathcal{Q}_j}{\partial q_l} + F_{q_k q_l} \right) \right\}$$

$$= \frac{\partial \mathcal{Q}_i}{\partial q_k} - \sum_{l=1}^{n} F_{q_k q_l} \frac{\partial \mathcal{Q}_i}{\partial p_l} + \sum_{j=1}^{n} F_{q_k Q_j} \cdot \left\{ \sum_{l=1}^{n} \frac{\partial \mathcal{Q}_i}{\partial q_l} \frac{\partial \mathcal{Q}_j}{\partial p_l} - \frac{\partial \mathcal{Q}_i}{\partial p_l} \frac{\partial \mathcal{Q}_j}{\partial q_l} \right\}$$

$$= 0 + \sum_{j=1}^{n} F_{q_k Q_j} \cdot [\mathcal{Q}_i, \mathcal{Q}_j] \quad \text{for all } i, k = 1, \ldots, n. \quad (1.260)$$

Because of Eq. (1.252), we have

$$[\mathcal{Q}_i, \mathcal{Q}_j] = 0 \quad \text{for } i, j = 1, \ldots, n. \quad (1.261)$$

Further, taking the Poisson bracket between $\mathcal{Q}_i$ and $\mathcal{P}_j$ as given by Eq. (1.251), and considering Eqs. (1.261) and (1.258), we have

$$[\mathcal{Q}_i, \mathcal{P}_j] = [\mathcal{Q}_i, F_{Q_j}(Q(\vec{q}, \vec{p}), \vec{q})]$$

$$= \sum_{k=1}^{n} \left\{ \frac{\partial \mathcal{Q}_i}{\partial q_k} \left(\sum_{l=1}^{n} F_{Q_j Q_l} \frac{\partial \mathcal{Q}_l}{\partial p_k} \right) - \frac{\partial \mathcal{Q}_i}{\partial p_k} \left(F_{Q_j q_k} + \sum_{l=1}^{n} F_{Q_j Q_l} \frac{\partial \mathcal{Q}_l}{\partial q_k} \right) \right\}$$

$$= -\sum_{k=1}^{n} F_{Q_j q_k} \frac{\partial \mathcal{Q}_i}{\partial p_k} + \sum_{l=1}^{n} F_{Q_j Q_l} \left(\sum_{k=1}^{n} \frac{\partial \mathcal{Q}_i}{\partial q_k} \frac{\partial \mathcal{Q}_l}{\partial p_k} - \frac{\partial \mathcal{Q}_i}{\partial p_k} \frac{\partial \mathcal{Q}_l}{\partial q_k} \right)$$

$$= \delta_{ij} + \sum_{l=1}^{n} F_{Q_j Q_l} \cdot [\mathcal{Q}_i, \mathcal{Q}_l] = \delta_{ij} \quad \text{for all } i, j = 1, \ldots, n \quad (1.262)$$

As the last step, we infer from Eq. (1.251) while utilizing Eq. (1.262) and (1.259) that

$$[\mathcal{P}_i, \mathcal{P}_j] = [\mathcal{P}_i, F_{Q_j}(\mathcal{Q}(\vec{q},\vec{p}),\vec{q})]$$

$$= \sum_{k=1}^{n} \left\{ \frac{\partial \mathcal{P}_i}{\partial q_k} \left(\sum_{l=1}^{n} F_{Q_j Q_l} \frac{\partial \mathcal{Q}_l}{\partial p_k} \right) - \frac{\partial \mathcal{P}_i}{\partial p_k} \left(F_{Q_j q_k} + \sum_{l=1}^{n} F_{Q_j Q_l} \frac{\partial \mathcal{Q}_l}{\partial q_k} \right) \right\}$$

$$= -\sum_{k=1}^{n} F_{Q_j q_k} \frac{\partial \mathcal{P}_i}{\partial p_k} + \sum_{l=1}^{n} F_{Q_j Q_l} \cdot \left(\sum_{k=1}^{n} \frac{\partial \mathcal{P}_i}{\partial q_k} \frac{\partial \mathcal{Q}_l}{\partial p_k} - \frac{\partial \mathcal{P}_i}{\partial p_k} \frac{\partial \mathcal{Q}_l}{\partial q_k} \right)$$

$$= -\sum_{k=1}^{n} F_{Q_j q_k} \frac{\partial \mathcal{P}_i}{\partial p_k} + \sum_{l=1}^{n} F_{Q_j Q_l} \cdot [\mathcal{P}_i, \mathcal{Q}_l]$$

$$= -\sum_{k=1}^{n} F_{Q_j q_k} \frac{\partial \mathcal{P}_i}{\partial p_k} - F_{Q_j Q_i} = 0 \quad \text{for all } i, j = 1, \ldots, n. \quad (1.263)$$

The three Poisson bracket relations (Eqs. (1.261), (1.262), and (1.263)) together assert that the transformation is indeed canonical.

Of course, the entire argument is not specific to the particular choice of the generating function $F(\vec{Q}, \vec{q})$ that we have made because through elementary canonical transformations (Eq. 1.154), it is possible to interchange both initial and final position and momentum variables. If one does this for all positions and momenta simultaneously, one obtains a total of four different generating functions that traditionally have been referred to as F_1–F_4 and have the form

$$F_1(\vec{q}, \vec{Q}), \quad F_2(\vec{q}, \vec{P}),$$
$$F_3(\vec{p}, \vec{Q}), \quad F_4(\vec{p}, \vec{P}). \quad (1.264)$$

Folding in the respective elementary canonical transformations exchanging positions and momenta, we obtain the following relationships:

$$\vec{p} = \frac{\partial F_1}{\partial \vec{q}}, \quad \vec{P} = -\frac{\partial F_1}{\partial \vec{Q}} \quad (1.265)$$

$$\vec{p} = \frac{\partial F_2}{\partial \vec{q}}, \quad \vec{Q} = \frac{\partial F_2}{\partial \vec{P}} \quad (1.266)$$

$$\vec{q} = -\frac{\partial F_3}{\partial \vec{p}}, \quad \vec{P} = -\frac{\partial F_3}{\partial \vec{Q}} \quad (1.267)$$

$$\vec{q} = -\frac{\partial F_4}{\partial \vec{p}}, \quad \vec{Q} = \frac{\partial F_4}{\partial \vec{P}}. \quad (1.268)$$

For each of the above generating functions F_i, its negative is also a valid generating function. This also entails that in each of the four equations (1.265) through

(1.268), it is possible to exchange both signs in front of the partials of F; and in fact, the generating function (1.248) is the negative of F_1. This fact helps memorization of the four equations; one now must remember only that the generating functions depending on position and momenta have equal signs, and those depending only on momenta or only on positions have unequal signs.

However, in light of our discussion it is clear that this selection of four types is indeed somewhat arbitrary, as it is not necessary to perform the elementary canonical transformations simultaneously for all components of the coordinates. So instead of only four cases shown here, there are **four choices for each component**. Moreover, one can subject the original function to other linear canonical transformations that are not merely combinations of elementary canonical transformations, increasing the multitude of possible generating functions even more.

Also, it is important to point out that **not every canonical transformation can be represented** through one of the four generators F_1–F_4, but rather, as seen in Section 1.4.6, at least a set of 4^n generators is needed.

1.4.9 Time-Dependent Canonical Transformations

The framework of canonical transformations is intrinsically time independent in the sense that the new coordinates and momenta depend on the old coordinates and momenta but not on time. In this respect, there is an apparent difference from the Lagrangian framework of transformations, which were allowed to be time dependent from the outset. However, this is not of fundamental concern since according to Eq. (1.107), in the $2(n+1)$-dimensional extended phase space $(\vec{q}, s, \vec{p}, p_s)$ with the extended Hamiltonian

$$\hat{H} = p_s + H(\vec{q}, \vec{p}, s), \tag{1.269}$$

time appears in the form of the dynamical variable s, and hence canonical transformation in extended phase naturally includes possible time dependence.

While fundamentally the issue is settled with this observation, practically the question often arises whether the transformed Hamiltonian system in extended phase space can again be expressed in terms of a nonautonomous system in conventional phase space. According to Eq. (1.107), this is the case if in the new variables, the Hamiltonian again has the form

$$\hat{H} = P_S + G(\vec{Q}, \vec{P}, S). \tag{1.270}$$

This requirement is easily met in the representation of canonical transformations through generating functions, and the interpretation of the resulting transformations in conventional phase space is straightforward. Let us consider the generating function in extended phase space

$$\hat{F}(\vec{q}, \vec{P}, s) + s \cdot P_S, \tag{1.271}$$

which produces a transformation satisfying

$$p_i = \frac{\partial \hat{F}}{\partial q_i}, \quad p_S = \frac{\partial \hat{F}}{\partial s} + P_S \tag{1.272}$$

$$Q_i = \frac{\partial \hat{F}}{\partial P_i}, \quad S = s.$$

We observe that in the first equation, the p_i do not depend on P_S since $\hat{F}$ is independent of P_S; likewise, after solving the second equation for q_i, these also do not depend on P_S. Therefore, inserting this transformation into the Hamiltonian (Eq. 1.269) yields

$$\hat{H}^{\text{new}} = P_S + \frac{\partial \hat{F}}{\partial s} + H^{\text{new}}(\vec{Q}, \vec{P}, S), \tag{1.273}$$

which is indeed of the form of Eq. (1.270). Viewed entirely within the nonextended phase space variables, the transformations are described by the time-dependent generating function

$$F(\vec{q}, \vec{P}, t); \tag{1.274}$$

the resulting transformation is not canonical in the conventional sense that the new Hamiltonian can be obtained by expressing old variables by new ones. Rather, we obtain a Hamiltonian system with the new Hamiltonian

$$H(\vec{Q}, \vec{P}, t) = \frac{\partial F}{\partial t} + H^{\text{new}}(\vec{Q}, \vec{P}, t). \tag{1.275}$$

1.4.10 The Hamilton–Jacobi Equation

The solution of a Hamiltonian system is simplified significantly if it is possible to find a canonical transformation such that the new Hamiltonian does not depend on some of the positions or momenta since by Hamilton's equations, the corresponding canonically conjugate quantities are constant. Certainly, the most extreme such case would be if the Hamiltonian in the new variables actually vanishes completely; then all new phase space variables are constant.

Apparently, this cannot be achieved with a conventional canonical transformation because if a Hamiltonian varies with the values of the old variables, it will also vary with the values of the new variables. However, for the case of time-dependent canonical transformations, it may be possible to achieve a completely vanishing Hamiltonian. For this purpose, we try a time-dependent generator $F_2(\vec{q}, \vec{P}, t)$, which, when viewed in conventional phase space according to the previous section, satisfies

$$\vec{p} = \frac{\partial F_2(\vec{q}, \vec{P}, t)}{\partial \vec{q}}, \quad \vec{Q} = \frac{\partial F_2(\vec{q}, \vec{P}, t)}{\partial \vec{P}}, \quad (1.276)$$

$$\tilde{H}(\vec{Q}, \vec{P}) = H(\vec{q}(\vec{Q}, \vec{P}), \vec{p}(\vec{Q}, \vec{P})) + \frac{\partial F_2(\vec{q}, \vec{P}, t)}{\partial t}. \quad (1.277)$$

Inserting the transformation equation for $\vec{p}$ shows the explicit form of the Hamiltonian in the new variables to be

$$\tilde{H}(\vec{Q}, \vec{P}) = H(\vec{q}(\vec{Q}, \vec{P}), \frac{\partial F_2(\vec{q}, \vec{P}, t)}{\partial \vec{q}}) + \frac{\partial F_2(\vec{q}, \vec{P}, t)}{\partial t}. \quad (1.278)$$

But now we utilize that the new Hamiltonian $\tilde{H}$ is supposed to be zero, which entails that the new momenta are all constants. Therefore the function F_2 is indeed only a function of the variables $\vec{q}$, and the momenta appear as the constants $\vec{\pi} = \vec{P}$. The previous equation then reduces to a condition on F_2, the **Hamilton–Jacobi equation**,

$$0 = H\left(\vec{q}, \frac{\partial F_2(\vec{q}, \vec{\pi}, t)}{\partial \vec{q}}\right) + \frac{\partial F_2(\vec{q}, \vec{\pi}, t)}{\partial t}. \quad (1.279)$$

This is an **implicit** partial differential equation for $F_2(\vec{q}, \vec{\pi})$ as a function of the n **variables $\vec{q}$ only**. Therefore, the number of unknowns has been reduced by half, at the expense of making the system implicit and turning it from an ODE to a partial differential equation (PDE).

If the Hamilton–Jacobi PDE can be solved for F_2 for each choice of the parameters $\vec{\pi}$, then it often allows for the complete solution of the Hamiltonian equations of motion. Indeed, if $\tilde{H} = 0$, then apparently the final positions are constant, and we have $\vec{Q} = \vec{\rho} =$ const. Then the equations for the generator F_2 in Eq. (1.276) read

$$\vec{p} = \frac{\partial F_2(\vec{q}, \vec{\pi}, t)}{\partial \vec{q}}, \quad \vec{\rho} = \frac{\partial F_2(\vec{q}, \vec{\pi}, t)}{\partial \vec{\pi}}. \quad (1.280)$$

The second vector equation is merely a set of n implicit algebraic equations in $\vec{q}$. If it can be solved for $\vec{q}$, we obtain $\vec{q}$ as a function of $\vec{\rho}$ and $\vec{\pi}$. The first equation then yields $\vec{p}$ directly by insertion of $\vec{q}(\vec{\rho}, \vec{\pi})$. Altogether, we obtain

$$\vec{q} = \vec{q}(\vec{\rho}, \vec{\pi}, t)$$
$$\vec{p} = \vec{p}(\vec{\rho}, \vec{\pi}, t). \quad (1.281)$$

The entire solution of the Hamiltonian equation can then be obtained by **algebraic inversion** of the relationship, if such inversion is possible.

Whether or not the Hamilton–Jacobi approach actually **simplifies** the solution of the Hamiltonian problem greatly depends on the circumstances, because in general the solution of the Hamilton–Jacobi PDE is by no means simple and straightforward. However, perhaps more important, it provides a **connection** between the theory of **PDEs and ODEs**, and sometimes also allows for the expression of PDE problems in terms of Hamiltonian dynamics.

An important approach that often allows one to solve or simplify PDEs is to assess whether the equation is **separable,** i.e., whether it can be broken down into separate parts of lower dimension. In our case, one makes the special Ansatz

$$F_2 = \sum_{i=1}^{n} F_{2,i}(q_i, \vec{\pi}, t) \tag{1.282}$$

and inserts into the Hamilton–Jacobi PDE. If it is then possible to split the PDE into n separate pieces of the form

$$H_i\left(q_i, \frac{\partial F_i}{\partial q_i}, c_1, \ldots, c_n\right) = c_i, \tag{1.283}$$

where the c_k are separation constants. Each of these equations merely requires the solution for $\partial F_i / \partial q_i$ and subsequent integration.

1.5 Fields and Potentials

In this section, we present an overview of concepts of electrodynamics, with a particular emphasis on deriving all the tools needed in the further development from the first principles. As in the remainder of the book, we will utilize the Système International d'unités (SI system), the well-known extension of the **MKSA** system with its units of meter, kilogram, second, and ampere. The symmetry between electric and magnetic phenomena manifests itself more dramatically in the **Gaussian system,** which because of this fact is often the preferred system for purely electrodynamic purposes. We will not dwell on the conversion rules that allow the description of one set of units in terms of another, but rather we refer the reader to the respective appendix in Jackson's book (Jackson, 1975) for details.

1.5.1 Maxwell's Equations

It is one of the beautiful aspects of electrodynamics that all the commonly treated phenomena can be derived from one set of four basic laws, the equations of Maxwell. In our theory, they have **axiomatic character** in the sense that they cannot be derived from any other laws, and they are assumed to be universally valid in all further developments. Of course, like other "good" axioms of physics,

FIELDS AND POTENTIALS 63

they can be checked directly experimentally, and all such checks so far have not shown any deviation from them.

The first of Maxwell's equations is **Coulomb's law**, which states that the quantity ρ, an elementary property of matter called the **charge density**, is the source of a field called the **electric flux density** (sometimes called electric displacement) denoted by $\vec{D}$:

$$\vec{\nabla} \cdot \vec{D} = \rho. \tag{1.284}$$

There is another field, the **magnetic flux density** (sometimes called magnetic induction) called $\vec{B}$, and this field does not have any sources:

$$\vec{\nabla} \cdot \vec{B} = 0. \tag{1.285}$$

In other words, there are no magnetic monopoles. **Faraday's law of induction** states that the curl of the electric field $\vec{E}$ is connected to change in the magnetic flux density $\vec{B}$:

$$\vec{\nabla} \times \vec{E} + \frac{\partial \vec{B}}{\partial t} = \vec{0}. \tag{1.286}$$

Finally, there is the **Ampère–Maxwell law**, which states that the curl of the magnetic field $\vec{H}$ is connected to change in the electric flux density $\vec{D}$ as well as another elementary property of matter called the current density $\vec{J}$:

$$\vec{\nabla} \times \vec{H} = \vec{J} + \frac{\partial \vec{D}}{\partial t}. \tag{1.287}$$

The electric and magnetic flux densities are related to the electric and magnetic fields, respectively, through two additional elementary properties of matter, **the electric and magnetic polarizations $\vec{P}$ and $\vec{M}$**:

$$\vec{D} = \varepsilon_0 \vec{E} + \vec{P}, \tag{1.288}$$
$$\vec{B} = \mu_0 \vec{H} + \vec{M}. \tag{1.289}$$

The magnetic polarization is often referred to as magnetization or magnetic moment density. The natural constants ε_0 and μ_0 are the electric permittivity constant or dielectric constant of vacuum and the magnetic permeability constant of vacuum, respectively. The relations which connect $\vec{E}$, $\vec{D}$, $\vec{B}$, and $\vec{H}$ are known as **constitutive relations**. In the case of many materials that are homogeneous and isotropic, the polarizations are proportional to the fields that are applied; in this case, the constitutive relations have the simplified forms

$$\vec{D} = \varepsilon \vec{E} \tag{1.290}$$
$$\vec{B} = \mu \vec{H}, \tag{1.291}$$

where the quantities ε and μ depend on the material at hand and are referred to as the relative electric permittivity or dielectric constant and the relative magnetic permeability of the material, respectively.

Similar to the constitutive relations, for many homogeneous and isotropic materials, there is a relation between current and electric field known as **Ohm's law**:

$$\vec{J} = \sigma \vec{E}, \tag{1.292}$$

where the quantity σ, the electric conductivity, is an elementary property of matter.

In many cases, it is of interest to study Maxwell's equations in the absence of matter at the point of interest. In this case, we have

$$\rho = 0, \qquad \vec{J} = \vec{0}. \tag{1.293}$$

Thus, Maxwell's equations take the simplified and very symmetric forms

$$\vec{\nabla} \cdot \vec{D} = 0, \quad \vec{\nabla} \times \vec{E} = -\frac{\partial \vec{B}}{\partial t} \tag{1.294}$$

$$\vec{\nabla} \cdot \vec{B} = 0, \quad \vec{\nabla} \times \vec{H} = \frac{\partial \vec{D}}{\partial t} \tag{1.295}$$

In many situations, it is also important to study the case of time independence, in which case all right-hand sides vanish:

$$\frac{\partial \vec{D}}{\partial t} = \vec{0}, \quad \frac{\partial \vec{B}}{\partial t} = \vec{0}. \tag{1.296}$$

An important consequence of Maxwell's equations is that the charge density and the current density are connected to each other through a relationship known as the **continuity equation**:

$$\vec{\nabla} \cdot \vec{J} + \frac{\partial \rho}{\partial t} = 0. \tag{1.297}$$

This equation is derived by applying the divergence operator on Ampere's law (Eq. 1.287) and combining it with Coulomb's law (Eq. 1.284); we have

$$\vec{\nabla} \cdot (\vec{\nabla} \times \vec{H}) = \vec{\nabla} \cdot \vec{J} + \vec{\nabla} \cdot \frac{\partial \vec{D}}{\partial t}$$

$$0 = \vec{\nabla} \cdot \vec{J} + \frac{\partial \rho}{\partial t},$$

where in the last step use has been made of the easily derivable vector formula $\vec{\nabla} \cdot (\vec{\nabla} \times \vec{v}) = 0$, which is also discussed later (Eq. 1.307). Similarly, the operation of the divergence on Eq. (1.286) shows agreement with Eq. (1.285):

$$\vec{\nabla} \cdot (\vec{\nabla} \times \vec{E}) + \vec{\nabla} \cdot \frac{\partial \vec{B}}{\partial t} = 0$$

$$\frac{\partial \vec{\nabla} \cdot \vec{B}}{\partial t} = 0.$$

There are two important theorems, the integral relations of **Stokes** and **Gauss**, that are of importance for many of the derivations that follow:

$$\oint_\gamma \vec{F} \cdot d\vec{s} = \int_A \vec{\nabla} \times \vec{F} \cdot \vec{n} \, dA \tag{1.298}$$

$$\int_A \vec{F} \cdot \vec{n} \, dA = \int_V \vec{\nabla} \cdot \vec{F} \, dV. \tag{1.299}$$

1.5.2 Scalar and Vector Potentials

In this section, we introduce a new set of scalar and vector fields called potentials that allow the determination of the electric and magnetic fields by differentiation and that allow a simplification of many electrodynamics problems. We begin with some definitions.

We call ψ a **scalar potential** for the n-dimensional vector field $\vec{F}$ if $\vec{F} = \vec{\nabla}\psi$. We call $\vec{A}$ a **vector potential** for the three-dimensional vector field $\vec{F}$ if $\vec{F} = \vec{\nabla} \times \vec{A}$.

The question now arises under what condition a given vector field $\vec{F}$ on R^n has a scalar potential or a vector potential. We first address the case of the scalar potential and answer this question for the general n-dimensional case. For the purposes of electrodynamics, the case $n = 3$ is usually sufficient, but for many questions connected to Lagrangian and Hamiltonian dynamics, the general case is needed.

In the following, we study the **existence and uniqueness of scalar potentials.** Let $\vec{F}$ be a continuously differentiable vector field on R^n.

1. If $\vec{F} = \vec{\nabla}\psi$, then

$$\partial F_i/\partial x_j = \partial F_j/\partial x_i \text{ for all } i, j = 1, \ldots, n. \tag{1.300}$$

2. If $\partial F_i/\partial x_j = \partial F_j/\partial x_i$ for all $i, j = 1, \ldots, n$, then there exists a scalar potential ψ such that

$$\vec{F} = \vec{\nabla}\psi. \tag{1.301}$$

3. A scalar potential ψ for the vector $\vec{F}$ is uniquely specified up to a constant.

Note the difference of (1) and (2), which are actually the converses of each other. We also note that in the three-dimensional case, the condition $\partial F_i/\partial x_j = \partial F_j/\partial x_i$ is equivalent to the more readily recognized condition $\vec{\nabla} \times \vec{F} = \vec{0}$. For the proof, we proceed as follows.

1. If there exists a twice continuously differentiable ψ such that $\vec{F} = \vec{\nabla}\psi$, then $F_i = \partial\psi/\partial x_i$ and $F_j = \partial\psi/\partial x_j$, and thus

$$\frac{\partial F_i}{\partial x_j} = \frac{\partial^2 \psi}{\partial x_j \partial x_i} = \frac{\partial^2 \psi}{\partial x_i \partial x_j} = \frac{\partial F_j}{\partial x_i}. \tag{1.302}$$

2. Now assume that $\vec{F}$ is continuously differentiable and

$$\frac{\partial F_i}{\partial x_j} = \frac{\partial F_j}{\partial x_i} \quad \text{for all } i, j = 1, \ldots, n. \tag{1.303}$$

Define ψ as a path integral of $\vec{F}$ from $(0, \ldots, 0)$ to $(x_1, x_2, \ldots, x_n)$ along a path that first is parallel to the x_1 axis from $(0, 0, \ldots, 0)$ to $(x_1, 0, \ldots, 0)$, then parallel to the x_2 axis from $(x_1, 0, \ldots, 0)$ to $(x_1, x_2, 0, \ldots, 0)$, and so on, and finally parallel to the x_n axis from $(x_1, x_2, \ldots, x_{n-1}, 0)$ to $(x_1, x_2, \ldots, x_{n-1}, x_n)$. Then we have

$$\begin{aligned}\psi &= \int_{(0,\ldots,0)}^{(x_1,\ldots,x_n)} \vec{F} \cdot d\vec{r} \\ &= \int_0^{x_1} F_1(x_1, 0, \ldots, 0)\, dx_1 + \int_0^{x_2} F_2(x_1, x_2, 0, \ldots, 0)\, dx_2 + \cdots \\ &\quad + \int_0^{x_n} F_n(x_1, x_2, \ldots, x_n)\, dx_n.\end{aligned} \tag{1.304}$$

In the following, we show that ψ indeed satisfies the requirement $\vec{F} = \vec{\nabla}\psi$. We first differentiate with respect to x_1 and obtain

$$\begin{aligned}\frac{\partial \psi}{\partial x_1} &= F_1(x_1, 0, 0, \ldots, 0) + \int_0^{x_2} \frac{\partial F_2(x_1, x_2, 0, \ldots, 0)}{\partial x_1}\, dx_2 + \cdots \\ &\quad + \int_0^{x_n} \frac{\partial F_n(x_1, x_2, \ldots, x_n)}{\partial x_1}\, dx_n \\ &= F_1(x_1, 0, 0, \ldots, 0) + \int_0^{x_2} \frac{\partial F_1(x_1, x_2, 0, \ldots, 0)}{\partial x_2}\, dx_2 + \cdots \\ &\quad + \int_0^{x_n} \frac{\partial F_1(x_1, x_2, \ldots, x_n)}{\partial x_n}\, dx_n\end{aligned}$$

$$= F_1(x_1, 0, 0, \ldots, 0)$$
$$+ [F_1(x_1, x_2, 0, \ldots, 0) - F_1(x_1, 0, 0, \ldots, 0)] + \cdots$$
$$+ [F_1(x_1, x_2, \ldots, x_{n-1}, x_n) - F_1(x_1, x_2, \ldots, x_{n-1}, 0)]$$
$$= F_1(x_1, x_2, \ldots, x_n).$$

where Eq. (1.303) was used for moving from the first to the second line. Similarly, we have

$$\frac{\partial \psi}{\partial x_2} = F_2(x_1, x_2, 0, \ldots, 0) + \int_0^{x_3} \frac{\partial F_3(x_1, x_2, x_3, 0, \ldots, 0)}{\partial x_2} dx_3 + \cdots$$
$$+ \int_0^{x_n} \frac{\partial F_n(x_1, x_2, \ldots, x_n)}{\partial x_2} dx_n$$
$$= F_2(x_1, x_2, 0, \ldots, 0) + \int_0^{x_3} \frac{\partial F_2(x_1, x_2, x_3, 0, \ldots, 0)}{\partial x_3} dx_3 + \cdots$$
$$+ \int_0^{x_n} \frac{\partial F_2(x_1, x_2, \ldots, x_n)}{\partial x_n} dx_n$$
$$= F_2(x_1, x_2, \ldots, x_n).$$

We continue in the same way for $\partial \psi / \partial x_3$ and $\partial \psi / \partial x_4$, and ultimately we derive the following for $\partial \psi / \partial x_n$:

$$\frac{\partial \psi}{\partial x_n} = F_n(x_1, x_2, \ldots, x_n), \tag{1.305}$$

which shows that indeed the scalar field ψ defined by Eq. (1.304) satisfies $\vec{F} = \vec{\nabla} \psi$.

3. Assume that there exist two scalars ψ_1 and ψ_2 which satisfy

$$\vec{F} = \vec{\nabla} \psi_1$$
$$\vec{F} = \vec{\nabla} \psi_2.$$

Then $\vec{\nabla}(\psi_1 - \psi_2) = \vec{0}$, which can only be the case if

$$\psi_1 - \psi_2 = \text{constant}.$$

Therefore, a scalar potential ψ is specified up to a constant, which concludes the proof.

We note that in the three-dimensional case, the peculiar-looking integration path in the definition of the potential ψ in Eq. (1.304) can be replaced by any other path connecting the origin and (x, y, z), because $\vec{\nabla} \times \vec{F} = 0$ ensures path independence of the integral according to the Stokes theorem (Eq. 1.298). Furthermore, the choice of the origin as the starting point is by no means mandatory; indeed, any other starting point $\vec{x}_0$ leads to the addition of a constant to the potential, because due to path independence the integral from $\vec{x}_0$ to $\vec{x}$ can be replaced by one from $\vec{x}_0$ to $\vec{0}$ and one from $\vec{0}$ to $\vec{x}$, and the first one always produces a constant contribution.

Next we study the **existence and uniqueness of vector potentials.** Let $\vec{F}$ be a continuously differentiable vector field.

1. If $\vec{F} = \vec{\nabla} \times \vec{A}$, then

$$\vec{\nabla} \cdot \vec{F} = 0. \tag{1.306}$$

2. If $\vec{\nabla} \cdot \vec{F} = 0$, there exists a vector potential $\vec{A}$ such that

$$\vec{F} = \vec{\nabla} \times \vec{A}. \tag{1.307}$$

3. A vector potential $\vec{A}$ for the vector $\vec{F}$ is uniquely determined up to a gradient of a scalar.

For the proof, we proceed as follows.

1. If there exists $\vec{A}$ such that $\vec{F} = \vec{\nabla} \times \vec{A}$, then

$$\vec{\nabla} \cdot \vec{F} = \vec{\nabla} \cdot (\vec{\nabla} \times \vec{A})$$
$$= 0,$$

as direct evaluation of the components reveals.

2. Since $\vec{\nabla} \cdot \vec{F} = 0$,

$$\frac{\partial F_x}{\partial x} + \frac{\partial F_y}{\partial y} + \frac{\partial F_z}{\partial z} = 0. \tag{1.308}$$

Considering

$$(\vec{\nabla} \times \vec{A})_x = \frac{\partial A_z}{\partial y} - \frac{\partial A_y}{\partial z}$$

$$(\vec{\nabla} \times \vec{A})_y = \frac{\partial A_x}{\partial z} - \frac{\partial A_z}{\partial x}$$

$$(\vec{\nabla} \times \vec{A})_z = \frac{\partial A_y}{\partial x} - \frac{\partial A_x}{\partial y},$$

define $\vec{A}$ as

$$A_x = \int_0^z F_y(x, y, z) \, dz$$
$$A_y = -\int_0^z F_x(x, y, z) \, dz + \int_0^x F_z(x, y, 0) \, dx$$
$$A_z = 0. \tag{1.309}$$

Then $\vec{A}$ indeed satisfies the requirement $\vec{F} = \vec{\nabla} \times \vec{A}$, as shown by computing the curl:

$$(\vec{\nabla} \times \vec{A})_x = F_x(x, y, z)$$
$$(\vec{\nabla} \times \vec{A})_y = F_y(x, y, z)$$
$$(\vec{\nabla} \times \vec{A})_z = -\int_0^z \frac{\partial F_x(x, y, z)}{\partial x} \, dz + F_z(x, y, 0) - \int_0^z \frac{\partial F_y(x, y, z)}{\partial y} \, dz$$
$$= -\int_0^z \left[\frac{\partial F_x(x, y, z)}{\partial x} + \frac{\partial F_y(x, y, z)}{\partial y} \right] dz + F_z(x, y, 0)$$
$$= \int_0^z \frac{\partial F_z(x, y, z)}{\partial z} \, dz + F_z(x, y, 0)$$
$$= F_z(x, y, z) - F_z(x, y, 0) + F_z(x, y, 0)$$
$$= F_z(x, y, z),$$

where Eq. (1.308) was used in moving from the second to the third line of the z component. Indeed, there exists a vector $\vec{A}$ which satisfies $\vec{F} = \vec{\nabla} \times \vec{A}$.

3. Assume that there exist two vectors $\vec{A}_1$ and $\vec{A}_2$ which satisfy

$$\vec{F} = \vec{\nabla} \times \vec{A}_1$$
$$\vec{F} = \vec{\nabla} \times \vec{A}_2,$$

then

$$\vec{\nabla} \times (\vec{A}_1 - \vec{A}_2) = \vec{0}.$$

Thus, according to the previous theorems about the existence of scalar potentials, there exists ξ which is unique up to a constant such that

$$\vec{A}_1 - \vec{A}_2 = \vec{\nabla}\xi.$$

So, the scalar potential $\vec{A}$ is unique up to the additional $\vec{\nabla}\xi$, which concludes the proof.

A closer study of the proof reveals that instead of having $A_z = 0$, by symmetry it is possible to construct vector potentials that satisfy either $A_x = 0$ or $A_y = 0$.

We now address the question of nonuniqueness of scalar and vector potentials in more detail. As was shown, different choices of the integration constant in the case of the scalar potential and specific choices of ξ in the case of the vector potential all lead to valid potentials. These transformations between potentials are simple examples of so-called **gauge transformations** which lead to the same physical system via different potentials. The various possible choices that yield the same field are called different gauges. We will discuss the matter in detail after developing a more complete understanding of the electrodynamics expressed in terms of potentials.

It is illuminating to note that the different gauges based on the non-uniqueness of the potentials, which at first may appear somewhat disturbing, can be accounted for very naturally with the concept of an **equivalence relation**. An equivalence relation is a relationship "$\sim$" between two elements of a given set that for all elements of the set satisfies

$$a \sim a,$$
$$a \sim b \Rightarrow b \sim a, \text{ and}$$
$$a \sim b, b \sim c \Rightarrow a \sim c. \tag{1.310}$$

For the case of **scalar potentials**, the relationship $\sim_s$ on the set of scalar functions is that $V_1 \sim_s V_2$ if $V_1 - V_2$ is a constant. For the case of **vector potentials**, the relationship $\sim_v$ on the set of vector functions is that $\vec{A}_1 \sim_v \vec{A}_2$ if and only if $\vec{A}_1 - \vec{A}_2$ can be written as the gradient of a scalar. In terms of gauges, in both cases potentials are related if one can be obtained from the other via a gauge transformation. One can quite readily verify that both these relations satisfy the condition in Eq. (1.310).

For a fixed element a, all elements related to it constitute the equivalence class of a, denoted by $[a]$; therefore,

$$[a] = \{x | x \sim a\}. \tag{1.311}$$

Apparently, the equivalence classes $[\ ]_s$ of scalar potentials and $[\ ]_v$ of vector potentials describe the collection of all functions ψ, $\vec{A}$ satisfying $\vec{F} = \vec{\nabla}\psi$ and $\vec{F} = \vec{\nabla} \times \vec{A}$, respectively, which are those that can be obtained through valid gauge transformations from one another. The concept of classes allows the description of scalar and vector potentials in a very natural way; indeed, we can say that the scalar potential ψ to a vector function $\vec{F}$ is really an equivalence class under the relationship $\sim_s$ of all those functions whose divergence is $\vec{F}$, and the vector potential $\vec{A}$ to $\vec{F}$ is the equivalence class under $\sim_v$ of all those whose curl is $\vec{F}$. Within this framework, both scalar and vector potentials are unique, while the underlying freedom of gauge is maintained and accounted for.

The theorems on the existence of **scalar and vector potentials** allow us to **simplify the equations of electrodynamics** and make them more compact. Suppose we are given $\vec{J}$ and ρ, and the task is to describe the system in its entirety. Utilizing Maxwell's equations, we can find $\vec{E}$ and $\vec{B}$ of the system, and for the description of the electrodynamic phenomena we need a total of six components.

By using scalar and vector potentials, we can decrease the number of variables in the electromagnetic system. The Maxwell equation $\vec{\nabla} \cdot \vec{B} = 0$ implies that there exists $\vec{A}$ such that $\vec{B} = \vec{\nabla} \times \vec{A}$. On the other hand, Faraday's law $\vec{\nabla} \times \vec{E} + \partial \vec{B}/\partial t = 0$ can then be written as

$$\vec{\nabla} \times \left(-\vec{E} - \frac{\partial \vec{A}}{\partial t}\right) = \vec{0}. \tag{1.312}$$

This ensures that there is a scalar potential Φ for $-\vec{E} - \partial \vec{A}/\partial t$ that satisfies

$$-\vec{E} - \frac{\partial \vec{A}}{\partial t} = \vec{\nabla}\Phi. \tag{1.313}$$

Thus, the two homogeneous Maxwell's equations are satisfied automatically by setting

$$\vec{B} = \vec{\nabla} \times \vec{A} \text{ and } \vec{E} = -\vec{\nabla}\Phi - \frac{\partial \vec{A}}{\partial t}. \tag{1.314}$$

The other two inhomogeneous Maxwell equations determine the space-time behavior of $\vec{A}$ and Φ. Using the constitutive relations, $\vec{D} = \varepsilon \vec{E}$ and $\vec{B} = \mu \vec{H}$, we express these two Maxwell equations in terms of $\vec{A}$ and Φ. The Ampère–Maxwell law takes the form

$$\vec{\nabla} \times \left(\frac{1}{\mu} \vec{\nabla} \times \vec{A}\right) = \vec{J} + \frac{\partial}{\partial t}\left[\varepsilon\left(-\vec{\nabla}\Phi - \frac{\partial \vec{A}}{\partial t}\right)\right]$$

$$\vec{\nabla}^2 \vec{A} - \varepsilon\mu \frac{\partial^2 \vec{A}}{\partial t^2} = \vec{\nabla}\left(\vec{\nabla} \cdot \vec{A} + \varepsilon\mu \frac{\partial \Phi}{\partial t}\right) - \mu \vec{J}. \tag{1.315}$$

On the other hand, Coulomb's law appears as

$$\vec{\nabla} \cdot \left[\varepsilon\left(-\vec{\nabla}\Phi - \frac{\partial \vec{A}}{\partial t}\right)\right] = \rho$$

$$\vec{\nabla}^2 \Phi + \frac{\partial}{\partial t}(\vec{\nabla} \cdot \vec{A}) = -\frac{\rho}{\varepsilon}. \tag{1.316}$$

Altogether, we have obtained a coupled set of two equations. To summarize, the following set of equations are equivalent to the set of Maxwell's equations:

$$\vec{B} = \vec{\nabla} \times \vec{A} \tag{1.317}$$

$$\vec{E} = -\vec{\nabla}\Phi - \frac{\partial \vec{A}}{\partial t} \tag{1.318}$$

$$\vec{\nabla}^2 \vec{A} - \varepsilon\mu \frac{\partial^2 \vec{A}}{\partial t^2} = \vec{\nabla}\left(\vec{\nabla} \cdot \vec{A} + \varepsilon\mu \frac{\partial \Phi}{\partial t}\right) - \mu \vec{J} \qquad (1.319)$$

$$\vec{\nabla}^2 \Phi + \frac{\partial}{\partial t}(\vec{\nabla} \cdot \vec{A}) = -\frac{\rho}{\varepsilon}. \qquad (1.320)$$

The fields $\vec{B}$ and $\vec{E}$ can be determined by the solutions $\vec{A}$ and Φ of Eqs. (1.319) and (1.320). However, as seen in the theorems in the beginning of this section, the potentials have the freedom of gauge in that $\vec{A}$ is unique up to the gradient $\vec{\nabla}u$ of a scalar field u, and Φ is unique up to a constant. For the coupled situation, it is possible to formulate a general gauge transformation that simultaneously affects $\vec{A}$ and Φ without influence on the fields. Indeed, if u is an arbitrary smooth scalar field depending on space and time, then the transformation

$$\vec{A} \to \vec{A}' = \vec{A} + \vec{\nabla}u \qquad (1.321)$$

$$\Phi \to \Phi' = \Phi - \frac{\partial u}{\partial t} \qquad (1.322)$$

does not affect the fields $\vec{E}$ and $\vec{B}$ and the two inhomogeneous equations. Equation (1.317) for the magnetic field has the form

$$\vec{B}' = \vec{\nabla} \times \vec{A}' = \vec{\nabla} \times (\vec{A} + \vec{\nabla}u)$$
$$= \vec{\nabla} \times \vec{A} + \vec{\nabla} \times \vec{\nabla}u = \vec{B},$$

whereas the corresponding one (Eq. 1.318) for the electric field has the form

$$\vec{E}' = -\vec{\nabla}\Phi' - \frac{\partial \vec{A}'}{\partial t} = -\vec{\nabla}\left(\Phi - \frac{\partial u}{\partial t}\right) - \frac{\partial}{\partial t}(\vec{A} + \vec{\nabla}u)$$
$$= -\vec{\nabla}\Phi - \frac{\partial \vec{A}}{\partial t} + \vec{\nabla}\frac{\partial u}{\partial t} - \frac{\partial \vec{\nabla}u}{\partial t} = \vec{E}.$$

Furthermore, Eq. (1.319) transforms as

$$\vec{\nabla}^2 \vec{A}' - \varepsilon\mu \frac{\partial^2 \vec{A}'}{\partial t^2} - \vec{\nabla}\left(\vec{\nabla} \cdot \vec{A}' + \varepsilon\mu \frac{\partial \Phi'}{\partial t}\right) + \mu \vec{J}$$
$$= \vec{\nabla}^2 \vec{A} + \vec{\nabla}^2(\vec{\nabla}u) - \varepsilon\mu \frac{\partial^2 \vec{A}}{\partial t^2} - \varepsilon\mu \frac{\partial^2 \vec{\nabla}u}{\partial t^2}$$
$$- \vec{\nabla}\left(\vec{\nabla} \cdot \vec{A} + \vec{\nabla}^2 u + \varepsilon\mu \frac{\partial \Phi}{\partial t} - \varepsilon\mu \frac{\partial^2 u}{\partial t^2}\right) + \mu \vec{J}$$
$$= \vec{\nabla}^2 \vec{A} - \varepsilon\mu \frac{\partial^2 \vec{A}}{\partial t^2} - \vec{\nabla}\left(\vec{\nabla} \cdot \vec{A} + \varepsilon\mu \frac{\partial \Phi}{\partial t}\right) + \mu \vec{J},$$

and Eq. (1.320) reads

$$\vec{\nabla}^2\Phi' + \frac{\partial}{\partial t}(\vec{\nabla} \cdot \vec{A}') + \frac{\rho}{\varepsilon}$$

$$= \vec{\nabla}^2\Phi - \vec{\nabla}^2\frac{\partial u}{\partial t} + \frac{\partial}{\partial t}(\vec{\nabla} \cdot \vec{A}) + \frac{\partial}{\partial t}\vec{\nabla}^2 u + \frac{\rho}{\varepsilon}$$

$$= \vec{\nabla}^2\Phi + \frac{\partial}{\partial t}(\vec{\nabla} \cdot \vec{A}) + \frac{\rho}{\varepsilon},$$

and altogether the situation is the same as before.

In the gauge transformation (Eqs. 1.321 and 1.322), the freedom of choosing u is left for the convenience depending on the problem. When $\vec{A}$ and Φ satisfy the so-called Lorentz condition,

$$\vec{\nabla} \cdot \vec{A} + \varepsilon\mu\frac{\partial \Phi}{\partial t} = 0, \qquad (1.323)$$

then the gauge is called the **Lorentz gauge**. Suppose there are solutions $\vec{A}_0$ and Φ_0 to Eqs. (1.319) and (1.320), then

$$\vec{A}_L = \vec{A}_0 + \vec{\nabla} u_L \qquad (1.324)$$

$$\Phi_L = \Phi_0 - \frac{\partial u_L}{\partial t} \qquad (1.325)$$

are also solutions as shown previously. Now,

$$\vec{\nabla} \cdot \vec{A}_L + \varepsilon\mu\frac{\partial \Phi_L}{\partial t} = \vec{\nabla} \cdot (\vec{A}_0 + \vec{\nabla} u_L) + \varepsilon\mu\frac{\partial}{\partial t}\left(\Phi_0 - \frac{\partial u_L}{\partial t}\right)$$

$$= \vec{\nabla} \cdot \vec{A}_0 + \varepsilon\mu\frac{\partial \Phi_0}{\partial t} + \vec{\nabla}^2 u_L - \varepsilon\mu\frac{\partial^2 u_L}{\partial t^2}.$$

By choosing u_L as a solution of

$$\vec{\nabla}^2 u_L - \varepsilon\mu\frac{\partial^2 u_L}{\partial t^2} = -\left(\vec{\nabla} \cdot \vec{A}_0 + \varepsilon\mu\frac{\partial \Phi_0}{\partial t}\right), \qquad (1.326)$$

obviously the Lorentz gauge condition

$$\vec{\nabla} \cdot \vec{A}_L + \varepsilon\mu\frac{\partial \Phi_L}{\partial t} = 0 \qquad (1.327)$$

is satisfied. Then the two inhomogeneous equations are decoupled, and we obtain two symmetric inhomogeneous wave equations

$$\vec{\nabla}^2 \vec{A}_L - \varepsilon\mu \frac{\partial^2 \vec{A}_L}{\partial t^2} = -\mu \vec{J} \qquad (1.328)$$

$$\vec{\nabla}^2 \Phi_L - \varepsilon\mu \frac{\partial^2 \Phi_L}{\partial t^2} = -\frac{\rho}{\varepsilon}. \qquad (1.329)$$

Another useful gauge is the **Coulomb gauge**, which satisfies the Coulomb condition

$$\vec{\nabla} \cdot \vec{A} = 0. \qquad (1.330)$$

Supposing $\vec{A}_0$ and Φ_0 are solutions to Eqs. (1.319) and (1.320), then

$$\vec{A}_C = \vec{A}_0 + \vec{\nabla} u_C \qquad (1.331)$$

$$\Phi_C = \Phi_0 - \frac{\partial u_C}{\partial t} \qquad (1.332)$$

are also solutions. Now observe

$$\vec{\nabla} \cdot \vec{A}_C = \vec{\nabla} \cdot (\vec{A}_0 + \vec{\nabla} u_C)$$
$$= \vec{\nabla} \cdot \vec{A}_0 + \vec{\nabla}^2 u_C.$$

By choosing u_C as a solution of

$$\vec{\nabla}^2 u_C = -\vec{\nabla} \cdot \vec{A}_0, \qquad (1.333)$$

we obtain that

$$\vec{\nabla} \cdot \vec{A}_C = 0 \qquad (1.334)$$

holds. Then the two inhomogeneous equations read

$$\vec{\nabla}^2 \vec{A}_C - \varepsilon\mu \frac{\partial^2 \vec{A}_C}{\partial t^2} = \varepsilon\mu \frac{\partial \vec{\nabla} \Phi_C}{\partial t} - \mu \vec{J} \qquad (1.335)$$

$$\vec{\nabla}^2 \Phi_C = -\frac{\rho}{\varepsilon}, \qquad (1.336)$$

and while there is no symmetry, it is convenient that the scalar potential Φ_C is the "instantaneous" Coulomb potential due to the time-dependent charge density $\rho(\vec{x}, t)$.

Sometimes a gauge where A_z, the z component of $\vec{A}$, is set to 0, is useful. In the proof of the existence of vector potentials we saw that it is possible to choose $\vec{A}$ in such a way; we now assume $\vec{A}$ does not satisfy this condition outright, and we attempt to bring it to this form. The gauge condition in this case is

$$A_z = 0. \tag{1.337}$$

Supposing $\vec{A}_0$ and Φ_0 are solutions of Eqs. (1.319) and (1.320), then

$$\vec{A}_S = \vec{A}_0 + \vec{\nabla} u_S \tag{1.338}$$

$$\Phi_S = \Phi_0 - \frac{\partial u_S}{\partial t} \tag{1.339}$$

are also solutions. In particular, we have

$$A_{Sz} = A_{0z} + \frac{\partial u_S}{\partial z},$$

and by choosing u_S as a solution of

$$\frac{\partial u_S}{\partial z} = -A_{0z}, \tag{1.340}$$

we obtain that, as needed,

$$A_{Sz} = 0. \tag{1.341}$$

In a similar way we can of course also construct vector potentials in which the x or y component vanishes.

We conclude the discussion of potentials with the special case of time-independent free space, in which the whole argument becomes very simple. With the constitutive relations $\vec{D} = \varepsilon_0 \vec{E}$ and $\vec{B} = \mu_0 \vec{H}$, Maxwell's equations are

$$\vec{\nabla} \cdot \vec{E} = 0, \quad \vec{\nabla} \times \vec{E} = \vec{0} \tag{1.342}$$

$$\vec{\nabla} \cdot \vec{B} = 0, \quad \vec{\nabla} \times \vec{B} = \vec{0}. \tag{1.343}$$

From the curl equations we infer that there are scalar potentials Φ and Φ_M for $\vec{E}$ and $\vec{B}$, respectively, such that we have

$$\vec{E} = -\vec{\nabla}\Phi \tag{1.344}$$

$$\vec{B} = -\vec{\nabla}\Phi_M. \tag{1.345}$$

Applying the divergence equations leads to Laplace equations for the electric and magnetic scalar potential:

$$\vec{\nabla}^2 \Phi = 0 \tag{1.346}$$

$$\vec{\nabla}^2 \Phi_M = 0. \tag{1.347}$$

1.5.3 Boundary Value Problems

The Maxwell theory yields a variety of partial differential equations specifying the fields and potentials, and the practical determination of these requires methods to solve **PDEs**. One of the important cases is the solution of the **Poisson equation**

$$\vec{\nabla}^2 \Phi = -\frac{\rho}{\varepsilon}, \tag{1.348}$$

which in the region where no charge exists reduces to the **Laplace equation**

$$\vec{\nabla}^2 \Phi = 0. \tag{1.349}$$

It is important to study under what conditions these **PDEs have solutions** and in what situations these solutions are **unique**. This question is similar in nature to the question of existence and uniqueness of solutions of ordinary differential equations, in which case even if existence is ensured, the specification of initial conditions is usually needed to assert uniqueness. However, in the case of PDEs, the situation is more complicated since usually conditions have to be specified not only at a point but also over extended regions. For example, a solution of the Poisson equation for $\rho = q = $ const. is given by

$$\Phi = \vec{a} \cdot (x, y, z) + \vec{b} \cdot (x^2, y^2, z^2) \tag{1.350}$$

which solves the equation as long as $b_1 + b_2 + b_3 = -q/2$; also, it is easy to construct more solutions. However, apparently it is not sufficient to specify the value of Φ at just one point to determine the exact values for $\vec{a}$ and $\vec{b}$; indeed, since there are five free parameters, at least five points would be needed.

Therefore, it is necessary to study in detail under what conditions solutions exist and in what situations they are unique. Of particular interest is the case of the boundary value problem, in which conditions are formulated along certain "boundaries," by which we usually mean simply connected sets that enclose the region of interest.

As we now show, if the values of the potential are specified over an entire boundary, then any solutions of the Poisson equation are unique. Let us assume we have given a region V which is enclosed by a simply connected boundary surface S. We consider **Green's first identity**,

$$\int_V (\phi \vec{\nabla}^2 \psi + \vec{\nabla}\phi \cdot \vec{\nabla}\psi) d^3x = \int_S \phi \frac{\partial \psi}{\partial n} da, \tag{1.351}$$

FIELDS AND POTENTIALS

where ϕ and ψ are scalar functions. Green's first identity is a direct consequence of the Gauss theorem,

$$\int_V \vec{\nabla} \cdot \vec{A} \, d^3x = \int_S \vec{A} \cdot \vec{n} \, da, \qquad (1.352)$$

where $\vec{A}$ is a vector function and $\vec{n}$ is the unit outward normal vector at the surface S. Choosing the special case of $\vec{A} = \phi \vec{\nabla} \psi$, and observing

$$\vec{\nabla} \cdot (\phi \vec{\nabla} \psi) = \phi \vec{\nabla}^2 \psi + \vec{\nabla} \phi \cdot \vec{\nabla} \psi$$

$$\phi \vec{\nabla} \psi \cdot \vec{n} = \phi \frac{\partial \psi}{\partial n},$$

yields Green's first identity.

Now let us assume there exist two solutions Φ_1 and Φ_2 for the same boundary value problem of the Poisson equation. Define the scalar function Ψ as the difference of Φ_2 and Φ_1,

$$\Psi = \Phi_1 - \Phi_2. \qquad (1.353)$$

Since Φ_1 and Φ_2 are both solutions,

$$\vec{\nabla}^2 \Phi_1 = -\frac{\rho}{\varepsilon} \text{ and } \vec{\nabla}^2 \Phi_2 = -\frac{\rho}{\varepsilon}$$

are satisfied inside the volume V, and we thus have

$$\vec{\nabla}^2 \Psi = 0 \text{ inside } V. \qquad (1.354)$$

Now apply Green's first identity (Eq. 1.351) to $\phi = \psi = \Psi$; we have

$$\int_V (\Psi \vec{\nabla}^2 \Psi + \vec{\nabla} \Psi \cdot \vec{\nabla} \Psi) d^3x = \int_S \Psi \frac{\partial \Psi}{\partial n} da \qquad (1.355)$$

$$\int_V |\vec{\nabla} \Psi|^2 d^3x = \int_S \Psi \frac{\partial \Psi}{\partial n} da, \qquad (1.356)$$

where (Eq. 1.354) is used in moving from the left-hand side of the first equation to the left-hand side of the second equation.

Now assume that at every point on the surface S, either $\Phi_1 = \Phi_2$ or $\partial \Phi_1 / \partial n = \partial \Phi_2 / \partial n$ holds. This entails that the right-hand side vanishes, and hence $\vec{\nabla} \Psi = $ const., even everywhere inside V, and

$$\Phi_1 = \Phi_2 + \text{const. inside } V, \qquad (1.357)$$

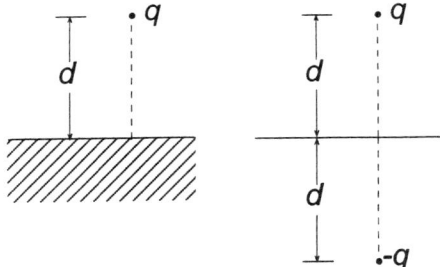

FIGURE 1.3. Method of images: A point charge put close to a conductor plane.

which means that the solution of the Poisson equation is unique up to a constant. Furthermore, if at least at one point on the boundary even $\Phi_1 = \Phi_2$, then $\Phi_1 = \Phi_2$ in all of V and hence the solution is completely unique.

The two most important special cases for the specifications of boundary conditions are those in which Φ is specified over the whole surface and $\partial \Phi_1 / \partial n$ is specified over the whole surface. The first case is usually referred to as the **Dirichlet boundary condition**, whereas the latter is usually called the **von Neumann boundary condition**. As discussed previously, however, a mixture of specification of value and derivative is sufficient for uniqueness, as long as at each point at least one of them is given.

In some situations, it is possible to obtain solutions to boundary value problems through simple symmetry arguments by suitably placing so-called **image charges** to obtain the proper potential on the boundary of the problem. A simple example is a point charge q put in the front of an infinite plane conductor with a distance d, as shown in Fig. 1.3.

The image charge $-q$, an equal and opposite charge, put at a distance d behind the plane ensures that the potential on the surface assumes the proper value, and thus that the unique potential has been found. Another example is a point charge q put in the region between two infinite conductor planes which form the angle π/k, as shown in Fig. 1.4.

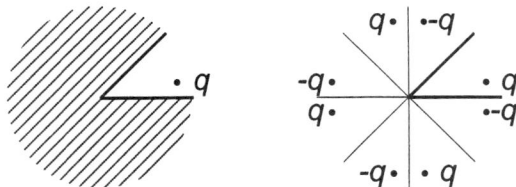

FIGURE 1.4. Method of images: a point charge put in a region between two conductor planes of angle π/k.

The problem is simulated by $2k - 1$ image charges put in the region out of interest to form mirror images. All $2k$ charges form potential 0 at each edge line in the picture.

As discussed in subsequent chapters, the method of images is often practically useful in order to numerically solve potential problems.

Chapter 2

Differential Algebraic Techniques

In this chapter, we discuss a technique that is at the core of further discussions in the following chapters. Differential algebraic techniques find their origin in the attempt to **solve analytic problems with algebraic means**. One of the initiators of the field was Liouville (Ritt, 1948) in connection with the problem of integration of functions and differential equations in finite terms. It was then significantly enhanced by Ritt (1932), who provided a complete algebraic theory of the solution of differential equations that are polynomials of the functions and their derivatives and that have meromorphic coefficients. Further development in the field is due to Kolchin (1973) and, with an eye on the algorithmic aspect, to Risch (1969, 1970, 1979).

Currently, the methods form the basis of many algorithms in modern formula manipulators, in which the treatment of differential equations and quadrature problems calls for the solution of analytic problems with algebraic means. Other important current work relying on differential algebraic methods is the practical study of differential equations under algebraic constraints, so-called differential algebraic equations (Ascher and Petzold, 1998; Griepentrog and Roswitha; Brenan et al., 1989; Matsuda, 1980).

For our purposes, we will concentrate on the use of differential algebraic techniques (Berz, 1986, 1987b, 1989) for the solution of differential equations and partial differential equations, in particular we discuss the efficient determination of **Taylor expansions of the flow** of differential equations in terms of initial conditions. The methods developed here have taken the perturbative treatment of flows $\vec{z}_f = \mathcal{M}(\vec{z}_i)$ of dynamical systems from the customary third (Brown, 1979a; Wollnik et al., 1987; Matsuo and Matsuda, 1976; Dragt et al., 1985) or fifth order (Berz et al., 1987) all the way to arbitrary order in a unified and straightforward way. Since its introduction, the method has been widely utilized in a large number of new map codes (Makino and Berz, 1996; Berz et al., 1996; Berz, 1992b, 1995a; Michelotti, 1990; Davis et al., 1993; van Zeijts and Neri, 1993; van Zeijts, 1993; Yan, 1993; Yan and Yan, 1990; Iselin, 1996).

2.1 FUNCTION SPACES AND THEIR ALGEBRAS

2.1.1 Floating Point Numbers and Intervals

The basic idea behind this method is to bring the treatment of **functions** and the operations on them to the computer in a similar way as the treatment of **num-**

bers. In a strict sense, neither functions (e.g., those that are C^∞, infinitely often differentiable) nor numbers (e.g., the real R) can be treated on a computer since neither can in general be represented by a finite amount of information; after all, a real number "really" (pun intended) is an equivalence class of bounded Cauchy sequences of rational numbers.

However, from the early days of computers we are used to dealing with numbers by **extracting information deemed relevant**, which in practice usually means the approximation by **floating point numbers** with finitely many digits. In a formal sense, this is possible since for every one of the operations on real numbers, such as addition and multiplication, we can craft an **adjoint** operation on the floating point numbers such that the following diagram commutes:

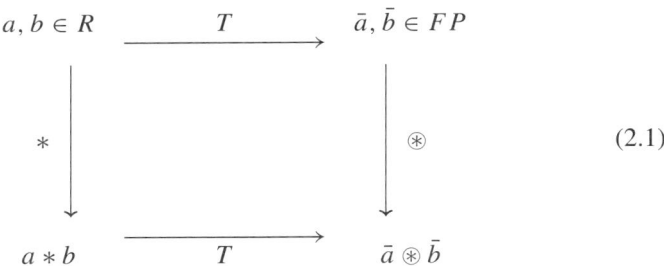

(2.1)

Of course, in reality the diagrams commute only "approximately," which typically makes the errors grow over time.

The approximate character of these arguments can be removed by representing a real number not by one floating point number but rather by **interval** floating point numbers providing a rigorous upper and lower bound (Hansen, 1969; Moore, 1979, 1988; Alefeld and Herzberger, 1983; Kaucher and Miranker, 1984; Kearfott and Kreinovich, 1996).

By rounding operations down for lower bounds and up for upper bounds, rigorous bounds can be determined for sums and products, and adjoint operations can be made such that diagram (2.1) commutes exactly. In practice, while always maintaining rigor, the method sometimes becomes somewhat pessimistic because over time the intervals often have a tendency to grow. This so-called dependency problem can be alleviated significantly in several ways, and a rather automated approach is discussed in (Makino and Berz, 1996).

2.1.2 Representations of Functions

Historically, the treatment of **functions** in numerics has been done based on the treatment of **numbers** and, as a result, virtually all classical numerical algorithms are based on the mere evaluation of functions at specific points. As a consequence, numerical methods for differentiation, which offer one way to attempt to compute Taylor representations of functions, are very cumbersome and prone to inaccuracies because of cancellation of digits, and they are not useful in practice for our purposes.

FUNCTION SPACES AND THEIR ALGEBRAS

The success of the new methods is based on the observation that it is possible to extract more information about a function than its mere values. Indeed, attempting to extend the commuting diagram in Eq. (2.1) to functions, one can demand the operation T to be the extraction of the Taylor coefficients of a prespecified order n of the function. In mathematical terms, T is an equivalence relation, and the application of T corresponds to the transition from the function to the **equivalence class** comprising all those functions with identical Taylor expansion to order n.

Since Taylor coefficients of order n for sums and products of functions as well as scalar products with reals can be computed from those of the summands and factors, it is clear that the diagram can be made to commute; indeed, except for the underlying inaccuracy of the floating point arithmetic, it will commute exactly. In mathematical terms, this means that the set of equivalence classes of functions can be endowed with well-defined operations, leading to the so-called **truncated power series algebra (TPSA)** (Berz, 1986, 1987b).

This fact was utilized in the first paper on the subject (Berz, 1987b), which led to a method to extract maps to any desired order from a computer algorithm that integrates orbits numerically. Similar to the need for algorithms within floating point arithmetic, the development of **algorithms for functions** followed, including methods to perform composition of functions, to invert them, to solve nonlinear systems explicitly, and to introduce the treatment of common elementary functions; many of these algorithms (Berz, 1998, 1991b) will be discussed later.

However, it became apparent quickly (Berz, 1988b, 1989) that this represents only a halfway point, and one should **proceed beyond mere arithmetic operations** on function spaces of addition and multiplication and consider their **analytic operations of differentiation and integration**. This resulted in the recognition of the underlying **differential algebraic structure** and its practical exploitation, based on the commuting diagrams for addition, multiplication, and differentiation and their inverses:

$$
\begin{array}{ccccccc}
f, g & \xrightarrow{T} & F, G & \quad & f, g & \xrightarrow{T} & F, G \\
{\scriptstyle +, -}\Big\downarrow & & \Big\downarrow{\scriptstyle \oplus, \ominus} & & {\scriptstyle \cdot, /}\Big\downarrow & & \Big\downarrow{\scriptstyle \odot, \oslash} \\
f \pm g & \xrightarrow{T} & F \overset{\oplus}{\ominus} G & & f/g & \xrightarrow{T} & F \overset{\odot}{\oslash} G \\
\end{array}
$$

$$
\begin{array}{ccc}
f & \xrightarrow{T} & F \\
{\scriptstyle \partial, \partial^{-1}}\Big\downarrow & & \Big\downarrow{\scriptstyle \partial_\bigcirc, \partial_\bigcirc^{-1}} \\
\partial f, \partial^{-1} f & \xrightarrow{T} & \partial_\bigcirc F, \partial_\bigcirc^{-1} F
\end{array}
\qquad (2.2)
$$

The theory of this approach will be developed in detail in this chapter.

Of course, the question of what constitutes "information deemed relevant" for functions does not necessarily have a unique answer. Formula manipulators, for example, attack the problem from a different perspective by attempting to algebraically express functions in terms of certain elementary functions linked by algebraic operations and composition. In practice, the Achilles' heel of this approach is the complexity that such representations can take after only a few operations. Compared to the mere Taylor expansion, however, they have the advantage of rigorously representing the function under consideration. Modern extensions of the Taylor method (Makino and Berz, 1996) have the ability to overcome this difficulty by providing fully rigorous bounds for remainder terms; but these methods go beyond the scope of this book.

2.1.3 Algebras and Differential Algebras

Before proceeding, it is worthwhile to put into perspective a variety of different concepts that were introduced to the field in connection with the previously discussed developments. We do this in parallel to establishing the scope of the further developments in which differential algebraic techniques will be applied extensively.

The first and simplest structure that was introduced (Berz, 1986, 1987b) is **TPSA**. This is the structure that results when the equivalence classes of functions are endowed with arithmetic such that the diagrams in Eq. (2.1) commute for the basic operations of addition, multiplication, and scalar multiplication. Addition and scalar multiplication lead to a **vector space**, and the multiplication operation turns it into a commutative **algebra.** In many respects, together with the polynomial algebras, this structure is an archetypal nontrivial algebra, and in fact it can be embedded into many larger and more interesting algebras.

TPSA can be equipped with an **order**, and then it contains differentials, i.e., infinitely small numbers. This fact triggered the study of such nonarchimedean structures in more detail and led to the introduction of a foundation of analysis (Berz, 1996, 1990a, 1994, 1999) on a larger and for such purposes much more useful structure, the Levi–Civita field discussed in Section 2.3.5. The Levi–Civita field is the smallest nonarchimedean extension of the real numbers that is algebraically and Cauchy complete, and many of the basic theorems of calculus can be proved in a similar way as in R. Furthermore, concepts such as Delta functions and the idea of derivatives as differential quotients can be formulated rigorously and integrated seamlessly into the theory. On the practical end, based on the latter concept, there are also several improvements regarding methods of computational differentiation (Shamseddine and Berz, 1996,1997).

The final concept that is connected to our methods and worth study is the technique of **automatic differentiation** (Berz et al., 1996; Griewank and Corliss, 1991; Berz, 1999). The purpose of this discipline is the automated transformation of existing code in such a way that derivatives of functional relationships between variables are calculated along with the original code. Besides the significantly increased computational accuracy compared to numerical differentiation, a striking advantage of this approach is the fact that in the so-called reverse mode it is actually possible in principle to calculate gradients in v variables in a **fixed amount of effort**; independent of v, in the optimal case the entire gradient can be obtained with a cost equalling only a few times the cost of the evaluation of the original functions, in stark contrast to numerical differentiation which requires $(v + 1)$ times the original cost.

In practice, multivariate automatic differentiation is almost exclusively restricted to **first order,** and as such is not directly useful for our purposes. One reason for this situation is the fact that conventional numerical algorithms avoid higher derivatives as much as possible because of the well-known difficulties when trying to obtain them via numerical differentiation, which for a long time represented the only available approach. On the other hand, the previously mentioned savings that are possible for linear derivatives are much harder to obtain in the same way for higher orders.

Altogether, the challenge in automatic differentiation is more **reminiscent of sparse matrix techniques** for management and manipulation of Jacobians than of a power series technique. It is perhaps also worth mentioning that because of the need for code restructuring in order to obtain performance, there is a certain reluctance in the community toward the use of the word "automatic." Mostly in order to avoid the impression of making false pretence, the technique has recently been referred to as **computational differentiation**.

Only very recently are other groups in computational differentiation picking up at least on second order (Abate *et al.*, 1997), but so far the only software for derivatives beyond order two listed in the automatic differentiation tool compendium (Bischof and Dilley,) is in fact the package DAFOR (Berz, 1991a, 1987a, 1990e) consisting of the FORTRAN precompiler DAPRE and the arbitrary order DA package that is also used as the power series engine in the code COSY INFINITY (Berz, 1997a; Berz *et al.*, ; Makino and Berz, 1999; Berz *et al.*, 1996; Makino and Berz, 1996).

As alluded to previously, the power of TPSA can be enhanced by the introduction of derivations ∂ and their inverses, corresponding to the differentiation and integration on the space of functions. The resulting structure, a **differential algebra (DA)**, allows the direct treatment of many questions connected with differentiation and integration of functions, including the solution of the ODEs $d\vec{x}/dt = \vec{f}(\vec{x}, t)$ describing the motion and PDEs describing the fields.

2.2 Taylor Differential Algebras

2.2.1 The Minimal Differential Algebra

We begin our study of differential algebraic structures with a particularly simple structure which is of historical significance and constitutes the simplest nontrivial differential algebra (Berz, 1989). In fact, the reals under their regular arithmetic with a derivation ∂ that vanishes identically trivially form a differential algebra, but it is not a very interesting one. To obtain the first nontrivial differential algebra, we have to move to R^2. Let us consider the set of all ordered pairs (q_0, q_1), where q_0 and q_1 are real numbers. We define addition, scalar multiplication and vector multiplication as follows:

$$(q_0, q_1) + (r_0, r_1) = (q_0 + r_0, q_1 + r_1) \tag{2.3}$$

$$t \cdot (q_0, q_1) = (t \cdot q_0, t \cdot q_1) \tag{2.4}$$

$$(q_0, q_1) \cdot (r_0, r_1) = (q_0 \cdot r_0, q_0 \cdot r_1 + q_1 \cdot r_0). \tag{2.5}$$

The ordered pairs with the arithmetic are called $_1D_1$. The first two operations are the familiar **vector space** structure of R^2. The multiplication, on the other hand, looks similar to that in the complex numbers; except here, as one sees easily, $(0, 1) \cdot (0, 1)$ does not equal $(-1, 0)$ but rather $(0, 0)$. Therefore, the element

$$d \stackrel{\text{def}}{=} (0, 1) \tag{2.6}$$

plays a quite different role than the imaginary unit i in the complex numbers. The multiplication of vectors is seen to have $(1, 0)$ as the unity element and to be commutative, i.e., $(q_0, q_1) \cdot (r_0, r_1) = (r_0, r_1) \cdot (q_0, q_1)$. It is also associative, i.e., $(q_0, q_1) \cdot \{(r_0, r_1) \cdot (s_0, s_1)\} = \{(q_0, q_1) \cdot (r_0, r_1)\} \cdot (s_0, s_1)$. It is also distributive with respect to addition, i.e., $(q_0, q_1) \cdot \{(r_0, r_1) + (s_0, s_1)\} = (q_0, q_1) \cdot (r_0, r_1) + (q_0, q_1) \cdot (s_0, s_1)$, and so the two operations $+$ and $\cdot$ form a (commutative) **ring**. Together, the three operations form an **algebra**. Like the complex numbers, they do form an **extension of the real numbers**; because $(r, 0) + (s, 0) = (r + s, 0)$ and $(r, 0) \cdot (s, 0) = (r \cdot s, 0)$, the pairs $(r, 0)$ behave like real numbers, and thus as in C, the reals can be embedded.

However $_1D_1$ is not a field. Indeed, it can be shown that $(q_0, q_1) \in {}_1D_1$ has a **multiplicative inverse** in $_1D_1$ if and only if $q_0 \neq 0$. If $q_0 \neq 0$ then

$$(q_0, q_1)^{-1} = \left(\frac{1}{q_0}, \frac{-q_1}{q_0^2} \right). \tag{2.7}$$

We also note that if q_0 is positive, then $(q_0, q_1) \in {}_1D_1$ has a **root**

$$\sqrt{(q_0, q_1)} = \left(\sqrt{q_0}, \frac{q_1}{2\sqrt{q_0}} \right), \tag{2.8}$$

as simple arithmetic shows.

The structure $_1D_1$ plays a quite important **role in the theory of real algebras**. All two-dimensional algebras, which are the smallest nontrivial algebras, are isomorphic to only three different kinds of algebras: the **complex numbers**, the so-called **dual numbers** in which $(0, 1) \cdot (0, 1) = (1, 0)$, and the **algebra** $_1D_1$.

In fact, this rather unique algebra has some interesting properties. One of them is that $_1D_1$ can be equipped with an **order** that is compatible with its algebraic operations. It is this requirement of compatibility of the order with addition and multiplication that makes R, the field of real numbers, an ordered field. On the other hand, while many orders can be introduced on the field of complex numbers, none can be made compatible with the field's arithmetic; we say that the complex numbers cannot be ordered.

Given two elements (q_0, q_1) and (r_0, r_1) in $_1D_1$, we define

$$(q_0, q_1) < (r_0, r_1) \text{ if } q_0 < r_0 \text{ or } (q_0 = r_0 \text{ and } q_1 < r_1)$$
$$(q_0, q_1) > (r_0, r_1) \text{ if } (r_0, r_1) < (q_0, q_1)$$
$$(q_0, q_1) = (r_0, r_1) \text{ if } q_0 = r_0 \text{ and } q_1 = r_1. \tag{2.9}$$

It follows immediately from this order definition that for any two elements (q_0, q_1) and (r_0, r_1) in $_1D_1$, one and only one of $(q_0, q_1) > (r_0, r_1)$, $(q_0, q_1) = (r_0, r_1)$ and $(q_0, q_1) < (r_0, r_1)$ holds. We say that the order is total; alternatively, $_1D_1$ is totally ordered. The order is compatible with addition and multiplication; for all (q_0, q_1), (r_0, r_1), $(s_0, s_1) \in {}_1D_1$, we have $(q_0, q_1) < (r_0, r_1) \implies (q_0, q_1) + (s_0, s_1) < (r_0, r_1) + (s_0, s_1)$, and $(q_0, q_1) < (r_0, r_1)$; and $(s_0, s_1) > (0, 0) = 0 \implies (q_0, q_1) \cdot (s_0, s_1) < (r_0, r_1) \cdot (s_0, s_1)$. We also see that the order on the reals embedded in $_1D_1$ is compatible with the order there.

The number d defined previously has the interesting property that it is positive but smaller than any positive real number; indeed, we have

$$(0, 0) < (0, 1) < (r, 0) = r. \tag{2.10}$$

We say that d is **infinitely small**. Alternatively, d is also called an **infinitesimal** or a **differential**. In fact, the number d is so small that its square vanishes, as shown previously. Since for any $(q_0, q_1) \in {}_1D_1$ we have

$$(q_0, q_1) = (q_0, 0) + (0, q_1) = q_0 + d \cdot q_1, \tag{2.11}$$

we call the first component of (q_0, q_1) the real part and the second component the differential part.

The number d has two more interesting properties. It has neither a multiplicative inverse nor an nth root in $_1D_1$ for any $n > 1$. For any (q_0, q_1) in $_1D_1$, we have

$$(q_0, q_1) \cdot (0, 1) = (0, q_0) \neq (1, 0). \tag{2.12}$$

On the other hand, one easily shows by induction on n that

$$(q_0, q_1)^n = (q_0^n, n \cdot q_0^{n-1} q_1) \text{ for all } (q_0, q_1) \in {}_1D_1 \text{ and all } n > 1. \tag{2.13}$$

Therefore, if d has an nth root $(q_0, q_1) \in {}_1D_1$, then

$$(q_0^n, n \cdot q_0^{n-1} q_1) = (0, 1), \tag{2.14}$$

from which we have simultaneously $q_0 = 0$ and $n \cdot q_0^{n-1} q_1 = 1$—a contradiction, for $n > 1$. Therefore, d has no nth root in ${}_1D_1$.

We next introduce a map ∂ from ${}_1D_1$ into itself, which will prove to be a derivation and will turn the algebra ${}_1D_1$ into a differential algebra.

Define $\partial : {}_1D_1 \longmapsto {}_1D_1$ by

$$\partial(q_0, q_1) = (0, q_1). \tag{2.15}$$

Note that

$$\partial\{(q_0, q_1) + (r_0, r_1)\} = \partial(q_0 + r_0, q_1 + r_1) = (0, q_1 + r_1)$$
$$= (0, q_1) + (0, r_1) = \partial(q_0, q_1) + \partial(r_0, r_1) \tag{2.16}$$

and

$$\partial\{(q_0, q_1) \cdot (r_0, r_1)\} = \partial(q_0 \cdot r_0, q_0 \cdot r_1 + q_1 \cdot r_0) = (0, q_0 \cdot r_1 + q_1 \cdot r_0)$$
$$= (0, q_1) \cdot (r_0, r_1) + (q_0, q_1) \cdot (0, r_1)$$
$$= \{\partial(q_0, q_1)\} \cdot (r_0, r_1) + (q_0, q_1) \cdot \{\partial(r_0, r_1)\}. \tag{2.17}$$

This holds for all (q_0, q_1), $(r_0, r_1) \in {}_1D_1$. Therefore, ∂ is a derivation, and hence $({}_1D_1, \partial)$ is a differential algebra.

There is another way of introducing a derivation on ${}_1D_1$, which maps into the real numbers, by setting $\partial^{(r)}(q_0, q_1) = q_1$. Also $\partial^{(r)}$ satisfies the previous addition and multiplication rules; but because it does not map ${}_1D_1$ into itself, it does not introduce a differential algebraic structure on ${}_1D_1$. The derivation ∂ is apparently connected to $\partial^{(r)}$ by $\partial = d \cdot \partial^{(r)}$, and hence has a resemblance to a **vector field** (Eq. 1.7).

For the purposes of beam physics, the most important aspect of ${}_1D_1$ is that it can be used for the automated computation of **derivatives.** This is based on the following observation. Let us assume that we have given the values and derivatives of two functions f and g at the origin; we put these into the real and differential components of two vectors in ${}_1D_1$, which have the form $(f(0), f'(0))$ and $(g(0), g'(0))$. Let us assume we are interested in the derivative of the product $f \cdot g$, which is given by $f'(0) \cdot g(0) + f(0) \cdot g'(0)$. Apparently, this value appears in the second component of the product $(f(0), f'(0)) \cdot (g(0), g'(0))$, whereas the first component of the product happens to be $f(0) \cdot g(0)$. Therefore, if two vectors contain the values and derivative of two functions, their product contains the values and derivatives of the product function. Defining the operation [ ] from the space of differentiable functions to $_1D_1$ via

$$[f] = (f(0), f'(0)), \qquad (2.18)$$

we thus have

$$[f + g] = [f] + [g] \qquad (2.19)$$
$$[f \cdot g] = [f] \cdot [g], \qquad (2.20)$$

and hence the operation closes the two upper commuting diagrams in Eq. (2.2). Similarly, the addition in $_1D_1$ is compatible with the sum rule for derivatives.

In this light, the derivations of rules for multiplicative inverses (Eq. 2.7) and roots (Eq. 2.8) appear as a fully algebraic derivation of the quotient and root rules of calculus, without any explicit use of limits of differential quotients; this is a small example of generic differential algebraic reasoning.

This observation about derivatives can now be used to compute derivatives of many kinds of functions algebraically by merely applying arithmetic rules on $_1D_1$, beginning with the value and derivative of the identity function. We illustrate this with an example using the function

$$f(x) = \frac{1}{x + (1/x)}. \qquad (2.21)$$

The derivative of the function is

$$f'(x) = \frac{(1/x^2) - 1}{(x + (1/x))^2}. \qquad (2.22)$$

Suppose we are interested in the value of the function and its derivative at $x = 3$. We obtain

$$f(3) = \frac{3}{10}, \quad f'(3) = -\frac{2}{25}. \qquad (2.23)$$

Now take the definition of the function f in Eq. (2.21) and evaluate it at value and derivative of the identity function $(3, 1) = 3 + d$. We obtain

$$f((3, 1)) = \frac{1}{(3, 1) + 1/(3, 1)} = \frac{1}{(3, 1) + (1/3, -1/9)}$$
$$= \frac{1}{(10/3, 8/9)} = \left(\frac{3}{10}, -\frac{8}{9} \bigg/ \frac{100}{9}\right) = \left(\frac{3}{10}, -\frac{2}{25}\right). \qquad (2.24)$$

As can be seen, after the evaluation of the function the real part of the result is the value of the function at $x = 3$, whereas the differential part is the value of the derivative of the function at $x = 3$. This result is not a coincidence. Using Eq. (2.18), we readily obtain that for all g with $g(0) \neq 0$ in $_1D_1$,

$$[1/g] = [1]/[g] = 1/[g]. \tag{2.25}$$

Thus,

$$\begin{aligned}
{[f(x)]} &= \left[\frac{1}{x+1/x}\right] = \frac{1}{[x+1/x]} \\
&= \frac{1}{[x]+[1/x]} = \frac{1}{[x]+1/[x]} \\
&= f([x]).
\end{aligned} \tag{2.26}$$

Since for a real x, we have $[x] = (x, 1) = x + d$, and $[f(x)] = (f(x), f'(x))$, apparently

$$(f(3), f'(3)) = f((3+d)). \tag{2.27}$$

The method can be generalized to allow for the treatment of common intrinsic functions g_i like sin, exp, by setting $\sin(q_0, q_1) = (\sin(q_0), q_1 \cos(q_0))$, $\exp(q_0, q_1) = (\exp(q_0), q_1 \exp(q_0))$, or more generally,

$$\begin{aligned}
g_i([f]) &= [g_i(f)] \quad \text{or} \\
g_i((q_0, q_1)) &= (g_i(q_0), q_1 g'_i(q_0))
\end{aligned} \tag{2.28}$$

By virtue of Eqs. (2.5) and (2.28), we see that any function f representable by finitely many additions, subtractions, multiplications, divisions, and intrinsic functions on $_1D_1$ satisfies the important relationship

$$[f(x)] = f([x]). \tag{2.29}$$

Therefore, for all $r \in R \subset {_1D_1}$, we can write

$$(f(r), f'(r)) = f(r+d), \tag{2.30}$$

from which we infer that $f(r)$ and $f'(r)$ are equal to the real and differential parts of $f(r+d)$, respectively.

Note that Eq. (2.30) can be rewritten as

$$f(r+d) = f(r) + d \cdot f'(r). \tag{2.31}$$

This resembles $f(x + \Delta x) \approx f(x) + \Delta x \cdot f'(x)$, in which case the approximation becomes increasingly better for smaller Δx. Here, we choose an infinitely small

Δx, and the error turns out to be zero. In Section 2.3.5 we provide a more detailed analysis of this interesting phenomenon and at the same time obtain some interesting new calculus.

2.2.2 The Differential Algebra $_nD_v$

In this section, we introduce a differential algebra that allows the computation of derivatives up to order n of functions in v variables. Similar to before, it is based on taking the space $C^n(R^v)$, the collection of n times continuously differentiable functions on R^v. As a vector space over R, $C^n(R^v)$ **is infinite dimensional**; for example, the set of all monomials x_1^i is linearly independent.

On the space $C^n(R^v)$, we now introduce an equivalence relation. For f and g in $C^n(R^v)$, we say $f =_n g$ if and only if $f(0) = g(0)$ and all the partial derivatives of f and g agree at 0 up to order n. The newly introduced relation $=_n$ apparently satisfies

$$f =_n f \text{ for all } f \in C^n(R^v),$$
$$f =_n g \implies g =_n f \text{ for all } f, g \in C^n(R^v) \text{ and}$$
$$f =_n g \text{ and } g =_n h \implies f =_n h \text{ for all } f, g, h \in C^n(R^v). \quad (2.32)$$

Thus, $=_n$ is an **equivalence relation**. We now group all those elements that are related to f together in one set, the **equivalence class** $[f]$ of the function f. The resulting equivalence classes are often referred to as **DA vectors** or **DA numbers**. Intuitively, each of these classes is then specified by a particular collection of partial derivatives in all v variables up to order n. We call the collection of all these classes $_nD_v$.

Now we try to carry over the arithmetic on $C^n(R^v)$ into $_nD_v$. We observe that if we know the values and derivatives of two functions f and g, we can infer the corresponding values and derivatives of $f + g$ (by mere addition) and $f \cdot g$ (by virtue of the product rule). Therefore, we can introduce arithmetic on the classes in $_nD_v$ via

$$[f] + [g] = [f + g] \quad (2.33)$$
$$t \cdot [f] = [t \cdot f] \quad (2.34)$$
$$[f] \cdot [g] = [f \cdot g]. \quad (2.35)$$

Under these operations, $_nD_v$ becomes an **algebra**. For each $k \in \{1, \ldots, v\}$, define the map ∂_k from $_nD_v$ to $_nD_v$ for f via

$$\partial_k[f] = \left[p_k \cdot \frac{\partial f}{\partial x_k} \right], \quad (2.36)$$

where

$$p_k(x_1, \ldots, x_v) = x_k \qquad (2.37)$$

projects out the kth component of the identity function. Note that nth order derivatives of $p_k \cdot (\partial f / \partial x_k)$ involve the product of the value of p_k at the origin with $(n+1)$st order derivatives of f at the origin. Even though we do not know the values of these latter derivatives at the origin, we do know the results of their multiplications with the value of p_k at the origin, which is 0. It is easy to show that, for all $k = 1, \ldots, v$ and for all $[f], [g] \in {}_n D_v$, we have:

$$\partial_k([f] + [g]) = \partial_k[f] + \partial_k[g] \qquad (2.38)$$

$$\partial_k([f] \cdot [g]) = [f] \cdot (\partial_k[g]) + (\partial_k[f]) \cdot [g]. \qquad (2.39)$$

Therefore, ∂_k is a derivation for all k, and hence $({}_n D_v, \partial_1, \ldots, \partial_v)$ is a **differential algebra**. Indeed, ${}_n D_v$ is a generalization of ${}_1 D_1$, which is obtained as the special case $n = 1$ and $v = 1$.

2.2.3 Generators, Bases, and Order

Similar to the case of ${}_1 D_1$, the differential algebra ${}_n D_v$ also contains the real numbers; in fact, by mapping the real number r to the class $[r]$ of the constant function that assumes the value r everywhere, we see that the arithmetic on ${}_n D_v$ corresponds to the arithmetic on R.

Next, we want to assess the dimension of ${}_n D_v$. We define the special numbers d_k as follows:

$$d_k = [x_k]. \qquad (2.40)$$

We observe that f lies in the same class as its Taylor polynomial T_f of order n around the origin; they have the same function values and derivatives up to order n, and nothing else matters. Therefore,

$$[f] = [T_f]. \qquad (2.41)$$

Denoting the Taylor coefficients of the Taylor polynomial T_f of f as $c_{j_1,\ldots,j_v}$, we have

$$T_f(x_1, \ldots, x_v) = \sum_{j_1 + \cdots + j_v \leq n} c_{j_1,\ldots,j_v} \cdot x_1^{j_1} \cdot \cdots \cdot x_v^{j_v}, \qquad (2.42)$$

with

$$c_{j_1,\ldots,j_v} = \frac{1}{j_1! \cdot \cdots \cdot j_v!} \cdot \frac{\partial^{j_1 + \cdots + j_v} f}{\partial x_1^{j_1} \cdot \cdots \cdot \partial x_v^{j_v}} \qquad (2.43)$$

and thus

$$[f] = [T_f] = \left[\sum_{j_1+\cdots+j_v \leq n} c_{j_1,\ldots,j_v} \cdot x_1^{j_1} \cdots\cdots x_v^{j_v}\right]$$
$$= \sum_{j_1+\cdots+j_v \leq n} c_{j_1,\ldots,j_v} \cdot d_1^{j_1} \cdots\cdots d_v^{j_v}, \quad (2.44)$$

where, in the last step, use has been made of $[a+b] = [a]+[b]$ and $[a \cdot b] = [a] \cdot [b]$. Therefore, the set $\{1, d_k : k = 1, 2, \ldots, v\}$ generates $_nD_v$; that is, any element of $_nD_v$ can be obtained from 1 and the d_k's via addition and multiplication. Therefore, as an algebra, $_nD_v$ has $(v+1)$ **generators**. Furthermore, the terms $d_1^{j_1} \cdots\cdots d_v^{j_v}$ for $0 \leq j_1 + \cdots + j_v \leq n$ form a generating system for the vector space $_nD_v$. They are in fact also linearly independent since the only way to produce the zero class by summing up linear combinations of $d_1^{j_1} \cdots\cdots d_v^{j_v}$ is by choosing all their coefficients to be zero; so the $d_1^{j_1} \cdots\cdots d_v^{j_v}$ actually form a **basis** of $_nD_v$, called the **differential basis**, for reasons that will become clear later. We now want to study how many basis elements exist and thus find the dimension of $_nD_v$. To this end, we represent each v-tuple of numbers $j_1, \ldots, j_v$ that satisfy $0 \leq j_1 + \cdots + j_v \leq n$ by the following unique sequence of ones and zeros:

$$(\underbrace{1, 1, \ldots, 1}_{j_1 \text{ times}}, 0, \underbrace{1, 1, \ldots, 1}_{j_2 \text{ times}}, 0, \ldots, 0, \underbrace{1, 1, \ldots, 1}_{j_v \text{ times}}, 0, \underbrace{1, 1, \ldots, 1}_{n-j_1-\cdots-j_v \text{ times}}). \quad (2.45)$$

Apparently, the value 1 appears a total of n times, and the value 0 appears v times, for a total length of $n + v$. The total number of such vectors, which is then equal to the number of basis vectors, is given by the number of ways that 0 can be distributed in the $n + v$ slots; this number is given by

$$N(n, v) \stackrel{\text{def}}{=} \dim\,_nD_v = \binom{n+v}{v} = \frac{(n+v)!}{n!\,v!}. \quad (2.46)$$

For later use, we also note that a similar argument shows that the basis vectors belonging to the precise order n can be described by the sequence

$$(\underbrace{1, 1, \ldots, 1}_{j_1 \text{ times}}, 0, \underbrace{1, 1, \ldots, 1}_{j_2 \text{ times}}, 0, \ldots, 0, \underbrace{1, 1, \ldots, 1}_{j_v \text{ times}}), \quad (2.47)$$

which has length $n + v - 1$ and $v - 1$ zeros, yielding the following total number of possibilities:

$$M(n, v) = \binom{n+v-1}{v-1} = N(n, v-1). \quad (2.48)$$

TABLE I

THE DIMENSION OF $_nD_v$ AS A FUNCTION OF n AND v, FOR $n, v = 1, \ldots, 10$

	1	2	3	4	5	6	7	8	9	10
1	2	3	4	5	6	7	8	9	10	11
2	3	6	10	15	21	28	36	45	55	66
3	4	10	20	35	56	84	120	165	220	286
4	5	15	35	70	126	210	330	495	715	1,001
5	6	21	56	126	252	462	792	1,287	2,002	3,003
6	7	28	84	210	462	924	1,716	3,003	5,005	8,008
7	8	36	120	330	792	1,716	3,432	6,435	11,440	19,448
8	9	45	165	495	1,287	3,003	6,435	12,870	24,310	43,758
9	10	55	220	715	2,002	5,005	11,440	24,310	48,620	92,378
10	11	66	286	1,001	3,003	8,008	19,448	43,758	92,378	184,756

Since the number of terms up to order n equals the number of terms up to order $n - 1$ plus the number of terms of exact order n, we have

$$N(n, v) = N(n - 1, v) + N(n, v - 1), \qquad (2.49)$$

which also follows directly from the properties of binomial coefficients.

Table I gives the dimension of $_nD_v$ as a function of n and v, for $n, v = 1, \ldots, 10$. Note that the table clearly shows the recursive relation (Eq. 2.49). Moreover, the table is symmetric about the diagonal, which is a direct result of the fact that the dimension of $_nD_v$, $N(n, v)$, is symmetric in n and v, i.e., $N(n, v) = N(v, n)$.

Since $_nD_v$ is finite dimensional, it cannot be a field. Otherwise, this would contradict Zermelo's Theorem, which asserts that the only finite dimensional vector spaces over R which are fields are C and the quaternions (in which multiplication is not commutative). This will be confirmed again later by the existence of nilpotent elements (the infinitesimals), which can have no multiplicative inverse.

Similar to the structure $_1D_1$, $_nD_v$ can also be ordered (Berz, 1990b). Given a DA number, we look for the terms that have the lowest sum of exponents of the d_i that occurs. From all combinations of these exponents, we find the ones with the highest exponent in d_1, and from these the ones with the highest exponent of d_2, and so on. The resulting combination is called the **leading term** and its coefficient the **leading coefficient**. For example, if a DA vector has a constant part, then the constant part is the leading term; if the constant part vanishes but the coefficient of d_1 is nonzero, then d_1 will be the leading term. We now say the DA number is **positive** if its leading coefficient is positive.

We conclude that if a and b are positive, then so are both $a + b$ and $a \cdot b$. The leading term of $a + b$ is either the leading term of a or the leading term of b, and if the leading terms are equal, the leading coefficients cannot add up to zero since they are both positive. The leading term of $a \cdot b$ is the product of the leading term of a and the leading term of b, and the leading coefficient is the product of the leading coefficients of a and b.

For $a \neq b$ in ${}_nD_v$, we say

$$\begin{cases} a > b & \text{if } a - b \text{ is positive} \\ a < b & \text{otherwise} \end{cases}. \tag{2.50}$$

Then, for any a, b in ${}_nD_v$, exactly one of each $a = b$, $a > b$ and $a < b$ holds. Moreover, for any a, b, c in ${}_nD_v$, $a < b$ and $b < c$ entails that $a < c$. Therefore, the order is total; alternatively, we say that ${}_nD_v$ is totally ordered. Moreover, the order is compatible with addition and multiplication, i.e., for all a, b, c in ${}_nD_v$ we have that

$$\begin{cases} a < b & \Longrightarrow a + c < b + c \\ a < b, \ c > 0 & \Longrightarrow a \cdot c < b \cdot c \end{cases}. \tag{2.51}$$

We call the order **lexicographic** because when comparing two numbers a and b, we look at the leading coefficient of $a - b$. This is equivalent to starting by comparing the leftmost terms of a and b, and then moving right until we have found a term in which the two numbers disagree. We also introduce the **absolute value** via

$$|a| = \begin{cases} a & \text{if } a \geq 0 \\ -a & \text{else} \end{cases}. \tag{2.52}$$

Note that the absolute value maps from ${}_nD_v$ into itself, and not into R as in the norm introduced later.

Observe that for all $r > 0$ in R, and for all k in $\{1, \ldots, v\}$,

$$0 < d_k < r. \tag{2.53}$$

Therefore, all d_k are infinitely small or **infinitesimal**.

Now we can order the basis defined in Eq. (2.44) as follows:

$$M_1 = 1 > M_2 = d_1 > M_3 = d_2 > \cdots > M_{v+1} = d_v > \cdots$$
$$> M_{N-1} > M_N = d_v^n; \tag{2.54}$$

and an arbitrary element $a = [f] \in {}_nD_v$ can now be represented as an ordered N-tuple

$$a = (c_1, c_2, \ldots, c_N), \tag{2.55}$$

where the c_i's are the expansion coefficients of a in the basis $\{M_1, M_2, \ldots, M_N\}$ and are given by Eq. (2.43). Addition and multiplication rules (2.33), (2.34) and (2.35) in $_nD_v$ can be reformulated as follows. For $a = (a_1, a_2, \ldots, a_N)$, $b = (b_1, b_2, \ldots, b_N)$ in $_nD_v$, and for t in R, we have:

$$a + b = (a_1 + b_1, a_2 + b_2, \ldots, a_N + b_N) \tag{2.56}$$

$$t \cdot a = (t \cdot a_1, t \cdot a_2, \ldots, t \cdot a_N) \tag{2.57}$$

$$a \cdot b = (c_1, c_2, \ldots, c_N), \tag{2.58}$$

where

$$c_k = \sum_{M_i \cdot M_j = M_k} a_i \cdot b_j. \tag{2.59}$$

It follows that an element a of $_nD_v$ is an infinitesimal if and only if its first component a_1, the so-called **real part**, vanishes. To conclude the section, we also observe that the total number $P(n, v)$ of multiplications of the form $a_i \cdot b_i$ necessary to compute all coefficients c_k of the product is given by

$$P(n, v) = N(n, 2v) = \binom{n + 2v}{2v} = \frac{(n + 2v)!}{n!\, (2v)!}, \tag{2.60}$$

since the number of each combination $M_i \cdot M_j$ of monomials in $_nD_v$ in Eq. (2.59) corresponds uniquely to one monomial in $_nD_{2v}$.

2.2.4 The Tower of Ideals, Nilpotency, and Fixed Points

To any element $[f] \in {}_nD_v$, we define the **depth** $\lambda([f])$ as

$$\lambda([f]) = \begin{cases} \text{Order of first nonvanishing derivative of } f & [f] \neq 0 \\ n + 1 & [f] = 0 \end{cases}. \tag{2.61}$$

In particular, any function f that does not vanish at the origin has $\lambda([f]) = 0$. In a similar way, on the set $_nD_v^m$ that describes vector functions $\vec{f} = (f_1, \ldots, f_m)$ from R^v to R^m, we define

$$\lambda(([f_1], \ldots, [f_m])) = \min_{1 \leq k \leq m} \lambda([f_k]) \tag{2.62}$$

We observe that for any $a, b \in {}_nD_v$, we have

$$\lambda(a \cdot b) = \min(\lambda(a) + \lambda(b), n + 1)$$

and

$$\lambda(a + b) \geq \min(\lambda(a), \lambda(b)). \tag{2.63}$$

In fact, except that λ never exceeds $n + 1$, it behaves like a **valuation**. We now define the sets I_k as

$$I_k = \{a \in {}_nD_v | \lambda(a) \geq k\}. \tag{2.64}$$

We apparently have the relation

$${}_nD_v = I_0 \supset I_1 \supset \ldots \supset I_n \supset I_{n+1} = \{0\}. \tag{2.65}$$

All I_k's are **ideals**, i.e., for $a \in I_k$ and any $b \in {}_nD_v$, we have $a \cdot b \in I_k$, which readily follows from Eq. (2.63). Because of Eq. (2.65), the I_k's are often referred to as the "tower of ideals." The Ideal I_1 apparently is just the collection of those elements that are infinitely small or zero in absolute value. An interesting property of the elements in I_1 (and hence also those in I_2, I_3, ...) is that they are **nilpotent**. Following the multiplication rules of the depth, we see that if $a \in I_1$ and hence $\lambda(a) \geq 1$,

$$\lambda\left(a^{n+1}\right) \geq (n+1) \cdot \lambda(a) \geq n+1, \tag{2.66}$$

and so indeed

$$a^{n+1} = 0. \tag{2.67}$$

In particular, all differential generators d_k are nilpotent. On the other hand, all elements in ${}_nD_v \setminus I_1$ are not nilpotent, but their powers actually stay in ${}_nD_v \setminus I_1$; indeed, they all have $\lambda(a) = 0$, which also entails that $\lambda(a^k) = 0$.

We now introduce a **norm** of the vector a in ${}_nD_v$. Let a have the components $(a_1, \ldots, a_N)$ in the differential basis; then we set

$$||a|| = N(n, v) \cdot \max_{i=1,\ldots,N} |a_i|. \tag{2.68}$$

We see that

$$||a|| = 0 \text{ if and only if } a = 0$$

$$||t \cdot a|| = |t| \cdot ||a|| \text{ for all real } t.$$

By considering $|a_i + b_i| \leq |a_i| + |b_i|$, we see that

$$||a + b|| \leq ||a|| + ||b|| \tag{2.69}$$

and hence $|| \ ||$ is indeed a norm. Let $c = (c_1, \ldots, c_N) = a \cdot b$. From the multiplication rule, we remember that each of the c_i's is made by summing over products of a_i and b_i. Since there cannot be more contributions to the sum as $N(n, v)$, we have $c_i \leq N(n, v) \cdot \max a_i \cdot \max b_i$. This entails that $N(n, v) \cdot \max |c_i| \leq N(n, v)^2 \max(a_i) \cdot \max(b_i)$ and hence

$$||a \cdot b|| \leq ||a|| \cdot ||b||, \tag{2.70}$$

so the norm satisfies a **Schwartz inequality**.

We proceed to the study of operators and prove a powerful fixed-point theorem, which will simplify many of the subsequent arguments. Let $\mathcal{O}$ be an operator on the set $M \subset {}_nD_v^m$. Then we say that $\mathcal{O}$ is **contracting** on M if for any $\vec{a}, \vec{b} \in M$ with $\vec{a} \neq \vec{b}$,

$$\lambda(\mathcal{O}(\vec{a}) - \mathcal{O}(\vec{b})) > \lambda(\vec{a} - \vec{b}) \tag{2.71}$$

Therefore, in practical terms this means that after application of $\mathcal{O}$, the derivatives in $\vec{a}$ and $\vec{b}$ agree to a higher order than they did before application of $\mathcal{O}$.

For example, consider the set $M = I_1$ and the operator $\mathcal{O}$ that satisfies $O(a) = a^2$. Then, $\mathcal{O}(a) - \mathcal{O}(b) = a^2 - b^2 = (a+b) \cdot (a-b)$, and thus $\lambda(\mathcal{O}(a) - \mathcal{O}(b)) = \min(\lambda(a+b) + \lambda(a-b), n+1) > \lambda(a-b)$, since $a+b \in I_1$ and so $\lambda(a+b) > 0$.

We also see that if we have two operators $\mathcal{O}_1$ and $\mathcal{O}_2$ that are contracting on M, then their **sum is contracting** on M because, according to Eq. (2.63),

$$\lambda((\mathcal{O}_1(\vec{a}) + \mathcal{O}_2(\vec{a})) - (\mathcal{O}_1(\vec{b}) + \mathcal{O}_2(\vec{a})))$$
$$\geq \min(\lambda(\mathcal{O}_1(\vec{a}) - \mathcal{O}_1(\vec{b})), \lambda(\mathcal{O}_2(\vec{a}) - \mathcal{O}_2(\vec{b})))$$
$$> \min(\lambda(\vec{a} - \vec{b}), \lambda(\vec{a} - b)) = \lambda(\vec{a} - \vec{b}). \tag{2.72}$$

Furthermore, if the set M on which the operators act is a subset of I_1, then their product is also contracting.

Another important case of a contracting operator is the **antiderivation** ∂_k^{-1}. Indeed, assume $\lambda(\vec{a} - \vec{b}) = l$, then the first nonvanishing derivatives of $\vec{a} - \vec{b}$ are of order l. However, after integration with respect to any variable k, the derivatives of $\vec{a} - \vec{b}$ are of order $l + 1$, and thus, using the linearity of the antiderivation,

$$\lambda(\partial_k^{-1}\vec{a} - \partial_k^{-1}\vec{b}) = \lambda(\partial_k^{-1}(\vec{a} - \vec{b})) > \lambda(\vec{a} - \vec{b}) \tag{2.73}$$

and thus the antiderivation operator is contracting. This will prove eminently important in the study of differential equations following.

Contracting operators have a very important property—they satisfy a **fixed-point theorem**. Let $\mathcal{O}$ be a contracting operator on $M \subset {}_nD_v$ that maps M into M. Then $\mathcal{O}$ has a **unique** fixed point $a \in M$ that satisfies the **fixed-point problem**

$$a = \mathcal{O}(a). \tag{2.74}$$

Moreover, let a_0 be any element in M. Then the sequence

$$a_k = \mathcal{O}(a_{k-1}) \text{ for } k = 1, 2, \ldots \tag{2.75}$$

converges in **finitely many steps** [in fact, at most $(n + 1)$ steps] to the fixed point a. The fixed-point theorem is of great practical usefulness since it ensures the existence of a solution and, moreover, allows its exact determination in a very simple way in finitely many steps.

The **proof** of the fixed-point theorem is rather simple. Consider the sequence $a_{k+1} = \mathcal{O}(a_k)$. We have that

$$\lambda(a_{k+2} - a_{k+1}) = \lambda(\mathcal{O}(a_{k+1}) - \mathcal{O}(a_k))$$
$$> \lambda(a_{k+1} - a_k).$$

Because λ assumes only integer values, we thus even have $\lambda(a_{k+2} - a_{k+1}) \geq \lambda(a_{k+1} - a_k) + 1$, and hence by induction it follows that

$$\lambda(a_{n+2} - a_{n+1}) \geq \lambda(a_{n+1} - a_n) + 1 \geq \ldots$$
$$\geq \lambda(a_1 - a_0) + (n + 1) \geq n + 1.$$

Because of the definition (Eq. 2.61) of λ, this entails that $a_{n+2} = a_{n+1}$, or because $a_{n+2} = \mathcal{O}(a_{n+1})$,

$$a_{n+1} = \mathcal{O}(a_{n+1}).$$

Therefore, the fixed point has indeed been reached exactly after $(n + 1)$ steps. The fixed point is also **unique**; if both a and a^* were distinct fixed points, then $\lambda(a - a^*) = \lambda(\mathcal{O}(a) - \mathcal{O}(a^*)) > \lambda(a - a^*)$, which is a contradiction.

The fixed-point theorem, and also its proof, has many similarities to its famous counterpart due to Banach on Cauchy-complete normed spaces. In Banach's version, contractivity is measured not discretely via λ but continuously via the norm, and it is sufficient that there exists a real q with $0 \leq q < 1$ with $|\mathcal{O}(a_1) - \mathcal{O}(a_2)| < q \cdot |a_1 - a_2|$. Then the resulting sequence is shown to be Cauchy, and its limit is the solution. In the DA version of the fixed-point theorem, convergence happens very conveniently after finitely many steps.

Similar to the case of the Banach fixed-point theorem, the DA fixed-point theorem also has many useful applications. In particular, it allows for a rather straightforward solution of ODEs and PDEs by utilizing operators that contain derivations, as will be shown later.

For example, we want to utilize the fixed-point theorem to prove the **quotient rule** of differentiation: Given a function f and its partial derivatives of order n, that satisfies $f(0) \neq 0$. Then we want to determine all the partial derivatives of order n of the function $1/f$. In DA terminology, this is equivalent to finding a multiplicative inverse of the element $a = [f] \in I_0 \subset {}_n D_v$. Let $a_0 = f(0)$. Then we have $\lambda(a - a_0) > 0$. Let $b = [1/f]$, and as demanded $a \cdot b = 1$. This can be rewritten as $(a - a_0) \cdot b + a_0 \cdot b = 1$ or

$$b = \frac{1}{a_0} \cdot \{1 - (a - a_0) \cdot b\}.$$

Defining the operator $\mathcal{O}$ by $\mathcal{O}(b) = \{1 - (a - a_0) \cdot b\}/a_0$, we have a fixed-point problem. Let $c_1 \neq c_2 \in {}_nD_v$. Then we have

$$\lambda\left(\mathcal{O}(c_1) - \mathcal{O}(c_2)\right) = \lambda\left(\frac{a - a_0}{a_0} \cdot (c_1 - c_2)\right)$$

$$= \min\left(\lambda\left(\frac{a - a_0}{a_0}\right) + \lambda(c_1 - c_2), n + 1\right)$$

$$> \lambda(c_1 - c_2),$$

and hence the operator is contracting. Thus, a unique fixed point exists, and it can be reached in $n + 1$ steps from any starting value. This method has particular advantages for the question of efficient **computation** of multiplicative inverses.

In a similar but somewhat more involved example, we derive the **root rule**, which allows the determination of the derivatives of $\sqrt{f}$ from those of f. We demand $f(0) > 0$. Let $a = [f]$, $b = [\sqrt{f}]$. Then $b^2 = a$. Setting $a_0 = f(0)$ and writing $a = a_0 + \tilde{a}$ and $b = \sqrt{a_0} + \tilde{b}$, we have $\lambda(\tilde{a}) > 0$ and $\lambda(\tilde{b}) > 0$. The condition $b^2 = a$ can then be rewritten as

$$\tilde{b} = \frac{\tilde{a} - \tilde{b}^2}{2\sqrt{a_0}} = \mathcal{O}(\tilde{b}). \tag{2.76}$$

The operator $\mathcal{O}$ is contracting on the set $M = \{\tilde{b} | \lambda(\tilde{b}) > 0\}$; if c_1 and c_2 are in M,

$$\lambda(\mathcal{O}(c_1) - \mathcal{O}(c_2)) = \lambda\left(\frac{1}{2\sqrt{a_0}}(c_2^2 - c_1^2)\right) = \lambda(c_2 - c_1) + \lambda(c_2 + c_1)$$

$$> \min(\lambda(c_2 - c_1), n + 1). \tag{2.77}$$

Hence, the operator is contracting, and the root can be obtained by iteration in finitely many steps. Again, the method is useful for the practical computation of roots.

2.3 ADVANCED METHODS

2.3.1 Composition and Inversion

Let us consider a function g on R^v in v variables that is at least n times differentiable. Given the set of derivatives $[\mathcal{M}]_n$ up to order n of a function $\mathcal{M}$ from R^v to R^v, we want to determine the derivatives up to order n of the **composition** $g \circ \mathcal{M} = g(\mathcal{M})$, and hence the class $[g \circ \mathcal{M}]_n$. According to the chain rule, the nth derivatives of $g \circ \mathcal{M}$ at the origin can be calculated from the respective derivatives of $\mathcal{M}$ at the origin and the derivatives of up to order n of g at the point

$\mathcal{M}(\vec{0})$. This implies that the mere knowledge of $[g]_n$ is sufficient for this purpose if $\mathcal{M}(\vec{0}) = \vec{0}$; on the other hand, if this is not the case, the knowledge of $[g]_n$ is not sufficient because this does not provide information about the derivatives of g at $\mathcal{M}(\vec{0})$.

Since the derivatives of g are only required to order n, for the practical purposes of calculating the class $[g \circ \mathcal{M}]_n$ it is actually possible to replace g by its Taylor polynomial T_g around $\mathcal{M}(\vec{0})$ since T_g and g have the same derivatives; therefore,

$$[g \circ \mathcal{M}]_n = [T_g \circ \mathcal{M}]_n. \tag{2.78}$$

However, since the polynomial evaluation requires only additions and multiplications, and both additions and multiplications commute with the bracket operation by virtue of Eqs. (2.33) and (2.34), i.e., $[a+b] = [a]+[b]$ and $[a \cdot b] = [a] \cdot [b]$, we have

$$[g \circ \mathcal{M}]_n = T_g([\mathcal{M}]_n). \tag{2.79}$$

This greatly simplifies the practical **computation** of the composition since it now merely requires the evaluation of the Taylor polynomial of g on the classes $[\mathcal{M}]_n$, i.e., purely DA arithmetic.

It now also follows readily how to handle compositions of **multidimensional** functions; replacing g by the function $\mathcal{N}$ mapping from R^v to R^v, and denoting the Taylor polynomial of $\mathcal{N}$ around $\mathcal{M}(\vec{0})$ by $\vec{T}_\mathcal{N}$, we have

$$[\mathcal{N} \circ \mathcal{M}]_n = \vec{T}_\mathcal{N}([\mathcal{M}]_n). \tag{2.80}$$

A particularly important case is the situation in which $\mathcal{M}(\vec{0}) = \vec{0}$; in this case, the Taylor polynomial $\vec{T}_\mathcal{N}$ of $\mathcal{N}$ around $\mathcal{M}(\vec{0})$ is completely specified by the class $[\mathcal{N}]_n$. This entails that we can define a composition operation in the ideal $I_1^v \subset {}_nD_v^v$ via

$$[\mathcal{N}]_n \circ [\mathcal{M}]_n \stackrel{\text{def}}{=} [\mathcal{N} \circ \mathcal{M}]_n. \tag{2.81}$$

Now we address the question of the determination of derivatives of **inverse** functions. Let us assume we are given a map $\mathcal{M}$ from R^v to R^v with $\mathcal{M}(\vec{0}) = \vec{0}$ that is n times differentiable, and let us assume that its linearization M around the origin is invertible. Then, according to the implicit function theorem, there is a neighborhood around $\vec{0}$ where the function is invertible, and so there is a function $\mathcal{M}^{-1}$ such that $\mathcal{M} \circ \mathcal{M}^{-1} = \mathcal{I}$. Moreover, $\mathcal{M}^{-1}$ is differentiable to order n.

The goal is to find the derivatives of $\mathcal{M}^{-1}$ from those of $\mathcal{M}$ in an automated way. In the DA representation, this seemingly difficult problem can be solved by an elegant and closed algorithm.

We begin by splitting the map $\mathcal{M}$ into its linear part M and its purely nonlinear part $\mathcal{N}$, so that we have

$$\mathcal{M} = M + \mathcal{N}. \tag{2.82}$$

Composing the inverse $\mathcal{M}^{-1}$ with $\mathcal{M}$, we obtain

$$\mathcal{M} \circ \mathcal{M}^{-1} = \mathcal{I}$$
$$M \circ \mathcal{M}^{-1} = \mathcal{I} - \mathcal{N} \circ \mathcal{M}^{-1}$$
$$\mathcal{M}^{-1} = M^{-1} \circ (\mathcal{I} - \mathcal{N} \circ \mathcal{M}^{-1}). \tag{2.83}$$

The resulting equation apparently is a **fixed-point problem**. Moving to equivalence classes, we see that the right-hand side is contracting, since

$$\left\{ M^{-1} \circ ([\mathcal{I}]_n - [\mathcal{N}]_n \circ [\mathcal{A}]_n) \right\} - \left\{ M^{-1} \circ ([\mathcal{I}]_n - [\mathcal{N}]_n \circ [\mathcal{B}]_n) \right\}$$
$$= M^{-1} \circ ([\mathcal{N}]_n \circ [\mathcal{B}]_n - [\mathcal{N}]_n \circ [\mathcal{A}]_n). \tag{2.84}$$

However, note that $[\mathcal{N}]_n \in I_2^v$, and so for any $[\mathcal{A}]_n$ and $[\mathcal{B}]_n \in I_1^v$, if $[\mathcal{A}]_n$ and $[\mathcal{B}]_n$ agree to order k, then $[\mathcal{N}]_n \circ [\mathcal{A}]_n$ and $[\mathcal{N}]_n \circ [\mathcal{B}]_n$ agree to order $k + 1$ because every term in $[\mathcal{A}]$ and $[\mathcal{B}]$ is multiplied by at least one nilpotent element (namely, other components from $[\mathcal{A}]$ and $[\mathcal{B}]$). According to the fixed-point theorem, this ensures the existence of a **unique fixed point**, which can be reached at most n iterative steps. Therefore, we have developed an algorithm to determine all derivatives of up to order n of $\mathcal{M}^{-1}$ in finitely many steps through mere iteration. Because it requires only iteration of a relatively simple operator, it is particularly useful for **computation**.

2.3.2 Important Elementary Functions

In this section, we introduce important functions on the DA $_nD_v$. In particular, this includes the functions typically available intrinsically in a computer environment, such as exp, sin, and log. As discussed in Section 2.3.1, for any function g that is at least n times differentiable, the chain rule in principle allows the computation of the derivatives of order n of $g \circ f$ at the origin, and hence $[g \circ f]$, from the mere knowledge of the derivatives of f, and hence $[f]$, as well as the derivatives of g at $f(\vec{0})$. This allows us to **define**

$$g([f]) \stackrel{\text{def}}{=} [g(f)]. \tag{2.85}$$

ADVANCED METHODS

In particular, we thus can obtain the exponential, the trigonometric functions, and a variety of other functions as

$$\exp([f]) \stackrel{\text{def}}{=} [\exp(f)], \quad \log([f]) \stackrel{\text{def}}{=} [\log(f)], \tag{2.86}$$

$$\sin([f]) \stackrel{\text{def}}{=} [\sin(f)], \quad \cos([f]) \stackrel{\text{def}}{=} [\cos(f)], \tag{2.87}$$

$$\sqrt{[f]} \stackrel{\text{def}}{=} [\sqrt{f}] \text{ for } f(0) \text{ positive}, \tag{2.88}$$

etc.

While this method formally settles the issue of the introduction of important functions on $_nD_v$ in a rather complete and straightforward way, it is **not useful for computation** of the previous functions in practice in an efficient and general way. In the following, we illustrate how this can be achieved.

We observe that by the very nature of the definition of the important functions via Eq. (2.85), many of the **original properties** of the functions translate to $_nD_v$. For example, functional relationships such as **addition theorems** continue to hold in $_nD_v$. To illustrate this fact, we show the addition theorem for the sin function:

$$\begin{aligned}
\sin([f] + [g]) &= [\sin(f + g)] = [\sin(f)\cos(g) + \cos(f)\sin(g)] \\
&= [\sin(f)\cos(g)] + [\cos(f)\sin(g)] \\
&= [\sin(f)][\cos(g)] + [\cos(f)][\sin(g)] \\
&= \sin([f])\cos([g]) + \cos([f])\sin([g]),
\end{aligned} \tag{2.89}$$

where we have made use of the definitions of the functions on $_nD_v$ as well as the fact that $[a \circledast b] = [a] \circledast [b]$ for all elementary operations such as $+$, $-$, and $*$. Obviously, other functional rules can be derived in a very analogous way.

The addition theorems for elementary functions allow us to derive **convenient computational tools** using only elementary operations on $_nD_v$. For this purpose, one rewrites the expression in such a way that elementary functions act only around points for which the coefficients of their Taylor expansion are readily available. As discussed in Section 2.3.1, according to Eq. (2.81) this allows substitution of the function itself by its nth order Taylor polynomial, which computationally requires only additions and multiplications. For example, consider the elementary function $g = \sin$. We split the argument function f in the class $[f]$ into two parts, its constant part $a_0 = f(0)$ and its differential (nilpotent) part $\tilde{f}$, so that

$$[f] = [a_0] + [\tilde{f}]. \tag{2.90}$$

We then conclude

$$\sin([f]) = \sin([a_0] + [\tilde{f}]) = \sin([a_0])\cos([\tilde{f}]) + \cos([a_0])\sin([\tilde{f}])$$
$$= [\sin(a_0)] \sum_{i=0}^{k(n)} (-1)^i \frac{[\tilde{f}]^{2i}}{(2i)!} + [\cos(a_0)] \sum_{i=0}^{k(n)} (-1)^i \frac{[\tilde{f}]^{2i+1}}{(2i+1)!}, \quad (2.91)$$

where $k(n)$ is the largest integer that does not exceed $n/2$.

The strategy for other functions is similar; for example, we use

$$\exp(a_0 + b) = \exp(a_0) \cdot \exp(b) \quad (2.92)$$

$$\sqrt{a_0 + b} = \sqrt{a_0} \cdot \sqrt{1 + \frac{b}{a_0}} \quad \text{(for } a_0 > 0\text{)} \quad (2.93)$$

$$\log(a_0 + b) = \log(a_0) + \log\left(1 + \frac{b}{a_0}\right) \quad \text{(for } a_0 > 0\text{)} \quad (2.94)$$

$$\frac{1}{a_0 + b} = \frac{1}{a_0} \cdot \frac{1}{1 + b/a_0} \quad \text{(for } a_0 > 0\text{)} \quad (2.95)$$

and the respective Taylor expansions around 0 or 1.

It is now possible to obtain derivatives of any order of a functional dependency that is given in terms of elementary operations and elementary functions by merely evaluating in the DA framework and observing

$$[f(x_1, \ldots, x_v)] = f([x_1], \ldots, [xv]) = f(d_1, \ldots, d_v). \quad (2.96)$$

2.3.3 Power Series on $_nD_v$

We say that a sequence (a_k) **converges** in $_nD_v$ if there exists a number a in $_nD_v$ such that $(\|a_k - a\|)$ converges to zero in R. That is, given $\epsilon > 0$ in R, there exists a positive integer K such that $\|a_k - a\| < \epsilon$ for all $k > K$.

With the previously defined norm, $_nD_v$ is a **Banach space**, i.e., a normed space which is Cauchy complete. A sequence (a_k) in $_nD_v$ is said to be Cauchy if for each $\epsilon > 0$ in R there exists a positive integer K such that $\|a_k - a_l\| < \epsilon$ whenever k and l are both $\geq K$. For the **proof** of the Cauchy completeness of $_nD_v$, we need to show that every Cauchy sequence in $_nD_v$ converges in $_nD_v$.

Let (a_k) be a Cauchy sequence in $_nD_v$ and let $\epsilon > 0$ in R be given. Then there exists a positive integer $K > 0$ such that

$$\|a_k - a_l\| < \epsilon \text{ for all } k, l \geq K. \quad (2.97)$$

According to the definition of the norm, it follows from Eq. (2.97) that

$$|a_{k,i} - a_{l,i}| < \epsilon \text{ for all } k, l \geq K \text{ and for all } i = 1, 2, \ldots, N, \quad (2.98)$$

where $a_{k,i}$ and $a_{l,i}$ denote the ith components of a_k and a_l, respectively. Thus, the sequence $(a_{k,i})$ is a Cauchy sequence in R for each i. Since R is Cauchy complete, the sequence $(a_{k,i})$ converges to some real number b_i for each i. Let

$$b = (b_1, b_2, \ldots, b_N). \tag{2.99}$$

We claim that (a_k) converges to b in ${}_nD_v$. Let $\epsilon > 0$ in R be given. For each i there exists a positive integer k_i such that $|a_{k,i} - b_i| < \epsilon/N$ for all $k \geq k_i$. Let $K = \max\{k_i : i = 1, 2, \ldots, N\}$. Then $|a_{k,i} - b_i| < \epsilon/N$ for all $i = 1, 2, \ldots, N$ and for all $k \geq K$. Therefore,

$$\|a_k - b\| < \epsilon \text{ for all } k \geq K, \tag{2.100}$$

and hence (a_k) converges to b in ${}_nD_v$ as claimed.

One particularly important class of sequences is that of **power series**. Let $\{a_k\}_{k=0}^{\infty}$ be a sequence of real numbers. Then $\sum_{k=0}^{\infty} a_k y^k$, $y \in {}_nD_v$ is called a power series in y (centered at 0). We say that a power series $\sum_{k=0}^{\infty} a_k y^k$ converges in ${}_nD_v$ if the sequence of partial sums, $A_m = \sum_{k=0}^{m} a_k y^k$, converges in ${}_nD_v$.

Now let $\sum_{k=0}^{\infty} a_k y^k$ be a power series in the real number y, with real coefficients and with real radius of convergence σ. We show that the series converges in ${}_nD_v$ if $|\Re(y)| < \sigma$, where $\Re(y)$ denotes the real part of y. From what we have done so far, it suffices to show that the sequence (A_m) is Cauchy in ${}_nD_v$, where

$$A_m = \sum_{k=0}^{m} a_k y^k. \tag{2.101}$$

Let $m > n$ be given. Write $y = y_0 + \tilde{y}$, where y_0 is the constant part of y, and hence $y - y_0 \in I_1$. Then,

$$\sum_{k=0}^{m} a_k (y_0 + \tilde{y})^k = \sum_{k=0}^{n} a_k (y_0 + \tilde{y})^k + \sum_{k=n+1}^{m} a_k (y_0 + \tilde{y})^k$$

$$= \sum_{k=0}^{n} a_k (y_0 + \tilde{y})^k + \sum_{k=n+1}^{m} a_k \sum_{i=0}^{k} \frac{k!}{i!(k-i)!} y_0^{k-i} \tilde{y}^i$$

$$= \sum_{k=0}^{n} a_k (y_0 + \tilde{y})^k$$

$$+ \sum_{k=n+1}^{m} a_k k \ldots (k-n) y_0^k \sum_{i=0}^{n} \left(\frac{\tilde{y}^i}{y_0^i} \frac{(k-n-1)!}{i!(k-i)!} \right),$$

$$\tag{2.102}$$

where use has been made of the fact that $\tilde{y}^i = 0$ for all $i > n$. Now let $l > m > n$ be given. Then,

$$A_l - A_m = \sum_{k=m+1}^{l} a_k \cdot k \cdot \ldots \cdot (k-n) y_0^k \sum_{i=0}^{n} \frac{\tilde{y}^i}{y_0^i} \frac{(k-n-1)!}{i!(k-i)!}. \quad (2.103)$$

Therefore,

$$\|A_l - A_m\| = \left\| \sum_{k=m+1}^{l} a_k \cdot k \ldots (k-n) y_0^k \sum_{i=0}^{n} \left(\frac{\tilde{y}^i}{y_0^i} \frac{(k-n-1)!}{i!(k-i)!} \right) \right\|$$

$$\leq \sum_{k=m+1}^{l} \|a_k \cdot k \ldots (k-n) y_0^k\| \cdot \left\| \sum_{i=0}^{n} \frac{\tilde{y}^i}{y_0^i} \frac{(k-n-1)!}{i!(k-i)!} \right\|$$

$$\leq \left(\sum_{k=m+1}^{l} |a_k \cdot k \ldots (k-n) y_0^k| \right) \cdot \left\| \sum_{i=0}^{n} \frac{\tilde{y}^i}{y_0^i i!(n+1-i)!} \right\|.$$
(2.104)

Since $k \cdot \ldots \cdot (k-n) \leq k^{n+1}$,

$$\lim_{k \to \infty} (k \cdot \ldots \cdot (k-n))^{1/k} \leq \lim_{k \to \infty} (k^{n+1})^{1/k} = 1. \quad (2.105)$$

Therefore, the first factor converges to 0 if $|y_0| < \sigma$. Since the second factor is finite, $\|A_l - A_m\|$ converges to 0. Thus, the sequence (A_m) is Cauchy in $_n D_v$ for $|y_0| < \sigma$. Therefore, the series **converges** if $|y_0| < \sigma$.

Now that we have seen that the power series actually converges, the question arises as to whether the DA vector of the power series applied to a function f is the same as the power series applied to the DA vector $[f]$, which we just proved to converge. Of course, if the power series has only finitely many terms, the question is settled because the elementary operations of addition and multiplication commute with the class operations, and we have that $[f + g] = [f] + [g]$, etc. However, in the case in which there are **infinitely many operations, commutativity is not ensured**. The answer to the question can be obtained with the help of the **derivative convergence theorem**.

Let $a < b$ in R and let $\{f_k\}_{k=0}^{\infty}$ be a sequence of real-valued functions that are differentiable on $[a, b]$. Suppose that $\{f_k\}_{k=0}^{\infty}$ converges uniformly on $[a, b]$ to a differentiable function f and that $\{f_k'\}_{k=0}^{\infty}$ converges uniformly on $[a, b]$ to some function g. Then $g(x) = f'(x)$ for all $x \in [a, b]$.

For the **proof**, let $x \in [a, b]$ and $\epsilon > 0$ in R be given. There exists a positive integer K such that

$$|f_k'(x) - g(x)| < \epsilon/4 \text{ for all } k \geq K \text{ and for } x \in [a, b], \quad (2.106)$$

from which we have

$$|f'_{k_1}(x) - f'_{k_2}(x)| < \epsilon/2 \text{ for all } x \in [a, b] \text{ and } k_1, k_2 \geq K. \qquad (2.107)$$

Let $k_0 \geq K$ be given. There exists $\delta_{k_0} > 0$ such that

$$\left| \frac{f_{k_0}(t) - f_{k_0}(x)}{t - x} - f'_{k_0}(x) \right| < \epsilon/4$$

whenever

$$t \in [a, b], \ t \neq x \text{ and } |t - x| < \delta_{k_0}. \qquad (2.108)$$

By the mean value theorem, for any integer $m \geq K$ and for any $t \neq x$ in $[a, b]$,

$$|(f_m - f_{k_0})(x) - (f_m - f_{k_0})(t)| = |(f_m - f_{k_0})'(c)| \cdot |x - t| \qquad (2.109)$$

for some c between x and t. Using Eq. (2.107) and the fact that $c \in [a, b]$ and $m, k_0 \geq K$,

$$|(f_m - f_{k_0})'(c)| < \epsilon/2. \qquad (2.110)$$

Combining Eqs. (2.110) and (2.109), we get

$$|(f_m - f_{k_0})(x) - (f_m - f_{k_0})(t)| < \epsilon/2 \cdot |x - t|, \qquad (2.111)$$

which we rewrite as

$$\left| \frac{f_m(t) - f_m(x)}{t - x} - \frac{f_{k_0}(t) - f_{k_0}(x)}{t - x} \right| < \epsilon/2. \qquad (2.112)$$

Letting $m \to \infty$ in Eq. (2.112), we obtain that

$$\left| \frac{f(t) - f(x)}{t - x} - \frac{f_{k_0}(t) - f_{k_0}(x)}{t - x} \right| < \epsilon/2. \qquad (2.113)$$

Using Eqs. (2.106), (2.108), and (2.113), we finally obtain that

$$\left| \frac{f(t) - f(x)}{t - x} - g(x) \right| \leq \left| \frac{f(t) - f(x)}{t - x} - \frac{f_{k_0}(t) - f_{k_0}(x)}{t - x} \right|$$

$$+ \left| \frac{f_{k_0}(t) - f_{k_0}(x)}{t - x} - f'_{k_0}(x) \right| + \left| f'_{k_0}(x) - g(x) \right|$$

$$< \epsilon/2 + \epsilon/4 + \epsilon/4 = \epsilon. \qquad (2.114)$$

Since Eq. (2.114) holds whenever $t \in [a, b]$ and $0 < |t - x| < \delta_{k_0}$, we conclude that $f'(x) = g(x)$. This finishes the proof of the theorem.

We next try to generalize the previous theorem to **higher order derivatives**: Let $\{f_k\}_{k=1}^{\infty}$ be a sequence of real-valued functions that are n times differentiable on $[a,b]$. Suppose that $\{f_k\}_{k=1}^{\infty}$ converges uniformly on $[a,b]$ to an n times differentiable function f and that $\{f_k^{(k)}\}_{k=1}^{\infty}$ converges uniformly on $[a,b]$ to some function g_k, $k=1,2,\ldots,n$. Then, $g_k(x) = f^{(k)}(x)$ for all $x \in [a,b]$ and for all $k=1,2,\ldots,n$.

The **proof** of the statement follows by induction on n. It is true for $n=1$ because it directly follows from the previous theorem. Assume it is true for $n=j$, and show that it will be true for $n=j+1$. Therefore, if $\{f_k\}_{k=1}^{\infty}$ is a sequence of real-valued functions that are j times differentiable on $[a,b]$, $\{f_k\}_{k=1}^{\infty}$ converges uniformly on $[a,b]$ to a j times differentiable function f, and $\{f_k^{(k)}\}_{k=1}^{\infty}$ converges uniformly on $[a,b]$ to some function g_k, $k=1,2,\ldots,j$, then $g_k(x) = f^{(k)}(x)$ for all $x \in [a,b]$ and for all $k=1,2,\ldots,j$. Now let $\{f_k\}_{k=1}^{\infty}$ be a sequence of real-valued functions that are $(j+1)$ times differentiable on $[a,b]$. Suppose that $\{f_k\}_{k=1}^{\infty}$ converges uniformly on $[a,b]$ to a $(j+1)$ times differentiable function f and that $\{f_k^{(k)}\}_{k=1}^{\infty}$ converges uniformly on $[a,b]$ to some function g_k, $k=1,2,\ldots,j+1$. Then $g_k(x) = f^{(k)}(x)$ for all $x \in [a,b]$ and for all $k=1,2,\ldots,j$ by the induction hypothesis. Furthermore, applying the previous theorem to the sequence $\{f_k^{(j)}\}_{k=1}^{\infty}$, we obtain that $g_{j+1}(x) = f^{(j+1)}(x)$ for all $x \in [a,b]$. Therefore, $g_k(x) = f^{(k)}(x)$ for all $x \in [a,b]$ and for all $k=1,2,\ldots,j+1$. Thus, the statement is true for $n=j+1$. Therefore, it is true for all n.

The two previous results are readily extendable to sequences of real-valued functions on R^v, in which derivatives are to be replaced by partial derivatives.

We can now apply the result to the question of convergence of power series of DA vectors to the DA vector of the power series. As shown previously, the power series of DA vectors converges in the DA norm within the classical radius of convergence. This entails that all its components, and hence the derivatives, converge uniformly to the values of the limit DA vector. Hence, according to the previous theorem, this limit contains the derivatives of the limit function.

2.3.4 ODE and PDE Solvers

Historically, the main impetus for the development of differential algebraic techniques were questions of quadrature and the solution of ordinary differential equations. In our case, the direct availability of the derivations ∂_i and their inverses ∂_i^{-1} allows to devise efficient numerical integrators of any order.

One such method is based on the common rewriting of the **ODE as a fixed-point problem** by use of

$$\vec{z} = A(\vec{z}) = \vec{z}_i + \int_{t_i}^{t} \vec{f}(\vec{z}, t') dt'. \tag{2.115}$$

Utilizing the operation ∂_{v+1}^{-1} for the integral, the problem readily translates into the differential algebra $_nD_{v+1}$, where besides the v position variables, there is an additional variable t'.

As shown in Eq. (2.73), the operation ∂_{v+1}^{-1} is contracting. Since

$$\lambda(\vec{f}(\vec{z}_1, t') - \vec{f}(\vec{z}_2, t')) \geq \lambda(\vec{z}_1 - \vec{z}_2), \quad (2.116)$$

it follows that

$$\lambda(A(\vec{z}_1) - A(\vec{z}_2)) > \lambda(\vec{z}_1 - \vec{z}_2), \quad (2.117)$$

and hence the operator A on $_nD_v$ is contracting. Thus, there exists a unique fixed point, which moreover can be reached in at most $n+1$ steps from any starting condition. Thus, iterating

$$\vec{z}_1 = \vec{0} \quad (2.118)$$
$$\vec{z}_{k+1} = A(\vec{z}_k) \text{ for } k = 1, 2, \ldots, n+1 \quad (2.119)$$

yields the **exact expansion** of $\vec{z}$ in the v variables and the time t to order n. The resulting polynomial can be used as an nth order numerical integrator by inserting an explicit time step.

It is interesting that the **computational expense** for this integrator is actually significantly lower than that of a conventional integrator. In the case of a conventional integrator, the right-hand side has to be evaluated repeatedly, and the number of these evaluations is at least equal to the time step order and sometimes nearly twice as much, depending on the scheme. The DA integrator requires $n+1$ iterations and hence $n+1$ evaluations of the right-hand side, but since in the $(k-1)$st step only the kth order will be determined exactly, it is sufficient to perform this step only at order k. Since the computational expense of DA operations increases quickly with order, by far the most important contribution to time comes from the last step. In typical cases, the last step consumes more than two-thirds of the total time, and the DA integrator is about **one order of magnitude faster** than a conventional integrator.

For ODEs that are simultaneously **autonomous** and **origin preserving**, i.e., time independent and admitting $\vec{z}(t) = \vec{0}$ as a solution, another even more efficient method can be applied. For a given function on phase space $g(\vec{z}, t)$, according to Eq. (1.7), the time derivative along the solution of any function g is given by

$$\frac{d}{dt} g(\vec{z}, t) = \vec{f} \cdot \vec{\nabla} g + \frac{\partial}{\partial t} g = L_{\vec{f}} g.$$

The operator $L_{\vec{f}}$, the **vector field** of the ODE, also allows computation of higher derivatives via $d^n/dt^n \, g = L_{\vec{f}}^n g$. This approach is well known (Conte and de Boor,

1980) and in fact is even sometimes used in practice to derive analytical low-order integration formulas for certain functions $\vec{f}$. The limitation is that unless $\vec{f}$ is very simple, it is usually impossible to compute the repeated action of $L_{\vec{f}}$ analytically, and for this reason this approach has not been very useful in practice. However, using the differential algebras $_nD_v$, and in particular the ∂_k operators which distinguish them from ordinary algebras, it is possible to perform the required operations easily. To this end, one just evaluates $\vec{f}$ in $_nD_v$ and uses the ∂_ks to compute the higher derivatives of g.

If the Taylor expansion of the solution in time t is possible, we have for the final coordinates $\vec{z}_f$

$$\vec{z}_f = \sum_{i=1}^{\infty} \frac{t^i \cdot L^i_{\vec{f}}}{i!} \mathcal{I}, \qquad (2.120)$$

where $\mathcal{I}$ is the identity function.

If g and f are not explicitly time dependent, the time derivative part vanishes for all iterations $L^n_{\vec{f}}$. Furthermore, if $\vec{f}$ is origin preserving, then the order that is lost by a plain application of ∂_k in the gradient is restored through the multiplication; for the derivatives up to order n of $f_k \cdot \partial_k g$, only the derivatives of up to order $n-1$ of $\partial_k g$ are important. This means that with one evaluation of $\vec{f}$, the powers of the operator $L_{\vec{f}}$ can be evaluated to any order directly within DA for any n. Using it for $g = z_k$, the components of the vector $\vec{z}$, we obtain an **integrator of adjustable order**.

While for general functions, it is not a priori clear whether the series in the so-called **propagator** converges, within the framework of $_nD_v$ this is always the case, as we will now show. Because of the Cauchy completeness of $_nD_v$ with respect to the norm on $_nD_v$, it is sufficient to show that

$$\sum_{i=k}^{k+l} \frac{L^i_{\vec{f}}}{i!} [g] \to 0 \text{ for } k \to \infty \text{ and for all } l. \qquad (2.121)$$

Using the definition of the norm (Eq. 2.68), the triangle inequality (Eq. 2.69), and the Schwarz inequality (Eq. 2.70), we first observe that

$$\| L_{\vec{f}}[g] \| = \left\| \sum_{k=1}^{v} f_k \cdot \partial_k[g] \right\| \leq \sum_{k=1}^{v} \| f_k \cdot \partial_k[g] \|$$

$$\leq \sum_{k=1}^{v} \| f_k \| \cdot \| [g] \| \leq C \cdot \| [g] \|, \qquad (2.122)$$

where the factor $C = \sum_{k=1}^{n} \|f_k\|$ is independent of $[g]$. Thus,

$$\left\| \sum_{i=k}^{k+l} \frac{L_{\vec{f}}^i}{i!} [g] \right\| \leq \sum_{i=k}^{k+l} \left\| \frac{L_{\vec{f}}^i}{i!} [g] \right\| \leq \|[g]\| \cdot \sum_{i=k}^{k+l} \frac{C^i}{i!}. \qquad (2.123)$$

However, since the real-number exponential series converges for any argument C, the partial sum $\sum_{i=k}^{k+l} C^i/i!$ tends to zero as $k \to \infty$ for all choices of l.

This method is the **computationally most advantageous approach** for autonomous, origin-preserving ODEs. In practice, one evaluation of the right-hand side $\vec{f}$ is required, and each new order requires merely one derivation and one multiplication per dimension. For complicated functions $\vec{f}$, the actual computational expense for each new order is only a small multiple of the cost of evaluation of $\vec{f}$, and hence this method is most efficient at high orders and large time steps. In practical use in the code COSY INFINITY, step sizes are fixed and orders adjusted dynamically, typically falling in a range between 25 and 30.

To conclude, we address the question of the **solution of PDEs**. Similar to the fixed point ODE solver, it is also possible to iteratively solve **PDEs** in finitely many steps by rephrasing them in terms of a fixed-point problem. The details depend on the PDE at hand, but the key idea is to eliminate differentiation with respect to one variable by integration. For example, consider the general PDE $a_1 \partial/\partial x (a_2 \partial/\partial x V) + b_1 \partial/\partial y (b_2 \partial/\partial y V) + c_1 \partial/\partial z (c_2 \partial/\partial z V) = 0$, where a_i, b_i, and c_i are functions of x, y, and z. The PDE is rewritten as

$$V = V|_{y=0} + \int_y \frac{1}{b_2} \left\{ \frac{\partial}{\partial y} V|_{y=0} \right.$$

$$\left. - \int_y \left(\frac{a_1}{b_1} \frac{\partial}{\partial x} \left(a_2 \frac{\partial}{\partial x} V \right) + \frac{c_1}{b_1} \frac{\partial}{\partial z} \left(c_2 \frac{\partial}{\partial z} V \right) \right) \right\}.$$

The equation is now in fixed-point form. Now assume the derivatives of V and $\partial V/\partial y$ with respect to x and z are known in the plane $y = 0$. Then the right-hand side is contracting, and the various orders in y can be iteratively calculated by mere iteration.

2.3.5 The Levi-Civita Field

We discuss in this section a very useful extension of DAs, a field that was discovered first by Levi-Civita, upon which concepts of analysis have recently been developed (Berz, 1994). The results obtained are very similar to the ones in **nonstandard analysis**; however, the number systems required here can be constructed directly and described on a computer, whereas the ones in nonstandard analysis are exceedingly large, nonconstructive (in the strict sense that the axiom of choice is used and in a practical sense), and require quite a machinery of formal logic for their formulation. For a detailed study of this field and its calculus, refer to (Berz, 1994, 1996). We start with the study of the algebraic structure of the field.

First, we introduce a family of special subsets of the rational numbers. A subset M of the rational numbers Q is said to be **left-finite** if below every (rational) bound there are only finitely many elements of M. We denote the family of all left-finite subsets of Q by $\mathcal{F}$. Elements of $\mathcal{F}$ satisfy the following properties. Let $M, N \in \mathcal{F}$ be given, then

$$M \neq \emptyset \implies M \text{ has a minimum} \quad (2.124)$$

$$X \subset M \implies X \in \mathcal{F} \quad (2.125)$$

$$M \cup N \in \mathcal{F} \text{ and } M \cap N \in \mathcal{F} \quad (2.126)$$

$$M + N = \{x + y | x \in M, y \in N\} \in \mathcal{F}. \quad (2.127)$$

$$x \in M + N \implies \exists \text{ only finitely many } (a, b) \in M \times N \text{ with } x = a + b. \quad (2.128)$$

We now define the new set of numbers $\mathcal{R}$:

$$\mathcal{R} = \{f : Q \to R \text{ such that } \operatorname{supp}(f) \in \mathcal{F}\}, \quad (2.129)$$

where

$$\operatorname{supp}(f) = \{q \in Q : f(q) \neq 0\} \quad (2.130)$$

Thus, $\mathcal{R}$ is the set of all real-valued functions on Q with left-finite support. From now on, we denote elements of $\mathcal{R}$ by $x, y, \ldots$, and their values at a given support point $q \in Q$ by $x[q], y[q], \ldots$. This will save some confusion when we talk about functions on $\mathcal{R}$.

According to Eq. (2.124), $\operatorname{supp}(x)$ has a minimum if $x \neq 0$. We define for $x \in \mathcal{R}$

$$\lambda(x) = \begin{cases} \min(\operatorname{supp}(x)) & \text{if } x \neq 0 \\ \infty & \text{if } x = 0 \end{cases}. \quad (2.131)$$

Comparing two elements x and y of $\mathcal{R}$, we say

$$x \sim y \iff \lambda(x) = \lambda(y). \quad (2.132)$$

$$x \approx y \iff \lambda(x) = \lambda(y) \text{ and } x[\lambda(x)] = y[\lambda(y)]. \quad (2.133)$$

$$x =_r y \iff x[q] = y[q] \text{ for all } q \leq r. \quad (2.134)$$

Apparently, the order is lexicographic in the sense that the functions x and y are compared from left to right. At this point, these definitions may look somewhat arbitrary, but after having introduced order on $\mathcal{R}$, we will see that λ describes "orders of infinite largeness or smallness", the relation "$\sim$" corresponds to agreement of order of magnitude, whereas the relation "$\approx$" corresponds to agreement up to an infinitely small relative error.

We define **addition** on $\mathcal{R}$ componentwise: For $x, y \in \mathcal{R}$, and $q \in Q$,

$$(x + y)[q] = x[q] + y[q]. \qquad (2.135)$$

We note that the support of $x + y$ is contained in the union of the supports of x and y and is thus itself left-finite by Eq. (2.126) and (2.125). Therefore, $x + y \in \mathcal{R}$, and thus $\mathcal{R}$ is closed under the addition defined previously.

Multiplication is defined as follows: For $x, y \in \mathcal{R}$, and $q \in Q$,

$$(x \cdot y)[q] = \sum_{\substack{q_x, q_y \in Q, \\ q_x + q_y = q}} x[q_x] \cdot y[q_y]. \qquad (2.136)$$

Note that $\operatorname{supp}(x \cdot y) \subset \operatorname{supp}(x) + \operatorname{supp}(y)$; hence according to Eq. (2.127), $\operatorname{supp}(x \cdot y)$ is left-finite. Thus $x \cdot y \in \mathcal{R}$. Note also that left-finiteness of the supports of $\mathcal{R}$ numbers is essential for the definition of multiplication; according to Eq. (2.128), for any given $q \in Q$, only finitely many terms contribute to the sum in the definition of the product.

With the addition and multiplication defined previously, $(\mathcal{R}, +, \cdot)$ is a field. The only nontrivial step in the proof of the previous statement is the proof of the existence of multiplicative inverses of nonzero elements. For this, we need to invoke the fixed-point theorem.

To show that the new field is an extension of the field of real numbers, define a map $g : R \longrightarrow \mathcal{R}$ by

$$g(x)[q] = \begin{cases} x & \text{if } q = 0 \\ 0 & \text{otherwise} \end{cases}. \qquad (2.137)$$

Then g "**embeds**" R into $\mathcal{R}$ (g is one-to-one, $g(x + y) = g(x) + g(y)$ and $g(x \cdot y) = g(x) \cdot g(y)$).

Regarded as a vector space over R, $\mathcal{R}$ is **infinite dimensional**. With the multiplication defined previously, $\mathcal{R}$ is an algebra. Furthermore, if we define an operation $\partial : \mathcal{R} \longrightarrow \mathcal{R}$ by

$$(\partial x)[q] = (q + 1) \cdot x[q + 1], \qquad (2.138)$$

then ∂ is a derivation and $(\mathcal{R}, +, \cdot, \partial)$ a **differential algebra** because ∂ can easily be shown to satisfy

$$\partial(x + y) = \partial x + \partial y$$

and

$$\partial(x \cdot y) = (\partial x) \cdot y + x \cdot (\partial y), \text{ for all } x, y \in \mathcal{R}. \tag{2.139}$$

We define an order on $\mathcal{R}$ by first introducing a set of positive elements, $\mathcal{R}^+$. Let $x \in \mathcal{R}$ be nonzero; then,

$$x \in \mathcal{R}^+ \iff x[\lambda(x)] > 0. \tag{2.140}$$

For x and y in $\mathcal{R}$,

$$x > y \iff x - y \in \mathcal{R}^+$$
$$x < y \iff y > x. \tag{2.141}$$

Then, the order is total and compatible with addition and multiplication in the usual way.

Having introduced an order on $\mathcal{R}$, we define the number d as follows:

$$d[q] = \begin{cases} 1 & \text{if } q = 1 \\ 0 & \text{if } q \neq 1 \end{cases}. \tag{2.142}$$

Then, according to the order defined previously,

$$0 < d < r \text{ for all positive } r \in R. \tag{2.143}$$

Hence, d is infinitely small. Moreover, it is easy to show that for all $t \in Q$,

$$d^t[q] = \begin{cases} 1 & \text{if } q = t \\ 0 & \text{if } q \neq t \end{cases}. \tag{2.144}$$

It follows directly from the previous equation that

$$d^t \text{ is infinitely small} \iff t > 0$$
$$d^t \text{ is infinitely large} \iff t < 0. \tag{2.145}$$

Having **infinitely small** and **infinitely large** numbers, $\mathcal{R}$ is nonarchimedean.

We introduce an **absolute value** on $\mathcal{R}$ in the natural way:

$$|x| = \begin{cases} x & \text{if } x \geq 0 \\ -x & \text{if } x < 0 \end{cases}. \tag{2.146}$$

$\mathcal{R}$ is **Cauchy complete** with respect to this absolute value, i.e., every Cauchy sequence in $\mathcal{R}$ converges in $\mathcal{R}$.

Let (x_n) be a Cauchy sequence in $\mathcal{R}$; let $q \in Q$ be given and let n be such that the terms of the Cauchy sequence do not differ by more than $\epsilon = d^{q+1}$ from n on. Define $x[q] := x_n[q]$. Then (x_n) converges to x. Since the limit x agrees with an element of the sequence to the left of any q, its support is left-finite and thus $x \in \mathcal{R}$.

We return to the **fixed-point theorem**, which proves to be a powerful mathematical tool for the detailed study of $\mathcal{R}$: Let $f : \mathcal{R} \to \mathcal{R}$ be a function defined on an interval M around the origin such that $f(M) \subset M$, and let f be contracting with an infinitely small contraction factor, then f has a unique fixed point in M.

The proof is very similar to the Banach space case. One begins with an arbitrary element x_0 and defines $x_{i+1} = f(x_i)$. The resulting sequence is Cauchy and converges by the **Cauchy completeness** of $\mathcal{R}$. Then, $x = \lim_{n \to \infty} x_n$ is the fixed point of f.

Now let $X \in \mathcal{R}$ be nonzero and show that X has a multiplicative inverse in $\mathcal{R}$. Write $X = x_0 \cdot d^r \cdot (1 + x)$, where x_0 is real, x is infinitely small, and $r = \lambda(X)$. Since $x_0 \cdot d^r$ has the inverse $x_0^{-1} \cdot d^{-r}$, it suffices to find an inverse to $(1 + x)$. Write the inverse as $(1 + y)$. Then we have

$$1 = (1 + x) \cdot (1 + y)$$
$$= 1 + x + y + x \cdot y.$$

Thus,

$$y = -x - x \cdot y. \qquad (2.147)$$

This is a fixed-point problem for y with the function $f(y) = -x - x \cdot y$ and $M = \{z \in \mathcal{R} : \lambda(z) > 0\}$. Since x is infinitely small, y is infinitely small by Eq. (2.147). Thus any solution y to Eq. (2.147) must lie in M. Also, since x is infinitely small, f is contracting with an infinitely small contraction factor. The fixed-point theorem then asserts the existence of a unique fixed point of f in M and thus in $\mathcal{R}$.

The fixed-point theorem can also be used to prove the existence of roots of positive numbers in $\mathcal{R}$ and to prove that the structure $\mathcal{C}$ obtained from $\mathcal{R}$ by adjoining the imaginary unit is algebraically closed (Berz, 1994).

The algebraic properties of $\mathcal{R}$ allow for a direct introduction of important **functions**, such as polynomials, rational functions, roots, and any combination thereof. Besides these conventional functions, $\mathcal{R}$ readily contains **delta functions**. For example, the function

$$f(x) = \frac{d}{d^2 + x^2} \qquad (2.148)$$

assumes the infinitely large value d^{-1} at the origin, falls off as $|x|$ gets larger, is infinitely small for any real x, and becomes even smaller for infinitely large (in magnitude) x.

For the scope of this book, however, it is more important to study the extensibility of the standard functions, in particular those representable by power series. For this purpose, we study two kinds of convergence. The first kind of convergence is associated with the absolute value defined in Eq. (2.146) and is called **strong convergence**. The second kind of convergence is called **weak convergence**; we say that the sequence (x_n) converges weakly to the limit $x \in \mathcal{R}$ if for all $q \in Q$, $x_r[q] \longrightarrow x[q]$ as $n \longrightarrow \infty$. Therefore, weak convergence is componentwise. It follows from the definitions that strong convergence implies weak convergence to the same limit. Weak convergence does not necessarily imply strong convergence, however. It is the weak convergence that will allow us to generalize power series to $\mathcal{R}$.

Let $\sum a_n x^n$ be a **power series** with real coefficients and conventional radius of convergence σ. Then $\sum a_n x^n$ converges weakly for all x with $|\Re(x)| < \sigma$ to an element of $\mathcal{R}$. This allows the automatic generalization of any power series within its radius of convergence to the field $\mathcal{R}$. Therefore, in a simple way, there is a very large class of functions readily available. In particular, this includes all the conventional **intrinsic functions** of a computer environment.

Let f be a function on a subset M of $\mathcal{R}$. We say f is **differentiable** with derivative $f'(x)$ at the point $x \in M$ if for any $\epsilon > 0 \in \mathcal{R}$ there exists a $\delta > 0 \in \mathcal{R}$ with δ/ϵ not infinitely small such that

$$\left| \frac{f(x + \Delta x) - f(x)}{\Delta x} - f'(x) \right| < \epsilon \qquad (2.149)$$

for all Δx with $x + \Delta x \in M$ and $|\Delta x| < \delta$. Therefore, this definition very much resembles the conventional differentiation, with an important difference being the requirement that δ not be too small. This restriction, which is automatically satisfied in archimedean structures (e.g., R), is crucial to making the concept of differentiation in $\mathcal{R}$ practically useful.

The usual rules for sums and products hold in the same way as in the real case, with the only exception that factors are not allowed to be infinitely large. Furthermore, it follows that if f coincides with a real function on all real points and is differentiable, then so is the real function, and the derivatives agree at any given real point up to an infinitely small error. This will allow the **computation of derivatives of real functions using techniques of** $\mathcal{R}$.

A very important consequence of the definition of derivatives is the fundamental result: **Derivatives are differential quotients** up to an infinitely small error.

Let $\Delta x \neq 0$ be a differential, i.e., an infinitely small number. Choose $\epsilon > 0$ infinitely small such that $|\Delta x|/\epsilon$ is also infinitely small. Because of differentiability, there is a $\delta > 0$ with δ/ϵ at most finite such that the difference quotient differs from the derivative by an infinitely small error less than ϵ for all Δx with $|\Delta x| < \delta$. However, since δ/ϵ is at most finite and $|\Delta x|/\epsilon$ is infinitely small, $|\Delta x|/\delta$ is infinitely small, and in particular $|\Delta x| < \delta$. Thus, Δx yields an infinitely small error in the difference quotient from the derivative.

This elegant method allows the compututation of the real derivative of any real function that has already been extended to the field $\mathcal{R}$ and is differentiable there. In particular, all real functions that can be expressed in terms of power series combined in finitely many operations can be conveniently differentiated. However, it also works for many other cases in which the DA methods fail. Furthermore, it is of historical interest since it retroactively justifies the ideas of the fathers of calculus of derivatives being differential quotients. It is worth pointing out that the computation of the derivatives as the real parts of difference quotients corresponds to the result in Eq. (2.31), except that division by d is impossible there, leading to the different form of the expression.

Equivalent results to the intermediate value theorem, Rolle's theorem, and Taylor's theorem can be proved to hold in $\mathcal{R}$. For more details, see (Berz, 1994, 1992a).

The last result we mention here is the **Cauchy point formula**. Let $f = \sum_{i=0}^{\infty} a_i (z - z_0)^i$ be a power series with real coefficients. Then the function is uniquely determined by its value at a point $z_0 + h$, where h is an arbitrary nonzero infinitely small number. In particular, if we choose $h = d$, we have

$$f(z_0 + d) = \sum_{i=0}^{\infty} a_i d^i, \qquad (2.150)$$

from which we readily obtain a formula to compute all the a_i's:

$$a_i = f(z_0 + d)[i]. \qquad (2.151)$$

This formula allows the computation of derivatives of any function which can be written as a power series with a nonzero radius of convergence; this includes all differentiable functions obtained in finitely many steps using arithmetic and intrinsic functions.

Besides allowing illuminating theoretical conclusions, the strength of the Levi-Civita numbers is that they can be used in practice in a **computer environment**. In this respect, they differ from the nonconstructive structures in nonstandard analysis.

Implementation of the Levi-Civita numbers is not as direct as that of the DAs discussed previously since $\mathcal{R}$ is infinite dimensional. However, since there are only finitely many support points below every bound, it is possible to pick any such bound and store all the values of a function to the left of it. Therefore, each Levi-Civita number is represented by these values and the value of the bound.

The sum of two such functions can then be computed for all values to the left of the minimum of the two bounds; therefore, the minimum of the bounds is the bound of the sum. In a similar way, it is possible to find a bound below which the product of two such numbers can be computed from the bounds of the two numbers. The bound to which each individual variable is known is carried along through all arithmetic.

Chapter 3

Fields

For the study of transfer maps of particle optical systems, it is first necessary to undertake a classification of the possible fields that can occur. All fields are governed by Maxwell's equations (1.284, 1.285, 1.286, 1.287), which in SI units have the form

$$\vec{\nabla}\vec{B} = 0, \; \vec{\nabla} \times \vec{H} = \vec{j} + \frac{\partial \vec{D}}{\partial t}$$

$$\vec{\nabla}\vec{D} = \rho, \; \vec{\nabla} \times \vec{E} = -\frac{\partial \vec{B}}{\partial t}$$

In the case of particle optics, we are mostly interested in cases in which there are no sources of the fields in the region where the beam is located, and so in this region we have $\rho = 0$ and $\vec{j} = \vec{0}$. Of course, any beam that is present would represent a ρ and a $\vec{j}$, but these effects are usually considered separately.

In the following, we restrict our discussion to time-independent situations and neglect the treatment of elements with quickly varying fields including cavities. This limitation in very good approximation also includes slowly time-varying fields such as the magnetic fields that are increased during the ramping of a synchrotron. In this case, Maxwell's equations simplify to

$$\begin{aligned}\vec{\nabla}\vec{B} = 0 \, , \; \vec{\nabla} \times \vec{H} = \vec{0} \\ \vec{\nabla}\vec{D} = 0 \, , \; \vec{\nabla} \times \vec{E} = \vec{0},\end{aligned} \quad (3.1)$$

where $\vec{B} = \mu_0 H$ and $\vec{D} = \varepsilon_0 \vec{E}$. Because of the vanishing curl, we infer that by virtue of Eq. (1.300), $\vec{B}$ and $\vec{E}$ have scalar potentials V_E and V_B such that

$$\vec{E} = -\vec{\nabla} V_E \text{ and } \vec{B} = -\vec{\nabla} V_B. \quad (3.2)$$

Note that here even the magnetic field is described by a scalar potential and not by the vector potential $\vec{A}$ that always exist. From the first and third equations, we infer that both scalar potentials V_E and V_B satisfy the **Laplace equation**, and thus

$$\Delta V_{E,B} = 0. \quad (3.3)$$

120   FIELDS

In order to study the solutions of Laplace's equations for the electric and magnetic scalar potentials, we discuss two special cases, each of which will be treated in a coordinate system most suitable for the problem.

3.1 ANALYTIC FIELD REPRESENTATION

3.1.1 Fields with Straight Reference Orbit

The first major case of systems is those that have a straight reference axis. In this case, there is no need to distinguish between particle optical coordinates and Cartesian coordinates, and in particular there is no need to transform Laplace's equation into a new set of coordinates. Many elements with a straight reference orbit possess a certain **rotational symmetry** around the axis of the reference orbit, and it is most advantageous to describe the potential in cylindrical coordinates with a "z axis" that coincides with the reference orbit. We first begin by expanding the r and ϕ components of the potential in Taylor and Fourier series, respectively; the dependence on the cylindrical "z" coordinate, which here coincides with the particle optical coordinate s, is not expanded. Therefore,

$$V = \sum_{k=0}^{\infty} \sum_{l=0}^{\infty} M_{k,l}(s) \cos\left(l\phi + \theta_{k,l}\right) r^k. \tag{3.4}$$

In cylindrical coordinates, the **Laplace Equation** has the form

$$\Delta V = \frac{1}{r} \frac{\partial}{\partial r} \left(r \frac{\partial V}{\partial r} \right) + \frac{1}{r^2} \frac{\partial^2 V}{\partial \phi^2} + \frac{\partial^2 V}{\partial s^2} = 0; \tag{3.5}$$

inserting the Fourier–Taylor expansion of the potential, we obtain

$$\Delta V = \frac{1}{r} \frac{\partial}{\partial r} \left(r \frac{\partial V}{\partial r} \right) + \frac{1}{r^2} \frac{\partial^2 V}{\partial \phi^2} + \frac{\partial^2 V}{\partial s^2}$$

$$= \frac{1}{r} \frac{\partial}{\partial r} \left\{ \sum_{k=1}^{\infty} \sum_{l=0}^{\infty} M_{k,l} \cos\left(l\phi + \theta_{k,l}\right) k r^k \right\}$$

$$+ \frac{1}{r^2} \sum_{k=0}^{\infty} \sum_{l=0}^{\infty} M_{k,l} \cos\left(l\phi + \theta_{k,l}\right) \left(-l^2\right) r^k$$

$$+ \sum_{k=0}^{\infty} \sum_{l=0}^{\infty} M''_{k,l}(s) \cos\left(l\phi + \theta_{k,l}\right) r^k$$

$$= \sum_{k=1}^{\infty}\sum_{l=0}^{\infty} M_{k,l} \cos(l\phi + \theta_{k,l}) k^2 r^{k-2}$$

$$- \sum_{k=0}^{\infty}\sum_{l=0}^{\infty} M_{k,l} \cos(l\phi + \theta_{k,l}) l^2 r^{k-2}$$

$$+ \sum_{k=2}^{\infty}\sum_{l=0}^{\infty} M''_{k-2,l}(s) \cos(l\phi + \theta_{k-2,l}) r^{k-2}$$

We note that in the first term, it is possible to let the sum start at $k = 0$ since there is no contribution because of the factor k^2. Furthermore, using the convention that the coefficient $M_{k,l}(s)$ vanishes for negative indices, we obtain

$$\Delta V = \sum_{k,l=0}^{\infty} \left\{ M_{k,l}(s) \cos(l\phi + \theta_{k,l}) \left(k^2 - l^2\right) \right.$$
$$\left. + M''_{k-2,l}(s) \cos(l\phi + \theta_{k-2,l}) \right\} r^{k-2}. \qquad (3.6)$$

We begin the analysis by studying the **case** $k = 0$. Apparently, $M_{0,0}$ and $\theta_{0,0}$ can be chosen freely because the factor $(k^2 - l^2)$ vanishes. Furthermore, since $M''_{k-2,l}(s)$ vanishes for all l because of the convention regarding negative indices, we infer $M_{0,l} = 0$ for $l \geq 1$.

By induction over k, we now show that $M_{k,l}(s) \equiv 0$ for all cases in which $k < l$. Apparently the statement is true for $k = 0$. Now let us assume that the statement is true up to $k - 1$. If $k < l$, then $k - 2 < l$, and thus $M''_{k-2,l}(s) = 0$. Since $(k^2 - l^2) \neq 0$ and $\cos(l\phi + \theta_{k,l}) \neq 0$ for some ϕ because $l \neq 0$, this requires $M_{k,l}(s) \equiv 0$ for $k < l$. Thus, the infinite matrix $M_{k,l}$ is strictly lower triangular.

We now study the situation for different values of l. We first notice that for all l, the choices of

$$M_{l,l}(s) \text{ and } \theta_{l,l} \text{ are free} \qquad (3.7)$$

because $M''_{l-2,l}(s) = 0$ by the previous observation, and $(k^2 - l^2) = 0$ because $k = l$. Next we observe that the value $M_{l+1,l}(s)$ must vanish because $(k^2 - l^2) \neq 0$, but $M''_{l-1,l}(s) \equiv 0$ because of the lower triangularity. Recursively, we even obtain that

$$M_{l+1,l}(s), M_{l+3,l}(s), \ldots \text{ vanish.} \qquad (3.8)$$

On the other hand, for $k = l + 2$, we obtain that $\theta_{l+2,l} = \theta_{l,l}$, and $M_{l+2,l}(s)$ is uniquely specified by $M_{l,l}(s)$. Applying recursion, we see that in general

$$\theta_{l,l} = \theta_{l+2,l}, \theta_{l+4,l},$$

and

$$M_{l+2n,l}(s) = \frac{M_{l,l}^{(2n)}(s)}{\prod_{\nu=1}^{n}\left((l)^2 - (l+2\nu)^2\right)}. \tag{3.9}$$

We now proceed with the physical interpretation of the result. The number l is called the **multipole order** because it describes how many oscillations the field will experience in one 2π sweep of ϕ. The free term $M_{l,l}(s)$ is called the **multipole strength**, and the term $\theta_{l,l}$ is called the **phase**. Apparently, frequency l and radial power k are coupled: The lowest order in r that appears is l, and if the multipole strength is s dependent the powers $l + 2, l + 4, \ldots$ will also appear.

For a multipole of order l, the potential has a total of $2l$ maxima and minima, and is often called a **2l-pole**. Often, Latin names are used for the $2l$ poles:

l	Leading term in V	Name
0	$M_{0,0}(s)\cos\left(\theta_{0,0}\right)$	
1	$M_{1,1}(s)\cos\left(\phi + \theta_{1,1}\right) r$	Dipole
2	$M_{2,2}(s)\cos\left(2\phi + \theta_{2,2}\right) r^2$	Quadrupole
3	$M_{3,3}(s)\cos\left(3\phi + \theta_{3,3}\right) r^3$	Sextupole
4	$M_{4,4}(s)\cos\left(4\phi + \theta_{4,4}\right) r^4$	Octupole
5	$M_{5,5}(s)\cos\left(5\phi + \theta_{5,5}\right) r^5$	Decapole

In many cases, it is very important to study the **Cartesian** (and hence also particle optical) form of the fields of the elements. The case $k = 1$ with $V = M_{1,1}\cos\left(\phi + \theta_{1,1}\right) r$ is quite trivial; for $\theta_{11} = 0$, we obtain $V = M_{1,1} \cdot x$, corresponding to a uniform field in x direction, and for another important subcase $\theta_{11} = -\pi/2$, we obtain $V = M_{1,1} \cdot y$, a uniform field in y direction.

The case $k = 2$ has $V = M_{2,2}\cos\left(2\phi + \theta_{22}\right) r^2$. Particularly important in practice will be the subcases $\theta_{22} = 0$ and $\theta_{22} = -\pi/2$. In the first case, we get

$$V = M_{2,2}\cos(2\phi)\, r^2 = M_{2,2}\left(\cos^2\phi - \sin^2\phi\right) r^2$$
$$= M_{2,2}\left(x^2 - y^2\right), \tag{3.10}$$

and in the second case we have

$$V = M_{2,2}\cos(2\phi - \pi/2)\, r^2 = M_{2,2}\left(2\sin\phi\cos\phi\right) r^2$$
$$= M_{2,2}\left(2xy\right). \tag{3.11}$$

All other angles θ_{22} lead to formulas that are more complicated; they can be obtained from the ones here by subjecting the x, y coordinates to a suitable rotation. This again leads to terms of purely second order.

Because the potential is quadratic, the resulting fields $\vec{E}$ or $\vec{B}$ are **linear**. Indeed, the quadrupole is the only s-independent element that leads to linear motion, similar to that in glass optics, and thus has great importance. In the electric case, one usually chooses $\theta_{2,2} = 0$, resulting in the fields

$$E_x = -2M_{2,2} \cdot x \tag{3.12}$$

$$E_y = 2M_{2,2} \cdot y. \tag{3.13}$$

Therefore, different from the case of glass optics, the motion cannot be rotationally symmetric: If there is focusing in the x direction, there is defocusing in the y direction and vice versa. This effect, completely due to Maxwell's equations, perhaps the biggest "nuisance" in beam physics: If one uses piecewise s-independent particle optical elements, the horizontal and vertical planes are always different from each other.

To make an electrostatic device that produces a quadrupole field, it is best to carve the electrodes along the equipotential surfaces and utilize the fact that if "enough" boundary conditions are specified, the field is uniquely determined and must hence be as specified by the formula used to determine the equipotential surfaces in the first place. Therefore in practice, the electrodes of an electric quadrupole often appear as shown in Fig. 3.1.

In the magnetic case, one indeed chooses $\theta_{2,2} = -\pi/2$, resulting in

$$B_x = -2M_{22} \cdot y \tag{3.14}$$

$$B_y = -2M_{22} \cdot x, \tag{3.15}$$

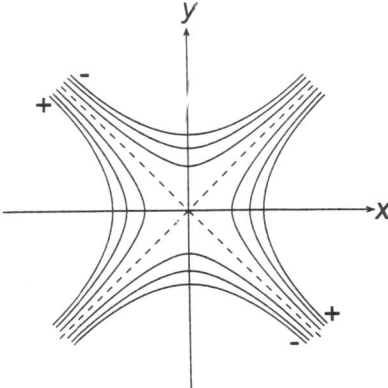

FIGURE 3.1. The equipotential surfaces of a quadrupole.

and examining the Lorentz forces that a particle moving mostly in s-direction experiences, we see that if there is focusing in x direction, there is defocusing in y direction and vice versa.

To study higher orders in k, let us consider the case $k = 3$. For $\theta_{3,3} = 0$, we obtain

$$V = M_{3,3} \cos(3\phi) r^3 = M_{3,3} (\cos\phi \cos 2\phi - \sin\phi \sin 2\phi) r^3$$
$$= M_{3,3} \left(xr^2 \cos 2\phi - yr^2 \sin 2\phi\right) = M_{3,3} \left(x^3 - 3xy^2\right). \quad (3.16)$$

In this case, the resulting forces are **quadratic** and are thus not suitable for affecting the linear motion; however, we later will show that they are indeed very convenient for the correction of nonlinear motion, and they even have the feature of having no influence on the linear part of the motion! Another important case for $\theta_{3,3}$ is $\theta_{3,3} = -\pi/2$, in which case one can perform a similar argument and again obtain cubic dependencies on the position.

For all the higher values of l, corresponding to octupoles, decapoles, duodecapoles, etc., the procedure is very similar. We begin with the addition theorem for $\cos(l\phi)$ or $\sin(l\phi)$, and by induction it can be seen that each of these consists of terms that have a product of precisely l cosines and sines. Since each of these terms is multiplied with r^l, each cosine multiplied with one r translates into an x, and each sine multiplied with one r translates into a y; the end result is always a **polynomial** in x and y of exact order l.

Because of their nonlinear field dependence, these elements will prove to have **no effect** on the motion up to order $l-1$ and thus allow one to selectively influence the higher orders of the motion without affecting the lower orders. Also, if it is the crux of particle optical motion that the horizontal and vertical linear motion cannot be affected simultaneously, it is its **blessing** that the nonlinear effects can be corrected order by order.

In the case in which there is no s dependence, the potential terms that we have derived are the only ones; under the **presence of s-dependence**, as shown in Eq. (3.9), to the given angular dependence there are higher order terms in r, the strengths of which are given by the s derivatives of the multipole strength $M_{l,l}$. The computation of their Cartesian form is very easy once the Cartesian form of the leading term is known because each additional term differs from the previous one by only the factor of $r^2 = (x^2 + y^2)$.

In practice, of course, s dependence is **unavoidable**: The field of any particle optical element has to begin and end somewhere, and it usually does this by rising and falling gently with s, entailing s dependence. This actually entails another crux of particle optics: Even the quadrupoles, the "linear" elements, have nonlinear effects at their edges, requiring higher order correction. The corrective elements in turn have higher order edge effects, possibly requiring even higher order correction, etc.

Although without s dependence the case $l = 0$ corresponding to full rotational symmetry was not very interesting, if we consider s dependence, it actually offers a remarkably useful effect. While there is no r dependence in the leading term, the contributions through the derivatives of $M_{0,0}(s)$ entail terms with an r dependence of the form $r^2, r^4, \ldots$. Of these, the r^2 terms will indeed produce linear, rotationally symmetric fields similar to those in the glass lens. Unfortunately, in practice these fields are restricted to the entrance and exit fringe field regions and are comparatively weak; furthermore, there are usually quite large nonlinearities, and these devices are usually used mostly for low-energy, small emittance beams, like those found in electron microscopes.

3.1.2 Fields with Planar Reference Orbit

In the case of the straight reference orbit, Maxwell's equations entail a very clean connection between rotational symmetry and radial potential. As one may expect, in the case of a nonstraight reference orbit, this is no longer the case; in this situation, Maxwell's equations have a rather different but not less interesting consequence as long as we restrict ourselves to the case in which the reference orbit stays in one **plane**.

Therefore, let us assume that the motion of the reference particle is in a plane, and that all orbits that are on this plane stay in it. Let $h(s)$ be the momentary curvature. As shown in (Berz et al., 1999), in these coordinates the **Laplacian** has the form

$$\Delta V = \frac{1}{1+hx} \frac{\partial}{\partial x}\left((1+hx)\frac{\partial V}{\partial x}\right) + \frac{1}{1+hx}\frac{\partial}{\partial s}\left(\frac{1}{1+hx}\frac{\partial V}{\partial s}\right) + \frac{\partial^2 V}{\partial y^2}. \tag{3.17}$$

For the **potential**, we again make an expansion in transversal coordinates and leave the longitudinal coordinates unexpanded. Since we are working with x and y, both expansions are Taylor, and we have

$$V = \sum_{k=0}^{\infty} \sum_{l=0}^{\infty} a_{k,l}(s) \frac{x^k y^l}{k! \cdot l!}. \tag{3.18}$$

This expansion must now be inserted into the Laplacian in particle optical coordinates. Besides the mere differentiation, we also have to Taylor expand $1/(1+hx) = 1 - (hx) + (hx)^2 - (hx)^3 + \ldots$. After gathering like terms and heavy arithmetic, and again using the convention that coefficients with negative indices are assumed to vanish, we obtain the recursion relation

$$\begin{aligned} a_{k,l+2} = &-a_{k,l}'' - kha_{k-1,l}'' + kh'a_{k-1,l}' - a_{k+2,l} - (3k+1)ha_{k+1,l} \\ &- 3kha_{k-1,l+2} - k(3k-1)h^2 a_{k,l} - 3k(k-1)h^2 a_{k-2,l+2} \\ &- k(k-1)^2 h^3 a_{k-1,l} - k(k-1)(k-2)h^3 a_{k-3,l+2}. \end{aligned} \tag{3.19}$$

Although admittedly horrible and unpleasant, the formula apparently allows a **recursive calculation** of coefficients. Indeed, the terms $a_{k,0}(s)$, $a_{k,1}(s)$ can be chosen freely, and all others are uniquely determined through them.

To study the significance of the free terms, let us consider the electric and magnetic cases separately. In the **electric** case, in order to ensure that orbits that were in the plane stay there, there must not be any field components in the y direction in the plane corresponding to $y = 0$. Computing the gradient of the potential, we have

$$E_x(x, y = 0) = -\sum_k a_{k,0} \frac{x^{k-1}}{(k-1)!}$$

$$E_y(x, y = 0) = -\sum_k a_{k,1} \frac{x^k}{k!} = 0,$$

and looking at E_y, we conclude that $a_{k,1} = 0$ for all k. Therefore, the terms $a_{k,0}$ alone specify the field. Looking at E_x, we see that these are the coefficients that specify the field within the plane, and so the **planar field determines the entire field**. Furthermore, looking at the details of the recursion relation, it becomes apparent that all second indices are either l or $l + 2$. This entails that as long as $a_{k,1}$ terms do not appear, $a_{k,3}, a_{k,5}, \ldots$ terms also do not appear. Indeed, the resulting potential is fully symmetric around the plane, and the resulting field lines above and below the plane are mirror images.

In the **magnetic** field, the argument is similar: Considering the fields in the plane, we have

$$B_y(x, y = 0) = -\sum_k a_{k,1} \frac{x^k}{k!}$$

$$B_x(x, y = 0) = -\sum_k a_{k,0} \frac{x^{k-1}}{(k-1)!} = 0.$$

In order for particles in the midplane to remain there, B_x must vanish in the midplane, which entails $a_{k,0} = 0$. Therefore, in the magnetic case, the coefficients $a_{k,1}$ specify everything. These coefficients, however, again describe the shape of the field in the plane, and so again the planar field determines the entire field. In the magnetic case, the potential is fully antisymmetric around the plane, and again the resulting field lines are mirror images of each other.

3.2 PRACTICAL UTILIZATION OF FIELD INFORMATION

In this section, we discuss techniques to provide the proper field data for the calculation of transfer maps. In particular, this requires the necessity to compute various high-order derivatives, for which differential algebra (DA) methods are very useful, but the case of measured data with a certain amount of noise dictates the use of special tools. We address both the utilization of local field measurements and the various global approaches for the description of data for field and geometry in three-dimensional space.

The following two sections address common scenarios in which measured field information exists as one- or two-dimensional data, in which case the problem is mostly connected to the proper utilization of local measurement data that require suitable smoothing and differentiation. The latter three sections discuss the question of global Maxwellian field representations for the case in which information about the field exists in a more global sense. Emphasis is put on robustness, differentiability, and suppression of measurement error on a global scale.

3.2.1 Multipole Measurements

The first method used for measured field data is used in the cases in which certain **multipole terms** $M_{l,l}$ have already been determined as a function of s through measurement or numerical techniques or combinations thereof. These terms can be used directly within the computation as long as it is possible to reasonably carry out the **differentiation in s direction** that is necessary to determine the terms in the recursion relation (Eq. 3.9). This is achieved with the use of an interpolation method that is sufficiently differentiable and that easily connects to the differential algebraic approach. We further desire that the interpolation be localized in the sense that for the interpolated fields at a certain point, only the fields at neighboring points contribute.

Both of these requirements are met by the wavelet techniques (Daubechies, 1992). Of the wide variety of choices, for our purposes it is sufficient to restrict the use to the **Gaussian Wavelet**, which ensures the required differentiability and locality.

Assume a set of data Y_i is given at N equidistant points x_i for $i = 1, \ldots, N$. Then, the interpolated value at a point x is expressed as

$$V(x) = \sum_{i=1}^{N} Y_i \frac{1}{\sqrt{\pi} S} \exp\left[-\frac{(x - x_i)^2}{\Delta x^2 S^2}\right],$$

where Δx is the distance of two neighboring points x_i and x_{i+1}, and S is the factor controlling the width of Gaussian wavelets.

Figure 3.2 and Table I show how the interpolation via a sum of Gaussian wavelets works for one-dimensional linear and Gaussian functions. The data values Y_i are obtained as the values of the function at the corresponding points x_i.

There are several other strategies for choosing the height of each Gaussian. One approach is to set up a local **least squares problem** to determine the heights that fit the data optimally. While this approach usually yields the best approximation, it does not preserve the integral of the function, which in many cases is very

 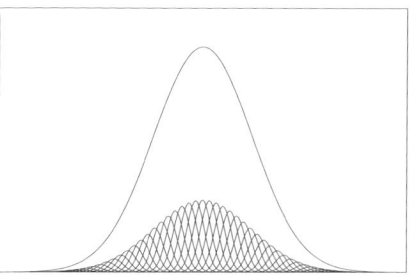

FIGURE 3.2. Gaussian wavelets representation for $f(x) = 1$ (left) and $f(x) = \exp(-x^2)$ (right).

desirable. Also, the method is not particularly well suited for the treatment of noisy data, which benefit from the **smoothing effect** achieved by wider Gaussians.

The resulting interpolated function is shown in Fig. 3.2, which represents the original functions very well. The value of S is chosen as 1.8 for all cases. Table I summarizes the accuracy of the method for those functions, although local accuracy for our purposes is of a lesser concern compared to smoothing and preservation of area.

The advantage of the Gaussian function and many other wavelets is that they fall off quickly. Thus, the potentially time-consuming summation over all wavelets can be replaced by the summation of only the neighboring Gaussian wavelets in the range of ±8S, which is in the vein of other wavelet transforms and greatly improves efficiency.

3.2.2 Midplane Field Measurements

A very common approach for the measurement of particle optical devices is the measurement of fields in the midplane, based on the observation that such information determines the field in all of space by virtue of Eq. (3.19). The difficulty with this approach is that while in principle the midplane data are indeed sufficient, in practice the determination of the out-of-plane field requires high-order **differentiation of the midplane data**.

This represents a fundamental **limitation** of the technique because of the subtlety of numerical differentiation of data, especially under the unavoidable pres-

TABLE I
ACCURACY OF THE GAUSSIAN WAVELETS REPRESENTATION FOR ONE-DIMENSIONAL FUNCTIONS

Test function	Number of Gaussian wavelets N	Average error	Maximum error
1	10	2.6×10^{-14}	2.6×10^{-14}
$\exp(-x^2)$	600	4.6×10^{-5}	1.6×10^{-4}

ence of measurement noise. In these circumstances the best approach appears to be the Gaussian wavelets representation discussed previously. Here, the favorable features of the Gaussian function of differentiability, locality, and adjustable smoothing help alleviate some of the inherent difficulty.

Of course, in the case of planar reference orbit, the procedure has to be extended to two dimensions; assume a set of data $B(i_x, i_z)$ is given at equidistant $N_x \times N_z$ grid points (x_{i_x}, z_{i_z}) for $i_x = 1, \ldots, N_x$ and $i_z = 1, \ldots, N_z$ in two-dimensional Cartesian coordinates. Then, the interpolated value at the point (x, z) is expressed as

$$B_y(x, z) = \sum_{i_x=1}^{N_x} \sum_{i_z=1}^{N_z} B(i_x, i_z) \frac{1}{\pi S^2} \exp\left[-\frac{(x - x_{i_x})^2}{\triangle x^2 S^2} - \frac{(z - z_{i_z})^2}{\triangle z^2 S^2}\right], \quad (3.20)$$

where $\triangle x$ and $\triangle z$ are the grid spacings in x and z directions, respectively, and S is the control factor for the width of Gaussian wavelets.

A suitable **choice for the control factor** S depends on the behavior of the original supplied data. If S is too small, the mountain structure of individual Gaussian wavelets is observed. On the other hand, if S is too large, the original value supplied by the data is washed out, which can also be used effectively for purposes of smoothing noisy data. For nearly constant fields, the suitable value S is about 1.8. For quickly varying fields, it should be as small as 1.0. In general, larger values of S usually provide more accurate evaluation of the derivatives; this is connected to their tendency to wash out local fluctuations of the data.

3.2.3 Electric Image Charge Methods

In this section, we address a global method to determine fields in the case of electrostatic geometries. It is based on a **global solution** of the boundary value problem by means of suitable image charges. The resulting fields are simultaneously Maxwellian and infinitely often differentiable, and because of their global nature, they provide **smoothing** of various errors. Using DA techniques, the method allows the direct and detailed computation of high-order multipoles for the geometry at hand.

We study the use of the method on the focusing elements of the **MBE-4** (Berz et al., 1991) Multiple Beam Experiment at Lawrence Berkeley Laboratory, in which four nonrelativistic, space-charge-dominated heavy ion beamlets travel simultaneously in a common acceleration channel. Because of significant interaction between the four beamlines, many additional multipole components in addition to the wanted quadrupole component exist. Although weak, the presence of these components, in conjunction with nonnegligible transverse beam displacements, will lead to undesirable nonlinear behavior during beam transport.

The presence of boundary surfaces with different potential values will lead to image charges on the conducting boundaries. When a complicated boundary shape is present, a conventional field solver using FFT or SOR techniques becomes computationally expensive because of the large number of mesh points required in three-dimensional calculations. This difficulty can be alleviated noticeably by calculating the induced **charge on boundary** surfaces directly rather than specifying the field boundary condition. Once the boundary charge distribution is known, the field at the location of a particle can be calculated straightforwardly. Since the number of required mesh nodes is confined to where the actual charges are located, i.e., at the boundary, **substantial reduction in total mesh points** is obtained.

In the boundary charge or capacitance matrix method, the electrodes are covered with test points $\vec{x}_i$, the so-called nodes. Charges Q_i located at the $\vec{x}_i$ are determined such that the potential induced on $\vec{x}_i$ by all the other nodes assumes the desired value. This reduces to solving the linear system

$$V_j = \sum_i G(\vec{x}_i, \vec{x}_j) \cdot Q_i, \qquad (3.21)$$

where $G(\vec{x}, \vec{x}_j)$ is the Green's function describing the effect of the potential of a charge at $\vec{x}_i$ to the point $\vec{x}_j$. The inverse to $G(\vec{x}_i, \vec{x}_j)$ is often called the **capacitance matrix** C_{ij}. Once the charges Q_i are known, the potential at any point in space can be computed as

$$V(\vec{x}) = \sum_i G(\vec{x}_i, \vec{x}) \cdot Q_i. \qquad (3.22)$$

In most cases, it is possible to use the simple Green's function

$$G(\vec{x}_i, \vec{x}_j) = \frac{1}{|\vec{x}_i - \vec{x}_j|} \text{ for } i \neq j \qquad (3.23)$$

for nonoverlapping test charges. When computing the self-potential ($i = j$) or when the test charge width σ_i exceeds an internode spacing, the test charge profile must be carefully considered. Frequently suitable choices are **triangular** or **Gaussian** charge distributions. The particular choice of the charge profile is somewhat arbitrary, but in practice numerical calculation shows that the determination of the multipole harmonics is not sensitive to the exact charge profile. The distribution width σ_i is typically set to the internode spacing, depending on the charge distribution used.

Once the boundary surface charge has been determined, the electric potential at an arbitrary point in space can be computed by the summation over the charges. The resulting approximate potential $V(\vec{x}) = \sum_i Q_i/|\vec{x}_i - \vec{x}|$ is infinitely differentiable and hence allows direct differentiation and multipole decomposition using DA methods.

In the case of the actual MBE-4 focusing lattice, there are a number of separate elements in the electrode as shown in Fig. 3.3. In particular, there are quadrupole

FIGURE 3.3. Geometry of the MBE-4 focusing element.

electrodes which are held at either ground or a negative voltage, flat aperture plates from which the electrodes are cantilevered and through whose holes the four ion beams travel, and a large "can" enclosure surrounding the entire focusing lattice.

The nodes were distributed uniformly over the surface of the electrode rods and with an $\sim 1/r$ dependence on the end plates. Typical node locations are plotted in Fig. 3.4. Beam holes in the end plates are represented by the absence of nodes in the interior and by a clustering around the boundary circles. The charge distribution width σ_i was set to a constant ($\approx$ 2.5 mm for a total node number $N = 5000$)—approximately the typical internode distance.

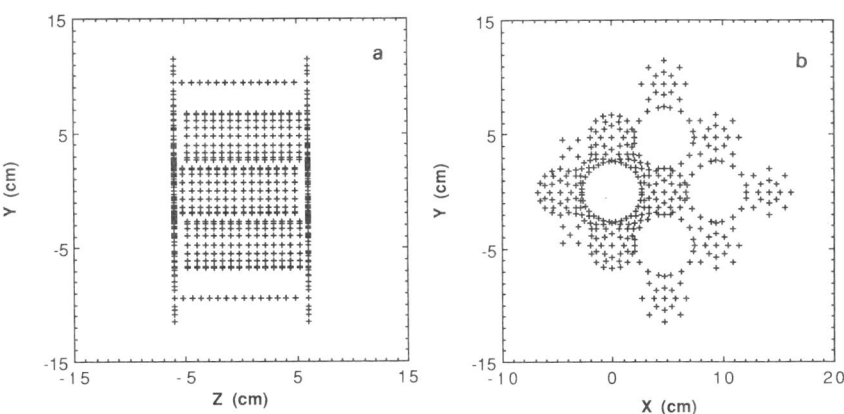

FIGURE 3.4. Distribution of nodes on the MBE-4 focusing element.

In order to evaluate expression (3.22) numerically, DA methods were used for the computation of the derivatives of the potential V. The $M_{k,l}$ were computed from 0th to 6th order and made dimensionless by scaling the potential by a factor of V_q and using a normalization length of the aperture radius a such that

$$V = V_q \cdot \sum_{k=0}^{6} \sum_{l=0}^{6} M_{k,l}(s) \left(\frac{r}{a}\right)^k \cos(l\phi). \qquad (3.24)$$

Figure 3.5 shows the results for the case of $N = 5000$. Compared with the conventional field-solver technique, in which the vacuum field is solved first and the multipole field decomposition is performed by some sort of fitting, the method described here is straightforward and more accurate, particularly when higher orders of harmonics are needed. Although the finite size of the charge distribution at a given node poses some uncertainty of the determination of the capacitance matrix, the effect can be minimized by optimally choosing the profile shape and width. Numerical experiments show that the decomposition is indeed insensitive to the charge distribution at the nodes as long as the internode distance is much smaller than the aperture radius.

3.2.4 Magnetic Image Charge Methods

A modification of the method discussed in the previous section is also applicable in the magnetic case, in which boundary conditions are not directly available. To this end, the magnetic field is generated by a suitable superposition of Gaussian charge distributions. In practice, image charges are placed on regular grids parallel to the midplane as illustrated in Fig. 3.6. For the determination of the strengths of the individual charges, a least square fit of the field values at the reference points is performed.

In order to reduce any fine structure due to the influence of individual image charges, we use extended distributions in the form of a three-dimensional **Gaussian charge distribution** (Berz, 1988a) of the form

$$\rho(r) = \rho_0 \cdot \exp\left(-\frac{r^2}{a^2}\right), \qquad (3.25)$$

where a is a measure for the width of the Gaussian, and ρ_0 determines the strength of the individual image charge. A magnetic field generated by a superposition of Gaussian charge distributions automatically satisfies Maxwell's equations.

By applying Gauss' law for the rotationally symmetric charge distribution (Eq. 3.25), the ith Gaussian positioned at the point (x_i, y_i, z_i) makes a contribution to

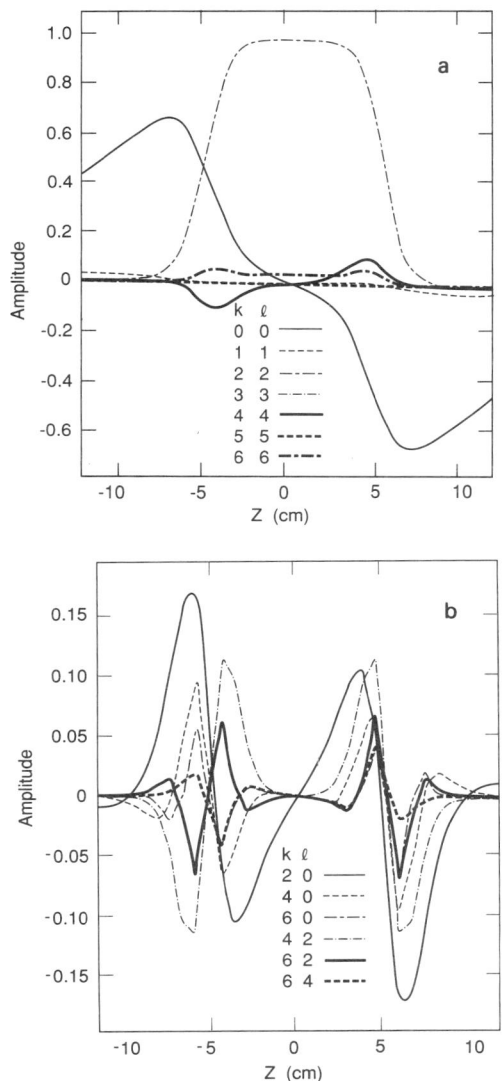

FIGURE 3.5. Multipole terms for MBE-4 using DA and the capacitance matrix method.

the y-component of the magnetic field of the form

$$H_{y,i}^C(x, y, z) = \frac{\rho_i(y - y_i)}{r_i^3} \cdot \left[-\frac{a_i^2 r_i}{2} \exp\left(-\left(\frac{r_i}{a_i}\right)^2\right) + \frac{\sqrt{\pi} a_i^3}{4} \operatorname{erf}\left(\frac{r_i}{a_i}\right), \right],$$

(3.26)

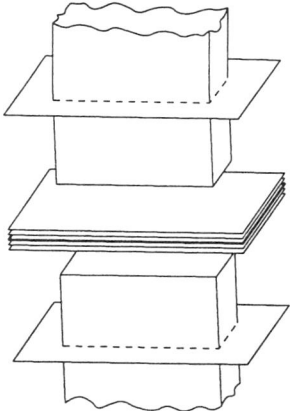

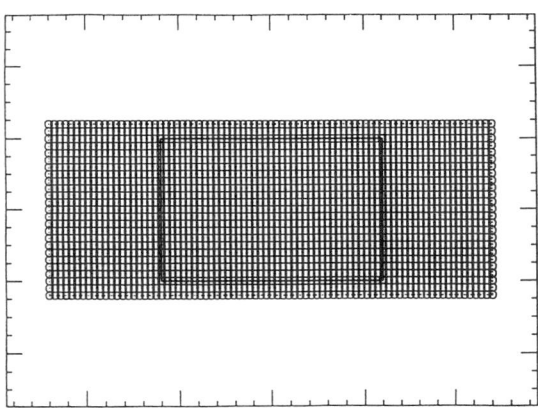

FIGURE 3.6. (Top): Schematic arrangement of the rectangular iron bars indicating in the gap the planes that contain the reference field data and two planes on which image charges are placed. (Bottom): Grid generated by Gaussian charge distributions. The bold inner rectangle indicates the cross section of the bars. Note the smaller step size in the horizontal direction in order to describe the fall off of the fringe field in more detail.

where $r_i = [(x - x_i)^2 + (y - y_i)^2 + (z - z_i)^2]^{1/2}$ and $\text{erf}(u) = 2/\sqrt{\pi} \int_0^u e^{-u'^2} du'$ is the error function. The total field $H_y^C(x, y, z)$ is obtained by summing over all individual Gaussians. Adjusting their width appropriately, the superposition of regularly distributed Gaussians has proven to result in a smooth global field distribution.

In order to assess the method, we study the case of a sample dipole for which an **analytical field** is known. We choose the magnetic field of rectangular iron bars with inner surfaces ($y = \pm y_0$) parallel to the midplane ($y = 0$). The geometry of these uniformly magnetized bars, which are assumed to be infinitely extended in the $\pm y$ directions, is defined by

$$-x_0 \leq x \leq x_0, \quad |y| \geq y_0, \quad -z_0 \leq z \leq z_0. \quad (3.27)$$

For this region of magnetization one obtains for the y-component of the magnetic bar field $H_y^B(x, y, z)$ an analytical solution of the form (Gordon and Taivassalo, 1986):

$$H_y^B(x, y, z) = \hat{H}_0^B \sum_{i,j} (-1)^{i+j} \left(\arctan\left(\frac{x_i \cdot z_j}{y_+ \cdot R_{ij}^+} \right) + \arctan\left(\frac{x_i \cdot z_j}{y_- \cdot R_{ij}^-} \right) \right), \quad (3.28)$$

with $i, j = 1, 2$ and where the abbreviations $x_1 = x - x_0$, $x_2 = x + x_0$, $z_1 = z - z_0$, $z_2 = z + z_0$, $y_- = y_0 - y$, $y_+ = y_0 + y$, and $R_{ij}^\pm = \sqrt{x_i^2 + y_\pm^2 + z_j^2}$ have been used. $\hat{H}_0^B$ determines the maximum of this field component which is reached at the bars.

In order to benchmark the performance of the method, we utilize the analytic formula of the reference field to determine data points modeling those that would be obtained by field measurements. From $H_y^B(x, y, z)$, we generate reference field points on a regular grid in the midplane (Fig. 3.7) as well as in planes above and below the midplane as indicated in Fig. 3.6.

One of the grids with Gaussian charge distributions is located in the present configuration at $y = \pm D$ and is larger than the cross section of the bars (see Fig. 3.6). The distance between the Gaussians is chosen to be approximately a. A second grid with $a = 1.5D$ (not shown in Fig. 3.6) is placed at the planes

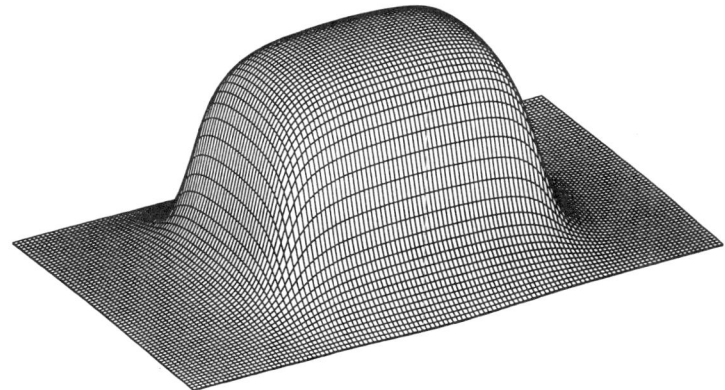

FIGURE 3.7. The analytical reference field.

$y = \pm 5D$ so that eight Gaussians cover almost the entire area shown in Fig. 3.6. Since the Gaussian distribution is very close to zero at a distance of $3a$, it is assumed that $\rho = 0$ at the reference planes, which are located at $y = 0, \pm 0.125D$, and $\pm 0.25D$. For the computations we also make use of the midplane symmetry of the magnet.

The least square algorithm results in a square **matrix** that has to be **inverted** in order to determine the strengths of the individual charges. In total, we placed $N_C = 535$ Gaussians on a quarter of the whole arrangement, taking advantage of the geometrical symmetry of the bars. The dimension of the matrix is given by the number of image charges and their strengths are determined in this case by $N_R = 10125$ reference field points.

We now concentrate on the calculation of the difference between the reference field and its approximation on a strip through the magnet in order to assess the applicability of our method to the description of sector magnets. In this case, our main focus is the proper representation of the entrance and exit fringe field region of the magnet. The strip covers 75% of the width of the bars and is twice as long as the bars. In this area, considered as relevant, the maximum difference between the y component of the two fields in the midplane $\Delta H_y(x, 0, z) = H_y^C(x, 0, z) - H_y^B(x, 0, z)$ normalized on $H_y^B(0, 0, 0)$ is smaller than 10^{-4}, as seen in the top of Fig. 3.8. The maximum differences occur in a limited region of the fringe field. The average error over the entire strip is estimated to be one order of magnitude smaller.

When experimentally obtained magnetic fields have to be approximated, local **measurement errors** are superimposed on the actual field data. In order to simulate the noise on measured field data, we add/subtract on every point of the reference field a field difference ΔH^N, which is randomly distributed in an interval $\pm \Delta H_{max}^N$. In the case $\Delta H_{max}^N / H_y^B(0, 0, 0) = 10^{-4}$, the precision with which we can approximate the reference field is essentially unchanged. Overall, the average approximation error increases slightly due to the noise. Nevertheless, as long as the amplitude of the noise is within the range of the precision of the charge density method, this kind of noise does not negatively affect the accuracy of the field approximation.

When the noise amplitude is increased by a factor of five, we find the expected beneficial result that the method provides **smoothing** of the noisy data. In this case the relative field difference $\Delta H_y(x, 0, z)/H_y^B(0, 0, 0)$ is smaller than 1.5×10^{-4}. The areas of maximal field differences are not restricted to the fringe field region anymore, resulting in a difference pattern that is dominated by the noise.

In order to calculate the **multipole content** of the analytical reference field and the one determined by the charge density method, we perform an expansion of the magnetic scalar potential as in Eq. (3.18). If the components of the fields or the magnetic scalar potential in the midplane are known in an analytical form, the decomposition into the individual **multipole moments** can immediately be calculated within a DA-based framework since the differentiations necessary

FIGURE 3.8. Normalized field difference $(H_y^C(x,0,z) - H_y^B(x,0,y))/H_y^B(0,0,0)$ on a strip through the bar covering 75% of its total width. Top, No noise is added to the reference field; Middle, With noise of amplitude $\Delta H_{max}^N / H_y^B(0,0,0) = 1 \cdot 10^{-4}$; Bottom, With noise of amplitude $\Delta H_{max}^N / H_y^B(0,0,0) = 5 \cdot 10^{-4}$.

in the theory developed in Section 3.1.2 can be performed directly and accurately. For particle optics calculations up to fifth order, the relevant coefficients are $a_{0,1}(s)$, $a_{2,1}(s)$, and $a_{4,1}(s)$. The coefficient $a_{0,1}(s)$ describes the field distribution whereas $a_{2,1}(s)$ and $a_{4,1}(s)$ determine the second and fourth derivatives with respect to x. In the case of the bar field, derivatives in x direction seem to be more sensitive than those for a homogenous sector magnet. Therefore, we assume that the accuracy which we obtain for the analytical reference field is a reasonable

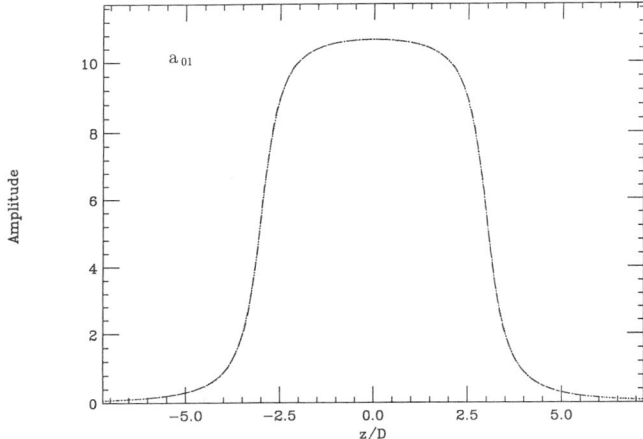

FIGURE 3.9. Multipole coefficient $a_{0,1}$ along the z axis through the magnet for the analytical reference field ($\cdots$) and its approximation by the charge density method (---).

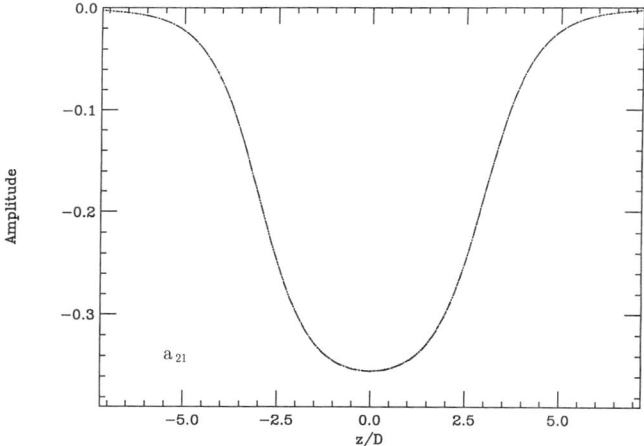

FIGURE 3.10. Multipole coefficient $a_{2,1(z)}$ (see the legend for Fig. 3.9).

estimate for the accuracy that we can expect in the case of homogeneous sector magnets including higher order derivatives.

Using DA methods, we calculated the distribution of the coefficients $a_{0,1}(s)$, $a_{2,1}(s)$, and $a_{4,1}(s)$ for the bar field and its approximated field. The results are shown in Figs. 3.9–3.11. It is apparent that the second derivatives agree very well with the analytic values, whereas a slight deviation is noticeable for the fourth derivative.

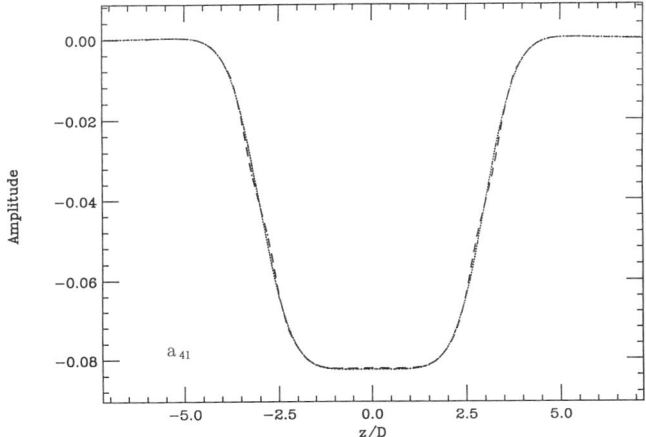

FIGURE 3.11. Multipole coefficient $a_{4,1(z)}$ (see the legend for Fig. 3.10).

3.2.5 The Method of Wire Currents

In the case of magnetic elements in which the field is dominated by the current coils the geometry of which is known, another useful method for the global description of the field is available. To this end, the entire field in space is calculated by summing up the fields of wire currents, an approach that is used in several field analysis programs including the widely used code ROXIE (Russenschuck, 1995). This method also results in a global field representation that is Maxwellian (as long as all coils close and there are no artificial sources and sinks of currents) and that is infinitely often differentiable and hence readily allows the use of DA methods. Due to the global character of the field description, local errors in the placement of coils, and so on are averaged out.

As a consequence of Ampere's law (Eq. 1.287) and $\vec{\nabla} B = 0$, the elementary magnetic flux density at a point $\vec{r}$ generated by a wire loop $\vec{l}(s)$ is given by the **Biot–Savart** formula

$$\vec{B}(\vec{r}) = \frac{\mu_0 I}{4\pi} \cdot \oint \frac{d\vec{l} \times (\vec{r} - \vec{l}(s))}{|\vec{r} - \vec{l}(s)|^3}.$$

To compute the magnetic field contribution generated by an extended straight-line current we parametrize the line by $\lambda \in [0, 1]$ and define $\vec{r}_0(\lambda) = \vec{r} + \vec{r}_s + \lambda \vec{l}$ and $\vec{r}_e = \vec{r}_s + \vec{l}$, where $\vec{r}_s$ and $\vec{r}_e$ represent the direction and the distance of the starting and ending points of the line current from the point $\vec{r}$. Integrating over the line, we obtain

$$\vec{B}_{\vec{l}} = -\frac{\mu_0 I}{4\pi} \cdot \int_0^1 \frac{\vec{l} \times (\vec{r}_s + \lambda\vec{l})}{|\vec{r}_s + \lambda\vec{l}|^3} d\lambda$$

$$= -\frac{\mu_0 I}{4\pi} \cdot (\vec{l} \times \vec{r}_s) \int_0^1 \frac{d\lambda}{|\vec{r}_s + \lambda\vec{l}|^3}.$$

Introducing the abbreviations $a = |\vec{r}_s|^2$, $b = 2\vec{r}_s \cdot \vec{l}$, and $c = |\vec{l}|^2$, the integral gives the result

$$\int_0^1 \frac{d\lambda}{(a + b\lambda + c\lambda^2)^{3/2}} = \frac{1}{b^2 - 4ac} \left(\frac{2b}{\sqrt{a}} - \frac{2b}{\sqrt{a+b+c}} - \frac{4c}{\sqrt{a+b+c}} \right).$$

While mathematically accurate, this formula exhibits several severe **numerical pitfalls** that restrict its direct practical use, in particular when high-order derivatives are to be computed. Indeed, first the formula apparently exhibits a problem of cancellation of close-by numbers if $b + c \ll a$. Introducing the quantity $\varepsilon = (b+c)/a$ yields

$$\vec{B} = -\frac{\mu_0 I}{4\pi} \cdot \frac{(\vec{l} \times \vec{r}_s)}{\sqrt{a}\,(b^2 - 4ac)} \left[2b \left(1 - \frac{1}{\sqrt{1+\varepsilon}}\right) - \frac{4c}{\sqrt{1+\varepsilon}} \right].$$

This problem can be substantially alleviated by observing that

$$1 - \frac{1}{\sqrt{1+\varepsilon}} = \frac{\varepsilon}{1 + \varepsilon + \sqrt{1+\varepsilon}},$$

which yields the field formula

$$\vec{B} = -\frac{\mu_0 I}{4\pi} \cdot \frac{(\vec{l} \times \vec{r}_s)}{\sqrt{a}\,(b^2 - 4ac)} \left[\frac{2b\varepsilon}{1 + \varepsilon + \sqrt{1+\varepsilon}} - \frac{4c}{\sqrt{1+\varepsilon}} \right].$$

However, there is a second numerical difficulty if the line current and the observation point are lying exactly or almost on the same line because in this case b^2 and $4ac$ assume similar values, which makes the evaluation of $b^2 - 4ac$ prone to numerical inaccuracy. To avoid this effect, we rewrite the formula in terms of the angle θ between $\vec{l}$ and $\vec{r}_s$. The relation among the angle and the products of vectors is

$$|\sin \theta| = \frac{|\vec{l} \times \vec{r}_s|}{|\vec{l}| \cdot |\vec{r}_s|} \quad \text{and} \quad \cos \theta = \frac{\vec{l} \cdot \vec{r}_s}{|\vec{l}| \cdot |\vec{r}_s|}.$$

This implies the relationships

$$b^2 - 4ac = -4|\vec{r}_s|^2|\vec{l}|^2 \sin^2\theta$$

$$\frac{2b\varepsilon}{1+\varepsilon+\sqrt{1+\varepsilon}} - \frac{4c}{\sqrt{1+\varepsilon}} = \frac{4|\vec{r}_s||\vec{l}|\cos\theta\left(2|\vec{r}_s||\vec{l}|\cos\theta + |\vec{l}|^2\right)}{|\vec{r}_e|(|\vec{r}_e|+|\vec{r}_s|)} - \frac{4|\vec{l}|^2|\vec{r}_s|}{|\vec{r}_e|}.$$

Finally, we obtain the magnetic field expressed in terms of $\vec{r}_s$ and $\vec{l}$ as

$$\vec{B} = \frac{\mu_0 I}{4\pi} \cdot \frac{(\vec{l} \times \vec{r}_s)}{|\vec{r}_s|^2|\vec{r}_s + \vec{l}|\left(|\vec{r}_s + \vec{l}| + |\vec{r}_s|\right)}$$

$$\times \left[-|\vec{r}_s| + \frac{|\vec{r}_s|\cos^2\theta + |\vec{l}|\cos\theta - |\vec{r}_s + \vec{l}|}{\sin^2\theta}\right].$$

Denoting $|\vec{r}_s|\cos^2\theta + |\vec{l}|\cos\theta = \alpha$ and $|\vec{r}_s + \vec{l}| = \beta$, we manage to eliminate the $\sin^2\theta$ term in the denominator with the help of the identity $\alpha - \beta = (\alpha^2 - \beta^2)/(\alpha + \beta)$. Direct calculation shows that $\alpha^2 - \beta^2 = -\sin^2\theta(|\vec{r}_s|^2\cos^2\theta + |\vec{r}_s + \vec{l}|^2)$. The following is the final result:

$$\vec{B} = -\frac{\mu_0 I}{4\pi} \cdot \frac{(\vec{l} \times \vec{r}_s)}{|\vec{r}_s|^2|\vec{r}_s + \vec{l}|\left(|\vec{r}_s + \vec{l}| + |\vec{r}_s|\right)}$$

$$\times \left[|\vec{r}_s| + \frac{|\vec{r}_s|^2\cos^2\theta + |\vec{r}_s + \vec{l}|^2}{|\vec{r}_s|\cos^2\theta + |\vec{l}|\cos\theta + |\vec{r}_s + \vec{l}|}\right].$$

The only case in which this is numerically unstable is when $|\vec{r}_s|\cos^2\theta + |\vec{l}|\cos\theta + |\vec{r}_s + \vec{l}|$ approaches zero—that is, $\theta = \pi$ and $|\vec{r}_s| < |\vec{l}|$; but this corresponds to a point in the close proximity of the wire.

To illustrate the use of the method in practice, we study the magnetic field of the High-Gradient Quadrupole (**HGQ**) of the LHC final focusing system. Figure 3.12 shows the layout of the coils of this magnet, including the lead ends emanating from the quadrupole. The wires were represented by 360,000 straight wire pieces, and the field was calculated in space and differentiated using DA methods. Because of the complicated geometry, an unusual s dependence of field distribution of the quadrupole results. Figure 3.13 shows the behavior of the quadrupole strength as well as its s-derivatives of orders 4, 8, and 12, all of which affect the nonlinear behavior of the system by virtue of (Eq. 3.9). Because of the breaking of quadrupole symmetry, there is also a noticeable skew quadrupole component, the value and derivatives of which are shown in Fig. 3.14.

Because of the almost fourfold symmetry of the system, the next important multipole is the duodecapole. Figure 3.15 shows the duodecapole strength as well as its fourth and eighth s derivative. Finally, Fig. 3.16 shows the strength of the 20-pole component of the quadrupole as well as its second and fourth s derivatives.

142 FIELDS

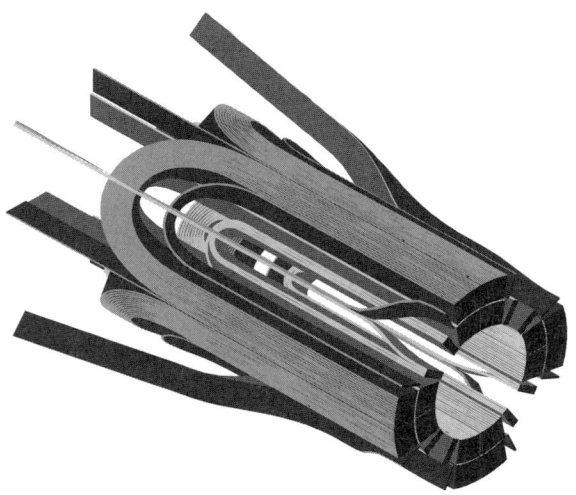

FIGURE 3.12. Schematic coil configuration of an LHC High-Gradient Quadrupole in the final focusing section.

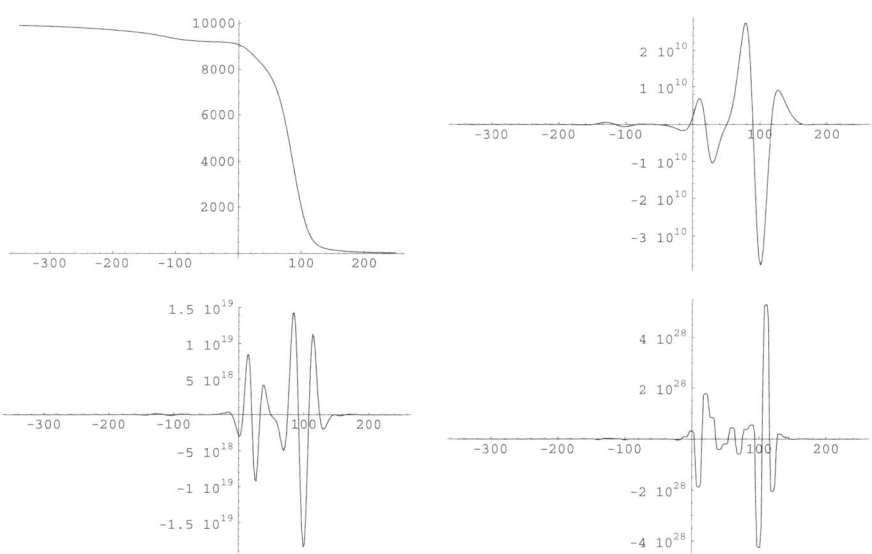

FIGURE 3.13. The value of the quadrupole strength as well as s derivatives of orders 4, 8, and 12 for the lead end of an LHC High-Gradient Quadrupole.

PRACTICAL UTILIZATION OF FIELD INFORMATION

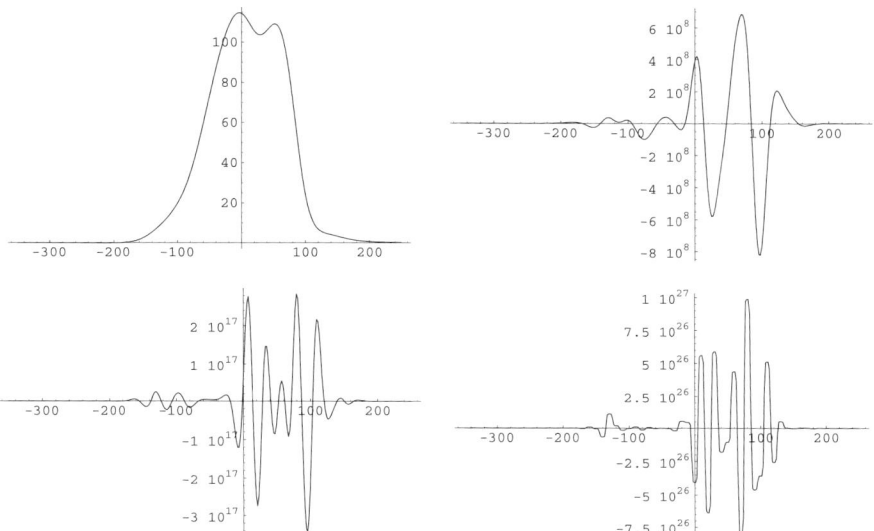

FIGURE 3.14. The value of the skew quadrupole strength as well as s derivatives of orders 4, 8, and 12 for the lead end of an LHC High-Gradient Quadrupole.

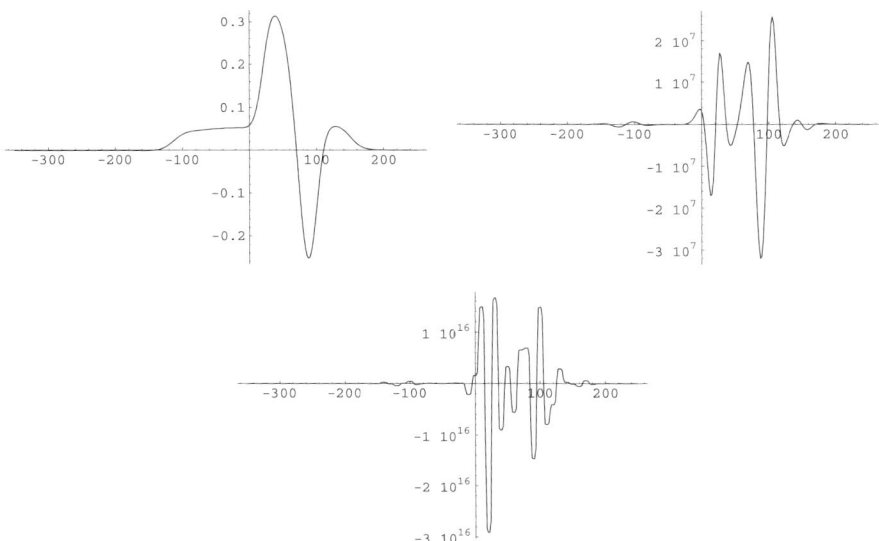

FIGURE 3.15. The value of the duodecapole strength as well as s derivatives of orders 4 and 8 for the lead end of an LHC High-Gradient Quadrupole.

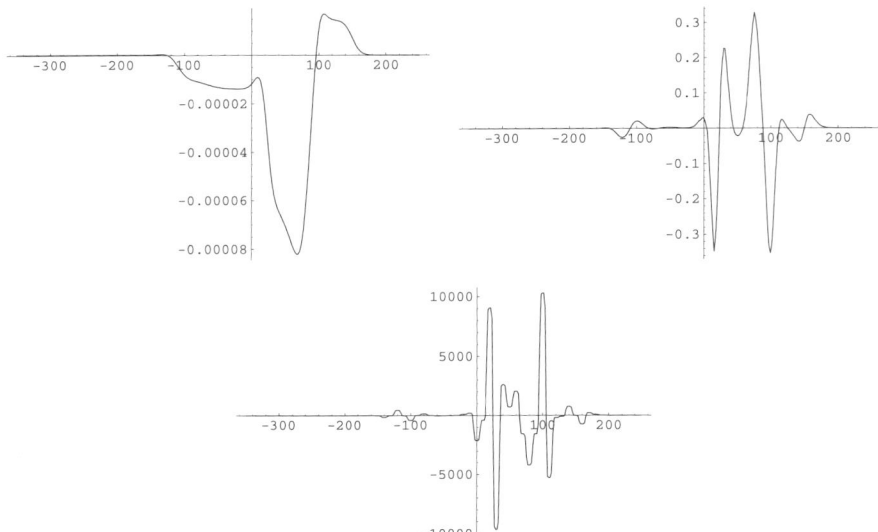

FIGURE 3.16. The value of the 20-pole strength as well as *s* derivatives of orders 2 and 4 for the lead end of an LHC High-Gradient Quadrupole.

Chapter 4

Maps: Properties

Chapters 4 and 5 will discuss methods for the determination and use of maps describing the flows for the ordinary differential equations occurring in beam physics. Specifically, we want to determine an accurate approximation of the functional dependency

$$\vec{z}_2 = \mathcal{M}(\vec{z}_1, \vec{\delta}), \tag{4.1}$$

relating initial conditions $\vec{z}_1$ and system parameters $\vec{\delta}$ at a time t_1 to final conditions $\vec{z}_2$ at a later time t_2. Because beams by their very definition (see the beginning of the book) occupy small regions of phase space in the proximity of a reference orbit, perturbative techniques are particularly suitable for this study. Specifically, we will use DA tools to develop methods that conveniently allow computation of the **Taylor expansion** or **class**

$$[\mathcal{M}]_n \tag{4.2}$$

of the map to any order n of interest. Furthermore, we will develop tools that allow the description and analysis of beam physics systems based on such maps.

A special and important case of transfer maps is the case of **linear transformations** between the coordinates where the maps preserve the origin, i.e., $\mathcal{M}(\vec{0}) = \vec{0}$. Such transfer maps arise as a flow of linear differential equations. In this case, the action of the map is described by a matrix:

$$\vec{z}_2 = \mathcal{M}(\vec{z}_1) = \begin{pmatrix} (z_1|z_1) & (z_1,z_2) & \cdots & (z_1,z_v) \\ (z_2|z_1) & (z_2,z_2) & \cdots & (z_2,z_v) \\ \vdots & \vdots & \vdots & \vdots \\ (z_v|z_1) & (z_v,z_2) & \cdots & (z_v,z_v) \end{pmatrix} \cdot \vec{z}_1. \tag{4.3}$$

If the map is linear and can be described by a matrix, the composition of maps is straightforward because it can be represented by matrix multiplication; if $\hat{M}_{1,2}$ and $\hat{M}_{2,3}$ describe the transformations from $\vec{z}_1$ to $\vec{z}_2$ and $\vec{z}_2$ to $\vec{z}_3$, respectively, then the transformation from $\vec{z}_1$ to $\vec{z}_3$ is given by

$$\hat{M}_{1,3} = \hat{M}_{2,3} \cdot \hat{M}_{1,2}. \tag{4.4}$$

In the case of a **nonlinear map**, it is often advantageous to consider its Jacobian matrix of partial derivatives as a linear approximation of the map. According to calculus rules, the Jacobian matrix $\hat{M}$ of the composition of maps $\mathcal{M}$ is the product of the Jacobians $\hat{M}_i$ of the individual maps $\mathcal{M}_i$. In the important case in which the individual maps are origin-preserving, i.e., $\mathcal{M}_i(\vec{0}) = \vec{0}$, all the Jacobians must be computed at the origin $\vec{0}$. In this case, the Jacobian of the total map can be computed as a product of the individual Jacobians evaluated at the origin.

This method fails if the sub-maps are not origin-preserving, i.e., $\mathcal{M}_i(\vec{0}) \neq \vec{0}$; in this case, the computation of the Jacobian of $\mathcal{M}_{i+1}$ must be computed around the value $\mathcal{M}_i(\vec{0})$, the computation of which requires the full knowledge of $\mathcal{M}_i$ and not only its Jacobian. Frequently, the Jacobian of the map can be used as a **linear approximation** of the map.

In the following, we will develop various tools for the manipulation and representation of maps, and in Chapter 5 develop methods to determine the maps for given systems.

4.1 MANIPULATIONS

In most cases, a beam optical system consists of more than one element and it is necessary to connect the maps of individual pieces. Often, the inverse of a map is needed. Sometimes one part of our system is the reversion of another part; therefore, it is time saving if the map of the reversed part can be obtained directly from that of the other part. All these map manipulations can be done elegantly using differential algebra (DA) techniques.

4.1.1 Composition and Inversion

One of the important aspects of the manipulations of maps is the combination of two maps of two consecutive pieces into one map describing both pieces together. This process corresponds to a **composition** of maps as described in Chapter 2, section 2.3.1, or as it is also often called, a **concatenation**.

Another application of concatenation is the **change of variables**, for example, to match those of other beam optics codes. For instance, while many codes work in the later discussed curvilinear canonical coordinates, others such as TRANSPORT (Brown, 1979a) and GIOS (Wollnik *et al.*, 1984) use the slope instead of the normalized momentum. Even others use Cartesian coordinates which are usually not very well suited for a discussion of the properties of beam lines. The transformation of a transfer map $\mathcal{M}_C$ to a different set of coordinates, which can be expressed as

$$\mathcal{M}_T = \mathcal{C}^{-1} \circ \mathcal{M}_C \circ \mathcal{C}, \tag{4.5}$$

is quite straightforward using DA techniques. One simply composes the transformation formulas between the two sets of coordinates, the map in one set of coordinates, and the inverse transformation in DA. Thus, it is easily possible to express the map in other sets of coordinates to arbitrary order.

Another important task for many practical questions is the **inversion** of transfer maps. As shown in Eq. (2.83), it is possible to obtain the inverse of a map $\mathcal{M}$ up to order n if and only if the linear part of $\mathcal{M}$, denoted by M, is invertible. If $\mathcal{M}$ is a symplectic transfer map, this is certainly the case since the determinant of M is unity, and we obtain the inverse by merely iterating the fixed-point equation

$$[\mathcal{M}^{-1}]_n = M^{-1}([\mathcal{I}]_n - [\mathcal{N}]_n \circ [\mathcal{M}^{-1}]_n), \tag{4.6}$$

where $\mathcal{N}$ is the purely nonlinear part of $\mathcal{M}$, i.e., $\mathcal{M} = M + \mathcal{N}$. The determination of M^{-1} is apparently just a matrix inversion.

4.1.2 Reversion

Besides inversion, another commonly needed technique is the so-called **reversion**. Throughout the development of beam optical systems, **mirror symmetry** has been frequently used, and it will be employed extensively in the following chapters. Indeed, among the various symmetry arrangements, the process of reversion of the direction of transversal of a certain arrangement of elements is perhaps the most commonly used. The reversed motion can be described by first reversing the independent variable s, which is tantamount to switching all the signs of p_x, p_y, and t, then transversing the inverse map, and finally re-reversing the independent variable s.

The time-reversal operations can be performed easily using two compositions before and after the inverse map. Specifically, reversion entails that if a particle enters the forward system at an initial point $(x_i, y_i, d_i, a_i, b_i, t_i)$ and exits at a final point $(x_f, y_f, d_f, a_f, b_f, t_f)$, it will exit the reversed system at $(x_i, y_i, d_i, -a_i, -b_i, -t_i)$ after entering this system at $(x_f, y_f, d_f, -a_f, -b_f, -t_f)$. This determines the reversion transformation to be described by the matrix

$$\hat{R} = \begin{pmatrix} 1 & 0 & 0 & 0 & 0 & 0 \\ 0 & 1 & 0 & 0 & 0 & 0 \\ 0 & 0 & 1 & 0 & 0 & 0 \\ 0 & 0 & 0 & -1 & 0 & 0 \\ 0 & 0 & 0 & 0 & -1 & 0 \\ 0 & 0 & 0 & 0 & 0 & -1 \end{pmatrix}; \tag{4.7}$$

hence, the map of a reversed system is given by

$$\mathcal{M}^R = \hat{R} \circ \mathcal{M}^{-1} \circ \hat{R}. \tag{4.8}$$

In DA representation, the nth order representation $\mathcal{M}_n$ is therefore obtained via concatenation operations as $\mathcal{M}_n = R \circ \mathcal{M}_n^{-1} \circ \hat{R}$. In fact, the second composition can be done by merely changing the signs of the a, and b and t components of the map, which reduces the computational effort.

An interesting point worth noting is that $\hat{R}$ is not symplectic. In fact, it satisfies the following relation:

$$\hat{R}^t \cdot \hat{J} \cdot \hat{R} = -\hat{J}, \tag{4.9}$$

which we call **antisymplecticity**. This plays an important role in the analytical theory of high-order achromats discussed in Chapter 6.

4.2 SYMMETRIES

Symmetry plays an important role in many fields in that it imposes restrictions on the system concerned. Specifically, symmetry in a dynamical system adds constraints to its transfer map. This is very helpful because those constraints reduce the number of unknowns and act as a powerful check of the correctness of the map. Most beam optical systems have either mid-plane symmetry or rotational symmetry. Most of them are Hamiltonian systems, which preserves symplecticity. These three symmetries will be discussed in the following.

4.2.1 Midplane Symmetry

Since it is convenient in practice to build and beam optical elements and systems such that they are symmetric about a certain plane, this approach is widely used for both individual particle optical elements and entire systems. Symmetry in geometry leads to symmetry in fields which in turn entails symmetry in particle motion. The symmetry of particle motion about a certain plane is called midplane symmetry and the plane is called the midplane. Midplane symmetry is probably the most important symmetry in beam optical systems due to its broad applications.

In a system with midplane symmetry, two particles that are symmetric about the midplane at the beginning **stay symmetric** throughout the system. Suppose that a particle is launched at $(x_i, y_i, d_i, a_i, b_i, t_i)$. After the map $\mathcal{M}$, its coordinates are

$$x_f = m_x(x_i, y_i, d_i, a_i, b_i, t_i),$$
$$y_f = m_y(x_i, y_i, d_i, a_i, b_i, t_i),$$
$$d_f = m_d(x_i, y_i, d_i, a_i, b_i, t_i),$$
$$a_f = m_a(x_i, y_i, d_i, a_i, b_i, t_i),$$
$$b_f = m_b(x_i, y_i, d_i, a_i, b_i, t_i),$$
$$t_f = m_t(x_i, y_i, d_i, a_i, b_i, t_i).$$

SYMMETRIES   149

Under the presence of midplane symmetry, a particle that starts at $(x_i, -y_i, d_i, a_i, -b_i, t_i)$ must end at $(x_f, -y_f, d_f a_f, -b_f, t_f)$, which implies

$$x_f = m_x(x_i, -y_i, d_i, a_i, -b_i, t_i), \tag{4.10}$$

$$-y_f = m_y(x_i, -y_i, d_i, a_i, -b_i, t_i), \tag{4.11}$$

$$d_f = m_d(x_i, -y_i, d_i, a_i, -b_i, t_i) \tag{4.12}$$

$$a_f = m_a(x_i, -y_i, d_i, a_i, -b_i, t_i), \tag{4.13}$$

$$-b_f = m_b(x_i, -y_i, d_i, a_i, -b_i, t_i), \tag{4.14}$$

$$t_f = m_t(x_i, -y_i, d_i, a_i, -b_i, t_i). \tag{4.15}$$

Thus, the Taylor coefficients of the map satisfy

$$\left(x | x_i^{i_x} y_i^{i_y} d_i^{i_d} a_i^{i_a} b_i^{i_b} t_i^{i_t}\right) = 0, \tag{4.16}$$

$$\left(d | x_i^{i_x} y_i^{i_y} d_i^{i_d} a_i^{i_a} b_i^{i_b} t_i^{i_t}\right) = 0, \tag{4.17}$$

$$\left(a | x_i^{i_x} y_i^{i_y} d_i^{i_d} a_i^{i_a} b_i^{i_b} t_i^{i_t}\right) = 0, \tag{4.18}$$

$$\left(t | x_i^{i_x} y_i^{i_y} d_i^{i_d} a_i^{i_a} b_i^{i_b} t_i^{i_t}\right) = 0 \tag{4.19}$$

for $i_y + i_b$ odd and

$$\left(y | x_i^{i_x} y_i^{i_y} d_i^{i_d} a_i^{i_a} b_i^{i_b} t_i^{i_t}\right) = 0, \tag{4.20}$$

$$\left(b | x_i^{i_x} y_i^{i_y} d_i^{i_d} a_i^{i_a} b_i^{i_b} t_i^{i_t}\right) = 0 \tag{4.21}$$

for $i_y + i_b$ even.

Since the terms that have to vanish in x_f and a_f are those that remain in y_f and b_f and vice versa, midplane symmetry cancels **exactly half** of the aberrations in the purely $x_f, y_f, a_f,$ and b_f.

As a special case, consider a linear matrix with midplane symmetry which has the form

$$M = \begin{pmatrix} (x|x) & 0 & (x|d) & (x|a) & 0 & (x|t) \\ 0 & (y|y) & 0 & 0 & (y|b) & 0 \\ (d|x) & 0 & (d|d) & (d|a) & 0 & (d|t) \\ (a|x) & 0 & (a|d) & (a|a) & 0 & (a|t) \\ 0 & (b|y) & 0 & 0 & (b|b) & 0 \\ (t|x) & 0 & (t|d) & (t|a) & 0 & (t|t) \end{pmatrix}.$$

Some elements, such as quadrupoles and octupoles, have two midplanes that are perpendicular to each other. We call this **double midplane symmetry**, which, in addition to the previous conditions, requires that

$$\left(y|x_i^{i_x} y_i^{i_y} d_i^{i_d} a_i^{i_a} b_i^{i_b} t_i^{i_t}\right) = 0, \tag{4.22}$$

$$\left(d|x_i^{i_x} y_i^{i_y} d_i^{i_d} a_i^{i_a} b_i^{i_b} t_i^{i_t}\right) = 0, \tag{4.23}$$

$$\left(b|x_i^{i_x} y_i^{i_y} d_i^{i_d} a_i^{i_a} b_i^{i_b} t_i^{i_t}\right) = 0, \tag{4.24}$$

$$\left(t|x_i^{i_x} y_i^{i_y} d_i^{i_d} a_i^{i_a} b_i^{i_b} t_i^{i_t}\right) = 0 \tag{4.25}$$

for $i_x + i_a$ odd and

$$\left(x|x_i^{i_x} y_i^{i_y} d_i^{i_d} a_i^{i_a} b_i^{i_b} t_i^{i_t}\right) = 0, \tag{4.26}$$

$$\left(a|x_i^{i_x} y_i^{i_y} d_i^{i_d} a_i^{i_a} b_i^{i_b} t_i^{i_t}\right) = 0 \tag{4.27}$$

for $i_x + i_a$ even. The additional midplane symmetry once again cancels half of the remaining aberrations in x_f, a_f, y_f, and b_f. Thus the number of aberrations in a system with double midplane symmetry is only **one-quarter** of that in a general system.

Applying the selection rules to a linear matrix, we obtain the structure

$$M = \begin{pmatrix} (x|x) & 0 & 0 & (x|a) & 0 & 0 \\ 0 & (y|y) & 0 & 0 & (y|b) & 0 \\ 0 & 0 & (d|d) & 0 & 0 & (d|t) \\ (a|x) & 0 & 0 & (a|a) & 0 & 0 \\ 0 & (b|y) & 0 & 0 & (b|b) & 0 \\ 0 & 0 & (t|d) & 0 & 0 & (t|t) \end{pmatrix}.$$

4.2.2 Rotational Symmetry

There are two types of rotational symmetry: One is characterized by the system being invariant under a rotation of **any angle,** which we call **continuous rotational symmetry**; the other is characterized by the system being invariant under a **fixed** angle, which we refer to as **discrete rotational symmetry**. The former is widely seen in light optics, in which almost all glass lenses are rotationally invariant, and in electron microscopes, in which solenoids are the primary focusing elements. The latter is preserved in quadrupoles and all higher multipoles.

For the analysis of both cases it is advantageous to use complex coordinates. Let us define

$$z = x + iy, \quad w = a + ib, \tag{4.28}$$

and hence $\bar{z} = x - iy$, $\bar{w} = a - ib$. After expressing x, a, y, and b in terms of z, $\bar{z}$, w, and $\bar{w}$, the transfer map is transformed into

$$\begin{pmatrix} z_f \\ w_f \end{pmatrix} = \begin{pmatrix} x_f + iy_f \\ a_f + ib_f \end{pmatrix} = \begin{pmatrix} F_z \\ F_w \end{pmatrix}(z_i, \bar{z}_i, w_i, \bar{w}_i, t_i, d_i), \tag{4.29}$$

where

$$\begin{pmatrix} F_z \\ F_w \end{pmatrix} = \sum_{j_1 j_2 j_3 j_4 j_t j_d} \begin{pmatrix} c_z \\ c_w \end{pmatrix}_{j_1 j_2 j_3 j_4 j_t j_d} z^{j_1} \bar{z}^{j_2} w^{j_3} \bar{w}^{j_4} t^{j_t} d^{j_d}. \tag{4.30}$$

Note that besides z and w, $\bar{z}$ and $\bar{w}$, also appear, contrary to the familiar Taylor expansion of analytic functions. This is due to the fact that while the original map may be Taylor expandable and hence analytic as a real function, it is not necessary that the resulting complex function be analytic in the sense of complex analysis.

Given the fact that a rotation by ϕ transforms z to $e^{i\phi}z$ and w to $e^{i\phi}w$, rotational symmetry requires that a rotation in initial coordinates results in the same transformation in the final coordinates, i.e., $z_f \to e^{i\phi}z_f$, $w_f \to e^{i\phi}w_f$. Inserting this yields

$$(j_1 - j_2 + j_3 - j_4 - 1)\phi = 2\pi n \tag{4.31}$$

for x, a, y, and b terms and

$$(j_1 - j_2 + j_3 - j_4)\phi = 2\pi n \tag{4.32}$$

for t and d terms, where n is an integer.

For continuous rotational symmetry, which means invariance for all ϕ, the j_i ($i = 1, 2, 3, 4$) should be independent of ϕ. Thus, we have

$$j_1 - j_2 + j_3 - j_4 = 1 \text{ for } z_f \text{ and } w_f \tag{4.33}$$

and

$$j_1 - j_2 + j_3 - j_4 = 0 \text{ for } t_f \text{ and } d_f, \tag{4.34}$$

which both eliminate many terms. First, all terms with $j_1 + j_2 + j_3 + j_4$ vanish because $j_1 + j_3$ and $j_2 + j_4$ always have the same parity, which means that $j_1 -$

$j_2 + j_3 - j_4$ is also even. This implies that, in rotationally symmetric systems, all even-order geometric aberrations disappear. As a summary, all remaining z and w terms up to order 3 are shown as follows:

Order	j_1	j_2	j_3	j_4
1	1	0	0	0
	0	0	1	0
2				
3	2	1	0	0
	2	0	0	1
	0	1	2	0
	0	0	2	1
	1	1	1	0
	1	0	1	1,

where the order represents the sum of the j_i. To illustrate the characteristic of such a map, let us derive the linear matrix from the previous conditions. First define

$$(z|z) = (c_z)_{100000},$$
$$(z|w) = (c_z)_{001000},$$
$$(w|z) = (c_w)_{100000},$$
$$(w|w) = (c_w)_{001000}.$$

The first-order map is then given by

$$x_f + iy_f = (z|z)(x_i + iy_i) + (z|w)(a_i + ib_i)$$
$$a_f + ib_f = (w|z)(x_i + iy_i) + (w|w)(a_i + ib_i),$$

which entails that the linear matrix is

$$M = \begin{pmatrix} \Re(z|z) & -\Im(z|z) & \Re(z|w) & -\Im(z|w) \\ \Im(z|z) & \Re(z|z) & \Im(z|w) & \Re(z|w) \\ \Re(w|z) & -\Im(w|z) & \Re(w|w) & -\Im(w|w) \\ \Im(w|z) & \Re(w|z) & \Im(w|w) & \Re(w|w) \end{pmatrix}. \quad (4.35)$$

As an example, we show the second-order map of a solenoid, which has rotational symmetry but exhibits a coupling between x and y and a and b. Table I shows that indeed all second-order geometric aberrations vanish, which is a consequence of the rotational symmetry.

Equation (4.35) also shows that a rotationally invariant system preserves mid-plane symmetry to first order when the first-order coefficients are real numbers. In

TABLE I
THE SECOND-ORDER MAP OF A SOLENOID

x	a	y	b	t	
0.9996624	-0.4083363E-03	-0.1866471E-01	0.7618839E-05	0.0000000E+00	100000
0.7998150	0.9996624	-0.1493335E-01	-0.1866471E-01	0.0000000E+00	010000
0.1866471E-01	-0.7618839E-05	0.9996624	-0.4083363E-03	0.0000000E+00	001000
0.1493335E-01	0.1866471E-01	0.7998150	0.9996624	0.0000000E+00	000100
0.0000000E+00	0.0000000E+00	0.0000000E+00	0.0000000E+00	1.000000	000010
0.0000000E+00	0.0000000E+00	0.0000000E+00	0.0000000E+00	0.1631010	000001
0.0000000E+00	0.0000000E+00	0.0000000E+00	0.0000000E+00	-0.1120070E-03	200000
0.0000000E+00	0.0000000E+00	0.0000000E+00	0.0000000E+00	-0.8764989E-09	110000
0.0000000E+00	0.0000000E+00	0.0000000E+00	0.0000000E+00	-0.2193876	020000
0.0000000E+00	0.0000000E+00	0.0000000E+00	0.0000000E+00	-0.1023936E-01	011000
0.0000000E+00	0.0000000E+00	0.0000000E+00	0.0000000E+00	-0.1120070E-03	002000
0.0000000E+00	0.0000000E+00	0.0000000E+00	0.0000000E+00	0.1023936E-01	100100
0.0000000E+00	0.0000000E+00	0.0000000E+00	0.0000000E+00	-0.8764989E-09	001100
0.3702835E-03	0.2238603E-03	0.1023256E-01	-0.8362261E-05	0.0000000E+00	100001
-0.4384743	0.3702835E-03	0.1637921E-01	0.1023256E-01	0.0000000E+00	010001
-0.1023256E-01	0.8362261E-05	0.3702835E-03	0.2238603E-03	0.0000000E+00	001001
0.0000000E+00	0.0000000E+00	0.0000000E+00	0.0000000E+00	-0.2193876	000200
-0.1637921E-01	-0.1023256E-01	-0.4384743	0.3702835E-03	0.0000000E+00	000101
0.0000000E+00	0.0000000E+00	0.0000000E+00	0.0000000E+00	-0.1341849	000002

fact, a simple argument shows that this is true even for higher orders. In complex coordinates, Eqs. (4.10)–(4.15) are transformed to

$$\begin{pmatrix} \bar{z}_f \\ \bar{w}_f \end{pmatrix} = \begin{pmatrix} F_z \\ F_w \end{pmatrix} (\bar{z}_i, z_i, \bar{w}_i, w_i, t_i, d_i)$$

$$\Rightarrow \begin{pmatrix} z_f \\ w_f \end{pmatrix} = \begin{pmatrix} \bar{F}_z \\ \bar{F}_w \end{pmatrix} (z_i, \bar{z}_i, w_i, \bar{w}_i, t_i, d_i)$$

$$= \sum_{j_1 j_2 j_3 j_4 j_t j_d} \begin{pmatrix} \bar{a}_z \\ \bar{a}_w \end{pmatrix}_{j_1 j_2 j_3 j_4 j_t j_d} z^{j_1} \bar{z}^{j_2} w^{j_3} \bar{w}^{j_4} t^{j_t} d^{j_d},$$

which shows that all coefficients have to be real numbers in order to preserve midplane symmetry. It is worth noting that this proof has nothing to do with rotational symmetry itself.

For discrete rotational symmetry, invariance occurs only when $\phi = 2\pi/k$, where k is an integer. Hence, the nonzero terms satisfy

$$j_1 - j_2 + j_3 - j_4 - 1 = nk. \tag{4.36}$$

In general, a $2k$ pole is invariant under rotation of $\phi = 2\pi/k$. For example, for a quadrupole, we have $k = 2$. Hence, the nonzero terms satisfy

$$j_1 - j_2 + j_3 - j_4 = 2n + 1.$$

Like round lenses, systems with quadrupole symmetry are also free of even-order geometric aberrations. The linear map of a quadrupole can be obtained from

$$j_1 - j_2 + j_3 - j_4 = \pm 1,$$

which is

$$x_f + iy_f = (z|z)(x_i + iy_i) + (z|w)(a_i + ib_i)$$
$$+ (z|\bar{z})(x_i - iy_i) + (z|\bar{w})(a_i - ib_i),$$
$$a_f + ib_f = (w|z)(x_i + iy_i) + (w|w)(a_i + ib_i)$$
$$+ (w|\bar{z})(x_i - iy_i) + (w|\bar{w})(a_i - ib_i).$$

Since a quadrupole has midplane symmetry, all the coefficients are real numbers. Thus its linear matrix is

$$\begin{pmatrix} (z|z) + (z|\bar{z}) & 0 & (z|w) + (z|\bar{w}) & 0 \\ 0 & (z|z) - (z|\bar{z}) & 0 & (z|w) - (z|\bar{w}) \\ (w|z) + (w|\bar{z}) & 0 & (w|w) + (w|\bar{w}) & 0 \\ 0 & (w|z) - (w|\bar{z}) & 0 & (w|w) - (w|\bar{w}) \end{pmatrix}. \quad (4.37)$$

For other multipoles, we have

$$j_1 - j_2 + j_3 - j_4 = nk + 1 = \begin{cases} \cdots \\ -k+1 \\ 1 \\ k+1 \\ \cdots \end{cases} \quad k = 3, 4, \ldots. \quad (4.38)$$

With midplane symmetry, the linear matrix of a $2k$ pole is

$$M = \begin{pmatrix} (z|z) & (z|w) & 0 & 0 \\ (w|z) & (w|w) & 0 & 0 \\ 0 & 0 & (z|z) & (z|w) \\ 0 & 0 & (w|z) & (w|w) \end{pmatrix}. \quad (4.39)$$

Since the linear matrix of a $2k$ pole ($k \leq 3$) is a drift, it satisfies Eq. (4.39). Equation (4.38) shows that the geometric aberrations appear only for orders of at least $k - 1$. The fact that multipoles do not have dispersion determines that the chromatic aberrations do not appear until order k. This can be easily seen from the equations of motion presented in Chapter 5, Section 5.1. Therefore, a $2k$ pole is a drift up to order $k - 2$.

4.2.3 Symplectic Symmetry

As shown in Section 1.4.7, the transfer map of any Hamiltonian system is symplectic, i.e., its Jacobian $\hat{M}$ satisfies

$$\hat{M}^t \cdot \hat{J} \cdot \hat{M} = \hat{J} \tag{4.40}$$

or any of the four equivalent forms discussed in section 1.4.7. Here, we study the consequences of the condition of symplecticity on the map, following the terminology of (Wollnik and Berz, 1985). Let the $2v$-dimensional vector $\vec{r}$ denote a point in phase space; a general map can then be written as

$$r_{if} = \sum_{j=1}^{2v} r_{jo} \left[(r_i|r_j) + \frac{1}{2} \sum_{k=1}^{2v} r_{ko} \left[(r_i|r_j r_k) + \frac{1}{3} \sum_{l=1}^{2v} r_{lo} \left[(r_i|r_j r_k r_l) + \cdots \right] \right] \right], \tag{4.41}$$

where we have used the abbreviations

$$(r_i|r_j) = \left. \frac{\partial r_{if}}{\partial r_{jo}} \right|_{\vec{r}=\vec{o}}, \quad (r_i|r_j r_k) = \left. \frac{\partial^2 r_{if}}{\partial r_{jo} \partial r_{ko}} \right|_{\vec{r}=\vec{o}}, \quad \text{etc.} \tag{4.42}$$

Note that the definitions differ by certain combinatorial factors from the conventionally used matrix elements.

Since for any $\hat{M}$, the product $\hat{M}^t \cdot \hat{J} \cdot \hat{M}$ is antisymmetric, Eq. (4.40) represents $2v(2v-1)/2$ independent conditions. Let m_{ij} be an element of $\hat{M}$; then, we have

$$m_{ij} = \frac{\partial r_{if}}{\partial r_{jo}} = (r_i|r_j) + \sum_{k=1}^{2v} r_{ko} \left[(r_i|r_j r_k) + \frac{1}{2} \sum_{l=1}^{2v} r_{lo} \left[(r_i|r_j r_k r_l) + \cdots \right] \right]$$

Let $\hat{V} = \hat{M}^t \cdot \hat{J} \cdot \hat{M}$. Then, we have

$$V_{ij} = \sum_{\mu=1}^{v} (m_{\mu,i} \cdot m_{\mu+v,j} - m_{\mu+v,i} \cdot m_{\mu,j}). \tag{4.43}$$

We introduce

$$D_\mu(M_i, M_j) = (r_\mu|M_i)(r_{\mu+v}|M_j) - (r_{\mu+v}|M_i)(r_\mu|M_j), \tag{4.44}$$

where M_i and M_j are monomials. After some straightforward arithmetic, V_{ij} can be simplified as follows:

$$V_{ij} = \sum_{\mu=1}^{v} \left[F_{0\mu} + \sum_{k_1=1}^{2v} r_{k_1 o} \left[F_{1\mu} + \frac{1}{2} \sum_{k_2=1}^{2v} r_{k_2 o} \left[F_{2\mu} + \cdots \right] \right] \right], \tag{4.45}$$

where

$$F_{\xi,\mu} = \binom{\xi}{0} \cdot D_\mu(r_i, r_{k_1} r_{k_2} \ldots r_{k_\xi} r_j) + \binom{\xi}{1} \cdot D_\mu(r_i r_{k_1}, r_{k_2} \ldots r_{k_\xi} r_j)$$
$$+ \binom{\xi}{2} \cdot D_\mu(r_i r_{k_1} r_{k_2}, \ldots r_{k_\xi} r_j) + \cdots + \binom{\xi}{\xi} \cdot D_\mu(r_i r_{k_1} r_{k_2} \ldots r_{k_\xi}, r_j)$$
(4.46)

It is worth noting that $F_{\xi,\mu}$ is the coefficient of monomial $r_{k_1} r_{k_2} \cdots r_{k_\xi}$ of order $\xi + 1$. For single particle motion, we have $2v = 6$ and $\vec{r} = (x, y, d, a, b, t)$. Therefore, the first-order conditions are

$$\sum_{\mu=1}^{3} F_{0\mu} = \sum_{\mu=1}^{3} D_\mu(r_i, r_j) = J_{ij}, \qquad (4.47)$$

where $i = (1, \ldots, 6)$ and $j = i + 1, \ldots, 6$. From the definition of $\hat{J}$, we have $J_{14} = J_{25} = J_{36} = 1$, and all other elements vanish. Altogether, there are 15 conditions.

Similarly, the second-order conditions are

$$\sum_{\mu=1}^{3} F_{1\mu} = \sum_{\mu=1}^{3} (D_\mu(r_i, r_{k_1} r_j) + D_\mu(r_i r_{k_1}, r_j)) = 0 \qquad (4.48)$$

for all k_1. Thus, we have 6×15 second-order conditions.

The third-order conditions are

$$\sum_{\mu=1}^{3} F_{2\mu} = \sum_{\mu=1}^{3} D_\mu(r_i, r_{k_1} r_{k_2} r_j) + 2D_\mu(r_i r_{k_1}, r_{k_2} r_j) + D_\mu(r_i r_{k_1} r_{k_2}, r_j) = 0$$
(4.49)

for all k_1 and k_2. Since there are $6(6+1)/2$ independent terms, we have 21×15 third-order conditions.

Now let us further assume that our system preserves midplane symmetry, its Hamiltonian is time independent of t, and the energy d is conserved, which is the most common case. As a result, the linear matrix is

$$M = \begin{pmatrix} (x|x) & 0 & (x|d) & (x|a) & 0 & 0 \\ 0 & (y|y) & 0 & 0 & (y|b) & 0 \\ 0 & 0 & 1 & 0 & 0 & 0 \\ (a|x) & 0 & (a|d) & (a|a) & 0 & 0 \\ 0 & (b|y) & 0 & 0 & (b|b) & 0 \\ (t|x) & 0 & (t|d) & (t|a) & 0 & 1 \end{pmatrix}.$$

Out of the 15 first-order conditions, 4 give nontrivial solutions. They are $(i, j) = (1, 4)$, $(i, j) = (2, 5)$, $(i, j) = (4, 6)$, and $(i, j) = (1, 6)$, which we will now study in detail.

Since for $i = 1$ and $j = 4$ we have $J_{14} = 1$, we obtain

$$\sum_{\mu=1}^{3} D_\mu(r_1, r_4) = 1. \tag{4.50}$$

D_μ ($\mu = 1, 2, 3$) can be obtained from Eq. (4.44) as

$$D_1(r_1, r_4) = (r_1|r_1)(r_4|r_4) - (r_4|r_1)(r_1|r_4) = (x|x)(a|a) - (a|x)(x|a),$$
$$D_2(r_1, r_4) = (r_2|r_1)(r_5|r_4) - (r_5|r_1)(r_2|r_4) = 0,$$
$$D_3(r_1, r_4) = (r_3|r_1)(r_6|r_4) - (r_6|r_1)(r_3|r_4) = 0.$$

Inserting the results into Eq. (4.50) yields

$$(x|x)(a|a) - (x|a)(a|x) = 1, \tag{4.51}$$

which shows that the determinant of the x–a linear matrix and hence that of the x–a Jacobian matrix equals 1.

Similarly, for $i = 2$ and $j = 5$, the symplectic condition yields

$$(y|y)(b|b) - (b|y)(y|b) = 1, \tag{4.52}$$

which is the same condition for the y–b block as for the x–a block. Together with the rest of the matrix, these two conditions show that the determinant of the Jacobian matrix equals 1. It is worth noting that the decoupled conditions on the x–a and y–b come from midplane symmetry.

For $i = 1$ and $j = 6$, we have

$$\sum_{\mu=1}^{3} D_\mu(r_1, r_3) = 0, \tag{4.53}$$

where D_μ ($\mu = 1, 2, 3$) are

$$D_1(r_1, r_3) = (r_1|r_1)(r_4|r_3) - (r_4|r_1)(r_1|r_3) = (x|x)(a|d) - (a|x)(x|d),$$
$$D_2(r_1, r_3) = (r_2|r_1)(r_5|r_3) - (r_5|r_1)(r_2|r_3) = 0,$$
$$D_3(r_1, r_3) = (r_3|r_1)(r_6|r_3) - (r_6|r_1)(r_3|r_3) = -(t|x).$$

Therefore, the condition yields

$$(x|x)(a|d) - (a|x)(x|d) - (t|x) = 0. \tag{4.54}$$

Finally, for $i = 4$ and $j = 6$, the condition yields

$$(x|a)(a|d) - (a|a)(x|d) - (t|a) = 0 \tag{4.55}$$

Equations (4.54) and (4.55) show that $(t|x)$ and $(t|a)$ are **completely determined** by the orbit motion. In particular, for achromatic systems in which $(x, d) = (a, d) = 0$, this means that $(t|x)$ and $(t|a)$. This concept has been used in designing time-of-flight mass spectrometers.

Now we examine some second-order examples. The first one is $i = 1$, $j = 4$, and $k_1 = 1$, which gives

$$(x|x)(a|xa) - (a|x)(x|xa) + (x|xx)(a|a) - (a|xx)(x|a) = 0 \tag{4.56}$$

This example shows an interdependence among second-order geometric aberrations.

The next example shows that $(t|xx)$ is a function of x and a terms. It is the case in which $i = 1$, $j = 6$, and $k_1 = 1$, and the result is

$$(x|x)(a|xd) - (a|x)(x|xd) + (x|xx)(a|d) - (a|xx)(x|d) - (t|xx) = 0. \tag{4.57}$$

Next we prove that as long as the kinetic energy is conserved, all t terms except $(t|d)$ for $n = 1, 2, \ldots$ are redundant. First, we show that $(t|d^n)$ is indeed independent.

Since all d terms are equal to zero except $(d|d)$, $(t|d^n)$ appears only in

$$D_3(r_3^n, r_3) = -(r_6|r_3^n)(r_3|r_3) = -(t|d^n)$$

and

$$D_3(r_3, r_3^n) = (r_3|r_3)(r_3|r_3^n) = (t|d^n).$$

Therefore, we have

$$\sum_{\mu=1}^{3} F_{n\mu} = D_3(r_3^n, r_3) + D_3(r_3, r_3^n) + \cdots$$

$$= 0 + \cdots,$$

which shows that $(t|d^n)$ does not appear in any symplectic condition and is therefore independent.

For other t terms, assume a general form $(t|r_{l_1} r_{l_2} \cdots r_{l_n})$. For $i = l_1$, $k_1 = l_2$, $\ldots$, $k_{n-1} = l_n$, and $j = 3$, $(t|r_{l_1} r_{l_2} \cdots r_{l_n})$ appears only in

$$D_3(r_{l_1} r_{l_2} \cdots r_{l_n}, r_3) = -(r_6|r_{l_1} r_{l_2} \cdots r_{l_n})(r_3|r_3)$$

$$= -(t|r_{l_1}r_{l_2}\cdots r_{l_n}),$$

which entails that $(t|r_{l_1}r_{l_2}\cdots r_{l_n})$ can be either zero or a function of x, a, y and b terms. In conclusion, $(t|r_{l_1}r_{l_2}\cdots r_{l_n})$ is redundant. As a consequence, an achromatic system in which all position and angle aberrations vanish is also free of all time-of-flight aberrations except $(t|d^n)$.

4.3 REPRESENTATIONS

In Section 4.2 it was shown that not all **aberrations** of a symplectic map are independent, but they are **connected through the condition of symplecticity**. In some situations it is helpful both theoretically and practically to represent a symplectic map with only independent coefficients. In the following sections, we introduce two such kinds of representations: the Lie factorizations and the generating functions.

4.3.1 Flow Factorizations

In this section, we develop techniques that allow the representation of maps via flows of autonomous dynamical systems. As a special case, we obtain factorizations into flow operators of Hamiltonian systems if the map is symplectic. Let a transfer map $\mathcal{M}$ be given. First, we perform a shift of the origin through the map $\mathcal{S}$ such that

$$\left(\mathcal{S}^{-1}\circ\mathcal{M}\right)(0) = 0. \qquad (4.58)$$

Next we factor out the linear map M of $\mathcal{M}$ and define the map $\mathcal{M}_1$ as

$$\mathcal{M}_1 = M^{-1}\circ\mathcal{S}^{-1}\circ\mathcal{M}. \qquad (4.59)$$

The map $\mathcal{M}_1$ is origin preserving, and up to first order equals the identity.

We now consider the operator $\exp(:L_{\vec{f}}:)$ in the DA ${}_nD_v$ for the case in which $\vec{f} =_1 0$, i.e., $\vec{f}$ has no constant and no linear part, and thus in the terminology of Eq. (2.64), $\vec{f} \in I_2^v$. If the operator $L_{\vec{f}}$ acts on a function $g \in I_1$, i.e., the ideal of functions that satisfy $g(0) = 0$, we have that all derivatives up to order 2 vanish. After the second application of $L_{\vec{f}}$, all derivatives of order 3 vanish, and so on. Altogether, we have that

$$L_{\vec{f}}g \in I_2,\ L_{\vec{f}}^2g \in I_3,\ \ldots,\ L_{\vec{f}}^n g \in I_{n+1} = \emptyset. \qquad (4.60)$$

In particular, this means that the operator $\exp(L_{\vec{f}})$ converges in finitely many steps in ${}_nD_v$.

We now choose any $\vec{f}_2$ that satisfies

$$\vec{f}_2 =_2 \mathcal{M}_1 - \mathcal{I} \tag{4.61}$$

where $\mathcal{I}$ is the identity map. In particular, we could choose the terms of $\vec{f}_2$ that are of order 3 and higher to vanish. Then we see that because $\mathcal{I} \in I_1^v$,

$$\exp\left(L_{\vec{f}_2}\right) \mathcal{I} =_2 \mathcal{I} + \vec{f}_2 \vec{\nabla} \mathcal{I} =_2 \mathcal{M}_1$$

Therefore, if we define

$$\mathcal{M}_2 = \left(\exp\left(L_{\vec{f}_2}\right) \mathcal{I}\right)^{-1} \circ \mathcal{M}_1, \tag{4.62}$$

then $\mathcal{M}_2$ satisfies

$$\mathcal{M}_2 =_2 \mathcal{I}. \tag{4.63}$$

Next, consider the operator $\exp(L_{\vec{f}})$ for the case of $\vec{f} \in I_3^v$. If the operator $L_{\vec{f}}$ acts on a function $g \in I_1$, all derivatives up to order 3 vanish. After the second application of $L_{\vec{f}}$, all derivatives of order 5 vanish, and so on. Altogether, we have that

$$L_{\vec{f}} g \in I_3, \ L_{\vec{f}}^2 g \in I_5, \ldots, \ L_{\vec{f}}^v g \in I_{2v+1}. \tag{4.64}$$

Now choose an $\vec{f}_3$ that satisfies $\vec{f}_3 =_3 \mathcal{M}_2 - \mathcal{I}$ and observe that

$$\exp(L_{\vec{f}_3}) \mathcal{I} =_3 \mathcal{I} + \vec{f}_3 \vec{\nabla} \mathcal{I} =_3 \mathcal{M}_2.$$

Therefore, if we define

$$\mathcal{M}_3 = \left(\exp\left(L_{\vec{f}_3}\right) \mathcal{I}\right)^{-1} \circ \mathcal{M}_2, \tag{4.65}$$

then $\mathcal{M}_3$ satisfies $\mathcal{M}_3 =_3 \mathcal{I}$. This procedure can be continued iteratively and finally ends with $\mathcal{M}_n =_n I$. Unrolling recursively the definitions of $\mathcal{M}_i$ then yields

$$\mathcal{M} =_n \mathcal{S} \circ M \circ \left(\exp\left(L_{\vec{f}_2}\right) \mathcal{I}\right) \circ \cdots \circ \left(\exp\left(L_{\vec{f}_n}\right) \mathcal{I}\right). \tag{4.66}$$

Several modifications of this approach are possible. First, we note that in our recursive definition of the $\mathcal{M}_i$, we always composed with the inverses of the respective maps from the left. If instead all such compositions are performed from the right, we obtain

$$\mathcal{M} =_n \mathcal{S} \circ M \circ \left(\exp\left(L_{\vec{f}_n}\right)\mathcal{I}\right) \circ \cdots \circ \left(\exp\left(L_{\vec{f}_2}\right)\mathcal{I}\right), \quad (4.67)$$

where of course the values of the $\vec{f}_i$ are different from what they were previously.

We also note that the appearance of the constant and linear parts $\mathcal{S}$ and M on the left is immaterial, and they can be made to appear on the right instead by Eq. (4.59), initially choosing $\mathcal{M}_1 = \mathcal{M} \circ \mathcal{S}^{-1} \circ M^{-1}$. In this case, however, $\mathcal{S}$ does not shift the coordinates by the amount $\mathcal{M}(\vec{0})$, but rather by an $\vec{x}$ that satisfies $\mathcal{M}(\vec{x}) = \vec{0}$.

Next we observe that in Eq. (4.66), all flow exponentials act on the identity map individually, and then all resulting maps are composed with each other. We now study a similar factorization where the exponentials act on each other successively. To this end, we first define $\mathcal{M}_1$ as in Eq. (4.59). Next, we choose $\vec{f}_2$ as before such that

$$\exp\left(L_{\vec{f}_2}\right)\mathcal{I} =_2 \mathcal{I} + \vec{f}_2 = \mathcal{M}_1 \quad (4.68)$$

holds. Different from before, we now define $\mathcal{M}_2$ as

$$\mathcal{M}_2 = \exp\left(L_{\vec{f}_2}\right)\mathcal{I}. \quad (4.69)$$

Next we try to find $\vec{f}_3$ such that

$$\mathcal{M}_1 =_3 \exp\left(L_{\vec{f}_3}\right)\mathcal{M}_2 =_3 \left(\mathcal{I} + \vec{f}_3 \vec{\nabla}\right)\mathcal{M}_2$$
$$=_3 \mathcal{M}_2 + \vec{f}_3, \quad (4.70)$$

where use has been made of the fact that $\mathcal{M}_2 =_1 \mathcal{I}$. Because $\mathcal{M}_2 =_2 \mathcal{M}_1$, this uniquely defines $\vec{f}_3$, and we now set

$$\mathcal{M}_3 = \exp\left(L_{\vec{f}_3}\right)\mathcal{M}_2. \quad (4.71)$$

We continue by trying to find $\vec{f}_4$ such that

$$\mathcal{M}_1 =_4 \exp\left(L_{\vec{f}_4}\right)\mathcal{M}_3 =_4 \left(\mathcal{I} + \vec{f}_4 \vec{\nabla}\right)\mathcal{M}_3$$
$$=_4 \mathcal{M}_3 + \vec{f}_4. \quad (4.72)$$

We now have used that $\mathcal{M}_3 =_1 \mathcal{I}$. Because $\mathcal{M}_1 =_3 \mathcal{M}_3$, this uniquely defines $\vec{f}_4$. The procedure continues, until in the last step we have $\mathcal{M}_1 =_n \exp(L_{\vec{f}_n})\mathcal{M}_{n-1}$.

162 MAPS: PROPERTIES

After unrolling the definitions of the $\mathcal{M}_i$, we arrive at the factorization

$$\mathcal{M} =_n \mathcal{S} \circ M \circ \left(\exp\left(L_{\vec{f}_n} \right) \ldots \exp\left(L_{\vec{f}_2} \right) \mathcal{I} \right). \tag{4.73}$$

Finally, we observe that according to Eq. (4.64), in the step in which $\vec{f}_3$ is chosen, we could not only choose $\vec{f}_3 = {}_3\mathcal{M}_2$, but also pick an $\vec{f}_{3,4}$ such that $\vec{f}_{3,4} = {}_4\mathcal{M}_2$. Therefore, in the second step, all terms of order 3 and 4 would be removed. In the following step, orders 5–8 could be removed and so on. In this case of a "**superconvergent**" factorization, many fewer terms are needed, and we would have

$$\mathcal{M} =_n \mathcal{S} \circ M \circ \left(\exp\left(L_{\vec{f}_2} \right) \mathcal{I} \right) \circ \left(\exp\left(L_{\vec{f}_{3,4}} \right) \mathcal{I} \right) \circ \ldots . \tag{4.74}$$

as well as the previous respective variations.

We now address the important case in which the map under consideration is **symplectic**. The question then arises whether the individually occurring flow operators could be chosen such that they represent a Hamiltonian system, i.e., according to Eq. (1.90), whether for each $\vec{f}_i$ there is an h_{i+1} such that

$$\vec{f}_i \begin{pmatrix} \vec{q} \\ \vec{p} \end{pmatrix} = \hat{J} \cdot \begin{pmatrix} \partial h_{i+1}/\partial \vec{q} \\ \partial h_{i+1}/\partial \vec{p} \end{pmatrix}.$$

In this case,

$$\exp\left(L_{\vec{f}_i} \right) = \exp(: h_{i+1} :), \tag{4.75}$$

where ": :" denotes a "Poisson bracket waiting to happen." According to Eq. (1.115), the h_{i+1} exist if and only if the potential problem

$$\begin{pmatrix} \partial h_{i+1}/\partial \vec{q} \\ \partial h_{i+1}/\partial \vec{p} \end{pmatrix} = -\hat{J} \cdot \vec{f}_i \begin{pmatrix} \vec{q} \\ \vec{p} \end{pmatrix} = -\hat{J} \cdot \begin{pmatrix} \vec{f}_{\vec{q}} \\ \vec{f}_{\vec{p}} \end{pmatrix} \begin{pmatrix} \vec{q} \\ \vec{p} \end{pmatrix} \tag{4.76}$$

can be solved. According to Eq. (1.300), this is the case if and only if the matrix

$$\hat{A} = -\hat{J} \cdot \begin{pmatrix} \partial \vec{f}_{\vec{q}}/\partial \vec{q} & \partial \vec{f}_{\vec{q}}/\partial \vec{p} \\ \partial \vec{f}_{\vec{p}}/\partial \vec{q} & \partial \vec{f}_{\vec{p}}/\partial \vec{p} \end{pmatrix}$$

is symmetric. We show that this is the case by induction over the order i. Therefore, we are given the symplectic map $\mathcal{M}_{i-1}$ which, according to the previous arguments, has the form

$$\mathcal{M}_{i-1} =_i \mathcal{I} + \vec{f}_i;$$

the symplectic condition for $\mathcal{M}_{i-1}$ now entails that

$$\hat{J} = \mathrm{Jac}(\mathcal{M}_{i-1}) \cdot \hat{J} \cdot \mathrm{Jac}(\mathcal{M}_{i-1})^t$$

$$= \left(\hat{I} + \begin{pmatrix} \partial \vec{f}_{\vec{q}}/\partial \vec{q} & \partial \vec{f}_{\vec{q}}/\partial \vec{p} \\ \partial \vec{f}_{\vec{p}}/\partial \vec{q} & \partial \vec{f}_{\vec{p}}/\partial \vec{p} \end{pmatrix} \right) \cdot \hat{J} \cdot \left(\hat{I} + \begin{pmatrix} \partial \vec{f}_{\vec{q}}/\partial \vec{q} & \partial \vec{f}_{\vec{q}}/\partial \vec{p} \\ \partial \vec{f}_{\vec{p}}/\partial \vec{q} & \partial \vec{f}_{\vec{p}}/\partial \vec{p} \end{pmatrix} \right)^t$$

$$=_i \hat{J} + \begin{pmatrix} \partial \vec{f}_{\vec{q}}/\partial \vec{q} & \partial \vec{f}_{\vec{q}}/\partial \vec{p} \\ \partial \vec{f}_{\vec{p}}/\partial \vec{q} & \partial \vec{f}_{\vec{p}}/\partial \vec{p} \end{pmatrix} \cdot \hat{J} + \hat{J} \cdot \begin{pmatrix} \partial \vec{f}_{\vec{q}}/\partial \vec{q} & \partial \vec{f}_{\vec{q}}/\partial \vec{p} \\ \partial \vec{f}_{\vec{p}}/\partial \vec{q} & \partial \vec{f}_{\vec{p}}/\partial \vec{p} \end{pmatrix}^t$$

$$= \hat{J} + \begin{pmatrix} \partial \vec{f}_{\vec{q}}/\partial \vec{q} & \partial \vec{f}_{\vec{q}}/\partial \vec{p} \\ \partial \vec{f}_{\vec{p}}/\partial \vec{q} & \partial \vec{f}_{\vec{p}}/\partial \vec{p} \end{pmatrix} \cdot \hat{J} - \left(\begin{pmatrix} \partial \vec{f}_{\vec{q}}/\partial \vec{q} & \partial \vec{f}_{\vec{q}}/\partial \vec{p} \\ \partial \vec{f}_{\vec{p}}/\partial \vec{q} & \partial \vec{f}_{\vec{p}}/\partial \vec{p} \end{pmatrix} \cdot \hat{J} \right)^t$$

$$= \hat{J} + \hat{J} \cdot \hat{A} \cdot \hat{J} - \left(\hat{J} \cdot \hat{A} \cdot \hat{J} \right)^t = \hat{J} + \hat{J} \cdot \left(\hat{A} - \hat{A}^t \right) \cdot \hat{J} \qquad (4.77)$$

from which we infer that $\hat{A}$ must be symmetric. According to Eq. (1.304), the polynomial h_{i+1} can then be obtained by integration along an arbitrary path as

$$h_{i+1} = \int_0^{\vec{r}} -\hat{J} \cdot \vec{f}_i(\vec{r}\,') \, d\vec{r}\,'. \qquad (4.78)$$

To conclude, we have proven Dragt's factorization theorem; furthermore, and perhaps more important, we have presented a relatively **straightforward algorithm** to obtain the f_i to arbitrary order with DA tools.

Next, we present a method to obtain a representation for the nonlinear part of the map in terms of a **single Lie exponent**, such that

$$\mathcal{M}_1 =_n \exp(:h:)\mathcal{I}, \qquad (4.79)$$

where $h = h_3 + h_4 + \cdots + h_{n+1}$ is a polynomial from order 3 to $n+1$. We begin by computing h_3 via

$$h_3 = \int_0^{\vec{r}} -\hat{J} \cdot \vec{f}_2(\vec{r}\,') \, d\vec{r}\,'.$$

To third order, we then have $\mathcal{M}_1 =_3 \mathcal{I} + \exp(:h_3:)\mathcal{I} + :h_4:\mathcal{I}$ and so

$$\vec{\nabla} h_4 =_3 -(\mathcal{M}_1 - \exp(:h_3:)\mathcal{I}) \cdot \hat{J}.$$

Proceeding along the same line, h_{m+1} can be computed from

$$\vec{\nabla} h_{m+1} =_m -(\mathcal{M}_1 - \exp(:h_3 + \cdots + h_m:)\mathcal{I}) \cdot \hat{J}. \qquad (4.80)$$

Dragt and coworkers (1988a) developed an extensive theory on how factorizations of Lie operators can be determined for a large class of particle optical elements and how two such factorizations can be combined into one. Because of the noncommutation of the operators in the respective exponentials, extensive use is made of the **Baker-Campbell-Hausdorff formula**. However, the effort required for this process increases rapidly with order and there is not much hope of streamlining the theory to work in an order-independent way. Therefore, results could be obtained only through order three, and in some simple cases to order five.

In conclusion, two observations are in order. First, we note that given any of the previous factorizations, it is possible to obtain the conventional **DA representation** of the map by mere evaluations of the propagators following the methods discussed in Chapter 2, Section 2.3.4. We also note that while in the differential algebra $_nD_v$ all flow operators converged in only finitely many steps and hence all arguments are straightforward, the **question of convergence**, whether various factorizations up to order n do indeed converge to the map as n goes to infinity, is difficult to answer in the general case.

4.3.2 Generating Functions

Historically, many important questions in Hamiltonian mechanics have been addressed using a generating function representation of the canonical transformations at hand. As shown in Chapter 1, Section 1.4.6, any canonical transformation and hence **any symplectic map** can locally be represented by **at least one generating function** that represents the transformation uniquely through implicit conditions. Four commonly used generating functions are given by

$$F_1(\vec{q}_i, \vec{q}_f), F_2(\vec{q}_i, \vec{p}_f), F_3(\vec{p}_i, \vec{q}_f), \text{ and } F_4(\vec{p}_i, \vec{p}_f). \tag{4.81}$$

which satisfy the following conditions:

$$(\vec{p}_i, \vec{p}_f) = \left(\vec{\nabla}_{\vec{q}_i} F_1, -\vec{\nabla}_{\vec{q}_f} F_1\right)$$
$$(\vec{p}_i, \vec{q}_f) = \left(\vec{\nabla}_{\vec{q}_i} F_2, \vec{\nabla}_{\vec{p}_f} F_2\right)$$
$$(\vec{q}_i, \vec{p}_f) = \left(-\vec{\nabla}_{\vec{p}_i} F_3, -\vec{\nabla}_{\vec{q}_f} F_3\right)$$
$$(\vec{q}_i, \vec{q}_f) = \left(-\vec{\nabla}_{\vec{p}_i} F_4, \vec{\nabla}_{\vec{p}_f} F_4\right). \tag{4.82}$$

As shown in Chapter 1, Section 1.4.6, **many more than these four exist**, and while none of the four may be suitable to represent the map, at least one of those introduced previously is suitable. The generating functions allow a **redundancy-free representation** of the map since as shown in Chapter 1, Section 1.4.8, every generating function produces a symplectic map.

As we shall now show, given an nth order Taylor expansion $[\mathcal{M}]_n$ of the symplectic map $\mathcal{M}$, there is a robust and straightforward method to obtain a polynomial generating function in such a way that the map produced by it is exactly $\mathcal{M}$ up to order n. Here we use the computation of an F_1-type generator and follow the reasoning leading to Eq. (1.210); it is immediately apparent how other generators can be determined.

Let $[\mathcal{Q}]_n$ and $[\mathcal{P}]_n$ denote the nth order maps of the position and momentum parts of $\mathcal{M}$, and let $\mathcal{I}_q$ and $\mathcal{I}_p$ denote the position and momentum parts of the identity map. Passing to classes, we have

$$\begin{pmatrix} \vec{q}_f \\ \vec{q}_i \end{pmatrix} = \begin{pmatrix} [\mathcal{Q}]_n \\ [\mathcal{I}_q]_n \end{pmatrix} \begin{pmatrix} \vec{q}_i \\ \vec{p}_i \end{pmatrix}. \quad (4.83)$$

If the linear part of the map $(\mathcal{Q}, \mathcal{I}_q)^t$ is regular, we thus have

$$\begin{pmatrix} \vec{q}_i \\ \vec{p}_i \end{pmatrix} = \begin{pmatrix} [\mathcal{Q}]_n \\ [\mathcal{I}_q]_n \end{pmatrix}^{-1} \begin{pmatrix} \vec{q}_f \\ \vec{q}_i \end{pmatrix}, \quad (4.84)$$

from which we then obtain the desired relationship

$$\begin{pmatrix} \vec{p}_f \\ \vec{p}_i \end{pmatrix} = \begin{pmatrix} [\mathcal{P}]_n \\ [\mathcal{I}_p]_n \end{pmatrix} \circ \begin{pmatrix} [\mathcal{Q}]_n \\ [\mathcal{I}_q]_n \end{pmatrix}^{-1} \begin{pmatrix} \vec{q}_f \\ \vec{q}_i \end{pmatrix}. \quad (4.85)$$

The $(n+1)$st-order expansion of the generator is obtained by integration over an arbitrary path. In the DA picture, the whole process consists merely of inversion and composition of DA maps. For the inversion to be possible, it is necessary and sufficient for the linear part $([\mathcal{Q}]_1, [\mathcal{I}_q]_1)^t$ of the map to be invertible, which is the requirement for the existence of the original generator. Hence, whenever the original generator exists, its Taylor expansion to $(n+1)$st order can be obtained in an automated way.

It is noteworthy that if $\mathcal{M}_n$ is not symplectic, the integration of $\mathcal{F}_n$ over an arbitrary path yields a generating function that represents a symplectic transfer map that is "near" the original one if $\mathcal{M}_n$ is "nearly" symplectic. In fact, the algorithm can also be used for **symplectification** of transfer maps that are not symplectic, for example, because of buildup of inaccuracies during the computation of $\mathcal{M}_n$. To this end, one computes the generating function and then follows the algorithm backwards, which merely requires the application of $\vec{\nabla}$ and inversion. The result is a new and now symplectic $\mathcal{M}_n$.

The process of obtaining the gradient of the generating function can be performed to arbitrary order using only the differential algebraic operations of composition, inversion, and antiderivation. The ease of computing a generating function with differential algebra is one of the strong points of the power series representation of the map. In the Lie representation, the computation of the generating function cannot be done in a straightforward pattern and gets increasingly cumbersome with high orders. We also note here that it is possible to solve for the generating function directly via the Hamilton-Jacobi equation (Eq. 1.279), without previously calculating a map. This has been demonstrated in (Pusch, 1990).

Chapter 5

Maps: Calculation

In this chapter we discuss the necessary tools for the computation of transfer maps or flows of particle optical systems that relate final coordinates to initial coordinates and parameters via

$$\vec{z}_f = \mathcal{M}(\vec{z}_i, \vec{\delta}). \tag{5.1}$$

Because the dynamics in these systems can generally be classified as being weakly nonlinear, such map methods usually rely on **Taylor expansion** representations of the map, assuming that this is possible. The first major code to employ such methods to second order was TRANSPORT (Brown *et al.*, 1964; Brown 1979a, 1982); later the methods were extended to third order in the codes TRIO (Matsuo and Matsuda, 1976) and the related GIOS (Wollnik *et al.*, 1988) and also in new versions of TRANSPORT (Carey, 1992). The Lie algebraic methods developed by Dragt and coworkers (Dragt and Finn, 1976; Dragt, 1982) were used for the third-order code MARYLIE (Dragt *et al.*, 1985). Using custom-made formula manipulators (Berz and Wollnik, 1987), an extension to fifth order was possible (Berz *et al.*, 1987).

All of the previous methods are based on finding **analytical representations** of formulas for aberrations of a select list of elements, which inevitably makes the use of these methods complicated for high-order or very complicated fields. The **differential algebraic methods** introduced in Chapter 2 that are based on transforming the three key function space operations of addition, multiplication, and differentiation, to a suitable space of equivalence classes, can circumvent this difficulty. They can be used to construct algorithms that allow the computation of maps to arbitrary orders, including parameter dependence, and as discussed in Chapter 4, they can be used in principle arbitrary fields. These methods were implemented in the code COSY INFINITY Berz, 1992b, 1993a, 1993b) as well as in a number of other recent codes (Yan, 1993; Yan and Yan, 1990; van Zeijts, 1993; van Zeijts and Neri, 1993; Michelotti, 1990; Davis *et al.*, 1993).

In this chapter, we first derive the equations of motion in particle optical coordinates using various canonical transformation tools of Hamiltonian mechanics and also derive equations of motion for other quantities. Then we present various differential algebra (DA)-based approaches for the determination of maps for specific cases.

5.1 THE PARTICLE OPTICAL EQUATIONS OF MOTION

In this section, we will derive the equations of motion of a particle in an electromagnetic field in the so-called **curvilinear coordinates**. For the purposes of this section, we will neglect radiation effects as well as any influence of the spin on the orbit motion, but it will be clear from the context how to include their treatment if so desired. The curvilinear coordinates are measured in a moving right-handed coordinate system that has one of its axes attached and parallel to a given reference curve in space; furthermore, usually the **time is replaced** as the independent variable **by the arc length** along the given reference curve.

While the transformation to curvilinear coordinates seems to complicate the description of the motion, it has several advantages. First, if the chosen reference curve in space is itself a valid orbit, then the resulting transfer map will be **origin preserving** because the origin corresponds to the reference curve. This then opens the door to the use of perturbative techniques for the analysis of the motion in order to study how small deviations from the reference curve propagate. In particular, if the system of interest is repetitive and the reference curve is closed, then the origin will be a **fixed point** of the motion; perturbative techniques around fixed points can be employed to study the one-turn transfer map, which here corresponds to the **Poincare map** of the motion.

Second, the method is also very practical in the sense that beams are usually rather **small in size**, while they often **cover large territory**. Therefore, it is more convenient to describe them in a local coordinate system following a reference particle instead of in a Cartesian coordinate system attached to the laboratory. Expressing the motion in terms of value of the transfer map at a given arc length very directly corresponds to the measurements by detectors, which usually determine particle coordinates at a fixed plane around the system instead of at a given time. Also, expressing the motion in terms of the **arc length** as an independent variable directly provides a natural scale since it is more natural to measure in meters along the system instead of in nano- or microseconds.

The following sections describe in detail the derivation of the motion in curvilinear coordinates. We will study the transformations of Maxwell's equations and the resulting fields and their potentials to the new coordinates, and then we derive the explicit forms of the Lagrangian and Hamiltonian with time as the independent variable. Finally, a special transformation on Hamiltonians is applied that replaces time as the independent variable by the arc length, while maintaining the Hamiltonian structure of the motion.

5.1.1 Curvilinear Coordinates

Let $\{\vec{e}_1, \vec{e}_2, \vec{e}_3\}$ denote a **Dreibein**, a right-handed set of fixed orthonormal basis vectors, which defines the so-called Cartesian coordinate systems. For any point in space, let (x_1, x_2, x_3) denote its Cartesian coordinates. In order to introduce the curvilinear coordinates, let $\vec{R}(s)$ be an infinitely often differentiable curve

THE PARTICLE OPTICAL EQUATIONS OF MOTION

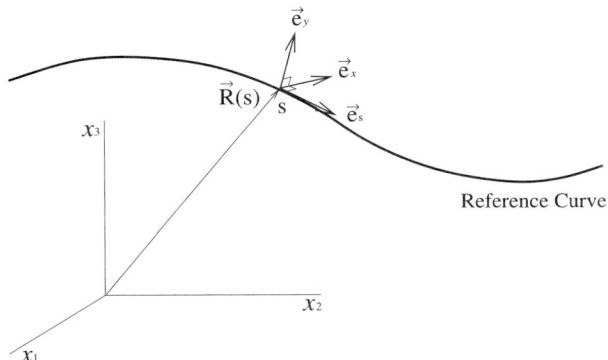

FIGURE 5.1. The reference curve and the locally attached Dreibeins.

parameterized in terms of its arc length s, the so-called **reference curve** (Fig. 5.1). For each value of s, let the vector $\vec{e}_s$ be parallel to the reference curve, i.e.,

$$\vec{e}_s(s) = \frac{d\vec{R}}{ds}. \tag{5.2}$$

We now choose the infinitely often differentiable vectors $\vec{e}_x(s)$ and $\vec{e}_y(s)$ such that for any value of s, the three vectors $\{\vec{e}_s, \vec{e}_x, \vec{e}_y\}$ form another Dreibein, a right-handed orthonormal system. For notational simplicity, in the following we also sometimes denote the curvilinear basis vectors $\{\vec{e}_s, \vec{e}_x, \vec{e}_y\}$ by $\{\vec{e}_1^C, \vec{e}_2^C, \vec{e}_3^C\}$.

Apparently, for a given curve $\vec{R}(s)$ there are a variety of choices for $\vec{e}_x(s)$ and $\vec{e}_y(s)$ that result in valid Dreibeins since $\vec{e}_x(s)$ and $\vec{e}_y(s)$ can be rotated around $\vec{e}_s$. A specific choice is often made such that additional requirements are satisfied; for example, if $\vec{R}(s)$ is never parallel to the vertical Cartesian coordinate $\vec{e}_3$, one may demand that $\vec{e}_x(s)$ always lie in the horizontal plane spanned by $\vec{e}_1$ and $\vec{e}_2$.

The functions $\vec{R}(s)$, $\vec{e}_x(s)$, and $\vec{e}_y(s)$ describe the so-called **curvilinear** coordinate system, in which a position is described in terms of s, x, and y via

$$\vec{r} = \vec{R}(s) + x\vec{e}_x + y\vec{e}_y. \tag{5.3}$$

Apparently, the position $\vec{r}$ in Cartesian coordinates is uniquely determined for any choice of (s, x, y). The converse, however, is not generally true: A point with given Cartesian coordinates $\vec{r}$ may lie in several different planes that are perpendicular to $\vec{R}(s)$, as shown in Fig. 5.2.

The situation can be remedied if the curvature $\kappa(s)$ of the reference curve $\vec{R}(s)$ never grows beyond a threshold, i.e., if

$$r_1 = 1/\max_s |\kappa(s)| \tag{5.4}$$

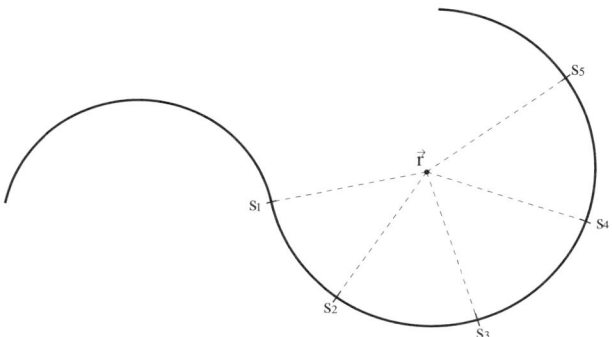

FIGURE 5.2. Non-uniqueness of curvilinear coordinates.

is finite. As Fig. 5.3 illustrates, if in this case we restrict ourselves to the inside of a tube of radius r_1 around $\vec{R}(s)$, for any vector within the tube, there is always one and only one set of coordinates (s, x, y) describing the point $\vec{r}$.

Let us now study the transformation matrix from the Cartesian basis $\{\vec{e}_1, \vec{e}_2, \vec{e}_3\}$ to the local basis of the curvilinear system $\{\vec{e}_s, \vec{e}_x, \vec{e}_y\} = \{\vec{e}_1^C, \vec{e}_2^C, \vec{e}_3^C\}$. The transformation between these basis vectors and the old ones is described by the matrix $\hat{O}(s)$, which has the form

$$\hat{O}(s) = \begin{pmatrix} \vec{e}_s(s) & \vec{e}_x(s) & \vec{e}_y(s) \end{pmatrix} = \begin{pmatrix} (\vec{e}_s \cdot \vec{e}_1) & (\vec{e}_x \cdot \vec{e}_1) & (\vec{e}_y \cdot \vec{e}_1) \\ (\vec{e}_s \cdot \vec{e}_2) & (\vec{e}_x \cdot \vec{e}_2) & (\vec{e}_y \cdot \vec{e}_2) \\ (\vec{e}_s \cdot \vec{e}_3) & (\vec{e}_x \cdot \vec{e}_3) & (\vec{e}_y \cdot \vec{e}_3) \end{pmatrix}.$$

(5.5)

Because the system $\{\vec{e}_s, \vec{e}_x, \vec{e}_y\}$ is orthonormal, so is $\hat{O}(s)$, and hence it satisfies

$$\hat{O}(s) \cdot \hat{O}(s)^t = \hat{I} \quad \text{and} \quad \hat{O}(s)^t \cdot \hat{O}(s) = \hat{I}.$$

(5.6)

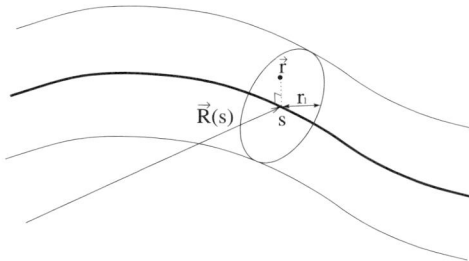

FIGURE 5.3. Uniqueness of curvilinear coordinates within a tube.

Since both the old and the new bases have the same handedness, we also have

$$\det\left(\hat{O}(s)\right) = 1, \tag{5.7}$$

and hence, altogether, $\hat{O}(s)$ belongs to the group $SO(3)$. It must be remembered that elements of $SO(3)$ preserve cross products, i.e., for $\hat{O} \in SO(3)$ and any vectors $\vec{a}, \vec{b}$, we have

$$(\hat{O}\vec{a}) \times (\hat{O}\vec{b}) = \hat{O}(\vec{a} \times \vec{b}). \tag{5.8}$$

One way to see this is to study the requirement of orthonormality on the matrix elements of $\hat{O}$. The elements of the matrix $\hat{O}$ describe the coordinates of the new parameter-dependent basis vectors in terms of the original Cartesian basis; explicitly, we have

$$[\vec{e}_s]_k = O_{k1}, \quad [\vec{e}_x]_k = O_{k2}, \quad [\vec{e}_y]_k = O_{k3}. \tag{5.9}$$

The demand of the right-handedness then reads

$$\vec{e}_l^C \times \vec{e}_m^C = \sum_{n=1}^{3} \epsilon_{lmn} \vec{e}_n^C,$$

where ϵ_{ijk} is the common totally antisymmetric tensor of rank three defined as

$$\epsilon_{ijk} = \begin{cases} 1 & \text{for } (i, j, k) = (1, 2, 3) \text{ and any cyclic permutation thereof} \\ -1 & \text{for other permutations of } (1, 2, 3) \\ 0 & \text{for two or more equal indices} \end{cases}$$

and reduces to a condition on the elements of the matrix $\hat{O}$,

$$\sum_{i,j=1}^{3} \epsilon_{ijk} O_{il} O_{jm} = \sum_{n=1}^{3} \epsilon_{lmn} O_{kn}. \tag{5.10}$$

It must be remembered that the symbol ϵ_{ijk} is very useful for the calculation of vector cross products; for vectors $\vec{a}$ and $\vec{b}$, we have

$$[\vec{a} \times \vec{b}]_k = \sum_{i,j=1}^{3} \epsilon_{ijk} a_i b_j.$$

Using the condition in Eq. (5.10), we readily obtain Eq. (5.8).

For the following discussion, it is useful to study how the transformation matrix $\hat{O}$ changes with s. Differentiating Eq. (5.6) with respect to the parameter s, we have

$$0 = \frac{d}{ds}(\hat{O}^t \cdot \hat{O}) = \frac{d\hat{O}^t}{ds}\hat{O} + \hat{O}^t\frac{d\hat{O}}{ds} = \left(\hat{O}^t\frac{d\hat{O}}{ds}\right)^t + \hat{O}^t\frac{d\hat{O}}{ds}.$$

Therefore, the matrix $\hat{T} = \hat{O}^t \cdot d\hat{O}/ds$ is antisymmetric; we describe it in terms of its three free elements via

$$\hat{O}^t \cdot \frac{d\hat{O}}{ds} = \hat{T} = \begin{pmatrix} 0 & -\tau_3 & \tau_2 \\ \tau_3 & 0 & -\tau_1 \\ -\tau_2 & \tau_1 & 0 \end{pmatrix}. \tag{5.11}$$

We group the three elements into the vector $\vec{\tau}$, which has the form

$$\vec{\tau} = \begin{pmatrix} \tau_1 \\ \tau_2 \\ \tau_3 \end{pmatrix}. \tag{5.12}$$

We observe that for any vector $\vec{a}$, we then have the relation

$$\hat{T} \cdot \vec{a} = \vec{\tau} \times \vec{a}.$$

The components of the vector $\vec{\tau}$, and hence the elements of the matrix $\hat{T}$, can be computed as

$$\tau_1 = \vec{e}_y \cdot \frac{d\vec{e}_x}{ds} = -\vec{e}_x \cdot \frac{d\vec{e}_y}{ds}$$

$$\tau_2 = \vec{e}_s \cdot \frac{d\vec{e}_y}{ds} = -\vec{e}_y \cdot \frac{d\vec{e}_s}{ds}$$

$$\tau_3 = \vec{e}_x \cdot \frac{d\vec{e}_s}{ds} = -\vec{e}_s \cdot \frac{d\vec{e}_x}{ds}. \tag{5.13}$$

These relationships give some practical meaning to the components of the vector $\vec{\tau}$: Apparently, τ_1 describes the current rate of rotation of the Dreibein around the reference curve $\vec{R}(s)$; τ_2 describes the current amount of curvature of $\vec{R}(s)$ in the plane spanned by $\vec{e}_y$ and $\vec{e}_s$; and τ_3 similarly describes the curvature of $\vec{R}(s)$ in the plane spanned by $\vec{e}_x$ and $\vec{e}_s$. In mathematical terms, because of

$$\vec{e}_s \cdot \frac{d\vec{e}_s}{ds} = 0, \quad \vec{e}_x \cdot \frac{d\vec{e}_x}{ds} = 0, \quad \vec{e}_y \cdot \frac{d\vec{e}_y}{ds} = 0, \tag{5.14}$$

we have

$$\frac{d\vec{e}_s}{ds} = \tau_3 \vec{e}_x - \tau_2 \vec{e}_y$$

$$\frac{d\vec{e}_x}{ds} = -\tau_3 \vec{e}_s + \tau_1 \vec{e}_y$$

$$\frac{d\vec{e}_y}{ds} = \tau_2 \vec{e}_s - \tau_1 \vec{e}_x, \tag{5.15}$$

as successive multiplication with $\vec{e}_s$, $\vec{e}_x$, and $\vec{e}_y$ and comparison with Eq. (5.13) reveals.

As the first step in the transformation of the Maxwell's equations and the equations of motion to the curvilinear coordinates, it is necessary to study the form of common differential operators in the new coordinates. From Eq. (5.9), which has the form

$$\vec{r} = \sum_{k=1}^{3} x_k \vec{e}_k = \sum_{k=1}^{3} \left\{ \vec{R} \cdot \vec{e}_k + x O_{k2} + y O_{k3} \right\} \vec{e}_k,$$

we see that the Cartesian components of $\vec{r}$ are

$$x_k = \vec{R} \cdot \vec{e}_k + x O_{k2} + y O_{k3} \quad \text{for } k = 1, 2, 3. \tag{5.16}$$

Through extended calculation, given in detail in (Berz et al., 1999), all common differential operators can be expressed in terms of curvilinear coordinates.

The final differential quantity we want to express in terms of curvilinear coordinates is the velocity vector $\vec{v}$. It is expressed as

$$\vec{v} = v_1 \vec{e}_1 + v_2 \vec{e}_2 + v_3 \vec{e}_3 = v_s \vec{e}_s + v_x \vec{e}_x + v_y \vec{e}_y.$$

We define

$$\vec{v}^{ct} = \begin{pmatrix} v_1 \\ v_2 \\ v_3 \end{pmatrix}, \quad \vec{v}^C = \begin{pmatrix} v_s \\ v_x \\ v_y \end{pmatrix},$$

and we have $\vec{v}^{ct} = \hat{O} \cdot \vec{v}^C$. To determine the velocity expressed in curvilinear coordinates, we differentiate the position vector $\vec{r}$ with respect to time t; from Eq. (5.16), we have

$$\vec{v}^{ct} = \frac{d\vec{r}}{dt} = \sum_{k=1}^{3} \frac{d}{dt} \{\vec{R} \cdot \vec{e}_k + x O_{k2} + y O_{k3}\} \vec{e}_k$$

$$= \sum_{k=1}^{3} \left\{ O_{k1} \dot{s} + O_{k2} \dot{x} + O_{k3} \dot{y} + \dot{s} \frac{d O_{k2}}{ds} x + \dot{s} \frac{d O_{k3}}{ds} y \right\} \vec{e}_k$$

$$= \hat{O} \cdot \begin{pmatrix} \dot{s} \\ \dot{x} \\ \dot{y} \end{pmatrix} + \dot{s} \frac{d\hat{O}}{ds} \cdot \begin{pmatrix} 0 \\ x \\ y \end{pmatrix} = \hat{O} \cdot \left\{ \begin{pmatrix} \dot{s} \\ \dot{x} \\ \dot{y} \end{pmatrix} + \dot{s} \hat{O}^t \cdot \frac{d\hat{O}}{ds} \cdot \begin{pmatrix} 0 \\ x \\ y \end{pmatrix} \right\}$$

$$= \hat{O} \cdot \left\{ \begin{pmatrix} \dot{s} \\ \dot{x} \\ \dot{y} \end{pmatrix} + \dot{s} \hat{T} \cdot \begin{pmatrix} 0 \\ x \\ y \end{pmatrix} \right\} = \hat{O} \cdot \begin{pmatrix} \dot{s} \, (1 - \tau_3 x + \tau_2 y) \\ \dot{x} - \dot{s} \tau_1 y \\ \dot{y} + \dot{s} \tau_1 x \end{pmatrix},$$

where Eq. (5.2) is used from the first line to the second line.

For later convenience, it is advantageous to introduce the abbreviation

$$\alpha = 1 - \tau_3 x + \tau_2 y. \tag{5.17}$$

We note that for x and y sufficiently close to zero, α does not vanish and is positive. Hence, besides the restriction for the motion to be inside a tube of radius r_1 imposed by the need for uniqueness of the transformation to curvilinear coordinates in Eq. (5.4), there is another condition; defining

$$r_2 = \frac{1}{2} \min_s \left(\left| \frac{1}{\tau_3} \right|, \left| \frac{1}{\tau_2} \right| \right). \tag{5.18}$$

If we restrict x, y to satisfy $|x|, |y| < r_2$, the quantity α never vanishes.

Utilizing $\vec{v}^{ct} = \hat{O} \cdot \vec{v}^C$, we conclude that the velocity expressed in terms of curvilinear coordinates is given by

$$\vec{v}^C = \begin{pmatrix} v_s \\ v_x \\ v_y \end{pmatrix} = \begin{pmatrix} \dot{s} \cdot (1 - \tau_3 x + \tau_2 y) \\ \dot{x} - \dot{s} \tau_1 y \\ \dot{y} + \dot{s} \tau_1 x \end{pmatrix} = \begin{pmatrix} \dot{s} \alpha \\ \dot{x} - \dot{s} \tau_1 y \\ \dot{y} + \dot{s} \tau_1 x \end{pmatrix}. \tag{5.19}$$

For future reference, we note that because of the orthonormality of $\hat{O}$, we also have the following relationships:

$$v^2 = \vec{v}^{ct} \cdot \vec{v}^{ct} = \vec{v}^C \cdot \vec{v}^C \tag{5.20}$$

$$\vec{v}^{ct} \cdot \vec{A}^{ct} = \vec{v}^C \cdot \vec{A}^C. \tag{5.21}$$

5.1.2 The Lagrangian and Lagrange's Equations in Curvilinear Coordinates

Now we are ready to develop Lagrangian and Hamiltonian methods in curvilinear coordinates. Following the transformation properties of Lagrangians, it is conceptually directly possible, albeit practically somewhat involved, to obtain the Lagrangian in curvilinear coordinates. To this end, we merely have to take the Lagrangian of a charged particle in an electromagnetic field in the Cartesian system,

$$L(x_1, x_2, x_3; \dot{x}_1, \dot{x}_2, \dot{x}_3; t) = -mc^2 \sqrt{1 - \frac{v^2}{c^2}} - e\Phi + e\vec{v}^{ct} \cdot \vec{A}^{ct},$$

and express all Cartesian quantities in terms of the curvilinear quantities. In this respect, it is very convenient that the scalar product of the velocity with itself and with $\vec{A}$ is the same in the Cartesian and curvilinear systems, according to Eqs. (5.20) and (5.21). Therefore, the Lagrangian in the curvilinear system is obtained completely straightforwardly as

$$L(s, x, y; \dot{s}, \dot{x}, \dot{y}; t) = -mc^2 \sqrt{1 - \frac{\vec{v}^{C2}}{c^2}} - e\Phi + e\vec{v}^C \cdot \vec{A}^C, \qquad (5.22)$$

where

$$\vec{v}^{C2} = v_s^2 + v_x^2 + v_y^2 \quad \text{and} \quad \vec{v}^C \cdot \vec{A}^C = v_s A_s + v_x A_x + v_y A_y.$$

Here, Φ and $\vec{A}^C$ are dependent on the position, i.e., $\{s, x, y\}$, and the time t. The quantities $\hat{O}$, $\hat{T}$, and hence τ_1, τ_2, τ_3 used in Eq. (5.23) are dependent on s.

The derivatives of v_s, v_x, v_y with respect to s, x, y, $\dot{s}$, $\dot{x}$, $\dot{y}$ are useful in order to determine the explicit form of Lagrange's equations:

$$\begin{aligned}
\frac{\partial v_s}{\partial \dot{s}} &= \alpha, & \frac{\partial v_s}{\partial \dot{x}} &= 0, & \frac{\partial v_s}{\partial \dot{y}} &= 0, \\
\frac{\partial v_x}{\partial \dot{s}} &= -\tau_1 y, & \frac{\partial v_x}{\partial \dot{x}} &= 1, & \frac{\partial v_x}{\partial \dot{y}} &= 0, \\
\frac{\partial v_y}{\partial \dot{s}} &= \tau_1 x, & \frac{\partial v_y}{\partial \dot{x}} &= 0, & \frac{\partial v_y}{\partial \dot{y}} &= 1, \\
\frac{\partial v_s}{\partial s} &= \dot{s}\left(-\frac{d\tau_3}{ds}x + \frac{d\tau_2}{ds}y\right), & \frac{\partial v_s}{\partial x} &= -\dot{s}\,\tau_3, & \frac{\partial v_s}{\partial y} &= \dot{s}\,\tau_2, \\
\frac{\partial v_x}{\partial s} &= -\dot{s}\frac{d\tau_1}{ds}y, & \frac{\partial v_x}{\partial x} &= 0, & \frac{\partial v_x}{\partial y} &= -\dot{s}\,\tau_1, \\
\frac{\partial v_y}{\partial s} &= \dot{s}\frac{d\tau_1}{ds}x, & \frac{\partial v_y}{\partial x} &= \dot{s}\,\tau_1, & \frac{\partial v_y}{\partial y} &= 0.
\end{aligned} \qquad (5.23)$$

The Lagrange equation for x is derived as follows. Using the derivatives of v_s, v_x, v_y in Eq. (5.23), we have

$$\frac{\partial v^2}{\partial \dot{x}} = 2v_s \frac{\partial v_s}{\partial \dot{x}} + 2v_x \frac{\partial v_x}{\partial \dot{x}} + 2v_y \frac{\partial v_y}{\partial \dot{x}} = 2v_x$$

$$\frac{\partial(\vec{v}^C \cdot \vec{A}^C)}{\partial \dot{x}} = A_s \frac{\partial v_s}{\partial \dot{x}} + A_x \frac{\partial v_x}{\partial \dot{x}} + A_y \frac{\partial v_y}{\partial \dot{x}} = A_x$$

$$\frac{\partial v^2}{\partial x} = 2v_s \frac{\partial v_s}{\partial x} + 2v_x \frac{\partial v_x}{\partial x} + 2v_y \frac{\partial v_y}{\partial x} = -2v_s \dot{s} \tau_3 + 2v_y \dot{s} \tau_1$$

$$= -2\dot{s}(\tau_3 v_s - \tau_1 v_y) = -2\dot{s}\,[\vec{\tau} \times \vec{v}^C]_2.$$

Altogether, we have

$$\frac{\partial L}{\partial \dot{x}} = \frac{m}{\sqrt{1 - v^2/c^2}} v_x + e A_x = p_x + e A_x, \tag{5.24}$$

where $\vec{p}^{ct} = m\vec{v}^{ct}/\sqrt{1 - v^2/c^2}$ and correspondingly $\vec{p}^C = m\vec{v}^C/\sqrt{1 - v^2/c^2}$ was used. We also have

$$\frac{\partial L}{\partial x} = -\frac{m}{\sqrt{1 - v^2/c^2}} \dot{s} \, [\vec{\tau} \times \vec{v}^C]_2 - e \frac{\partial}{\partial x}(\Phi - \vec{v}^C \cdot \vec{A}^C)$$

$$= -\dot{s} \, [\vec{\tau} \times \vec{p}^C]_2 - e \frac{\partial}{\partial x}(\Phi - \vec{v}^C \cdot \vec{A}^C).$$

Thus, the Lagrange equation for x is

$$\frac{dp_x}{dt} + \dot{s} \, [\vec{\tau} \times \vec{p}^C]_2 = e \left[-\frac{dA_x}{dt} - \frac{\partial}{\partial x}(\Phi - \vec{v}^C \cdot \vec{A}^C) \right]. \tag{5.25}$$

The Lagrange equation for y is derived in the same way, and it is

$$\frac{dp_y}{dt} + \dot{s} \, [\vec{\tau} \times \vec{p}^C]_3 = e \left[-\frac{dA_y}{dt} - \frac{\partial}{\partial y}(\Phi - \vec{v}^C \cdot \vec{A}^C) \right]. \tag{5.26}$$

It is more complicated to derive the Lagrange equation for s. Using the derivatives of v_s, v_x, v_y in Eq. (5.23), we obtain

$$\frac{\partial v^2}{\partial \dot{s}} = 2v_s \alpha - 2v_x \tau_1 y + 2v_y \tau_1 x$$

$$\frac{\partial (\vec{v}^C \cdot \vec{A}^C)}{\partial \dot{s}} = A_s \alpha - A_x \tau_1 y + A_y \tau_1 x$$

$$\frac{\partial v^2}{\partial s} = 2v_s \dot{s} \left(-\frac{d\tau_3}{ds} x + \frac{d\tau_2}{ds} y \right) - 2v_x \dot{s} \frac{d\tau_1}{ds} y + 2v_y \dot{s} \frac{d\tau_1}{ds} x$$

$$= -2\dot{s} x \left[\frac{d\vec{\tau}}{ds} \times \vec{v}^C \right]_2 - 2\dot{s} y \left[\frac{d\vec{\tau}}{ds} \times \vec{v}^C \right]_3,$$

and so

$$\frac{\partial L}{\partial \dot{s}} = \frac{m}{\sqrt{1 - v^2/c^2}} \cdot (v_s \alpha - v_x \tau_1 y + v_y \tau_1 x) + e (A_s \alpha - A_x \tau_1 y + A_y \tau_1 x)$$

$$= (p_s + e A_s) \alpha - (p_x + e A_x) \tau_1 y + (p_y + e A_y) \tau_1 x \tag{5.27}$$

THE PARTICLE OPTICAL EQUATIONS OF MOTION

as well as

$$\frac{\partial L}{\partial s} = \frac{m}{\sqrt{1-v^2/c^2}} \cdot \left\{ -\dot{s}x \left[\frac{d\vec{\tau}}{ds} \times \vec{v}^C\right]_2 - \dot{s}y \left[\frac{d\vec{\tau}}{ds} \times \vec{v}^C\right]_3 \right\}$$
$$- e \frac{\partial}{\partial s}(\Phi - \vec{v}^C \cdot \vec{A}^C)$$
$$= -\dot{s}x \left[\frac{d\vec{\tau}}{ds} \times \vec{p}^C\right]_2 - \dot{s}y \left[\frac{d\vec{\tau}}{ds} \times \vec{p}^C\right]_3 - e \frac{\partial}{\partial s}(\Phi - \vec{v}^C \cdot \vec{A}^C).$$

Thus, the Lagrange equation for s is

$$\frac{d}{dt}(p_s \alpha - p_x \tau_1 y + p_y \tau_1 x) + \dot{s}x \left[\frac{d\vec{\tau}}{ds} \times \vec{p}^C\right]_2 + \dot{s}y \left[\frac{d\vec{\tau}}{ds} \times \vec{p}^C\right]_3$$
$$= e \left[-\frac{d}{dt}(A_s \alpha - A_x \tau_1 y + A_y \tau_1 x) - \frac{\partial}{\partial s}(\Phi - \vec{v}^C \cdot \vec{A}^C) \right]. \quad (5.28)$$

The left-hand side is modified as follows

$$\alpha \frac{dp_s}{dt} - \tau_1 y \frac{dp_x}{dt} + \tau_1 x \frac{dp_y}{dt} + p_s(-\tau_3 \dot{x} + \tau_2 \dot{y}) - p_x \tau_1 \dot{y} + p_y \tau_1 \dot{x}$$
$$+ p_s \left(-\dot{s}\frac{d\tau_3}{ds}x + \dot{s}\frac{d\tau_2}{ds}y\right) - p_x \dot{s}\frac{d\tau_1}{ds}y + p_y \dot{s}\frac{d\tau_1}{ds}x$$
$$+ \dot{s}x \left[\frac{d\vec{\tau}}{ds} \times \vec{p}^C\right]_2 + \dot{s}y \left[\frac{d\vec{\tau}}{ds} \times \vec{p}^C\right]_3$$
$$= \alpha \frac{dp_s}{dt} - \tau_1 y \frac{dp_x}{dt} + \tau_1 x \frac{dp_y}{dt} - \dot{x}[\vec{\tau} \times \vec{p}^C]_2 - \dot{y}[\vec{\tau} \times \vec{p}^C]_3$$
$$= \alpha \left(\frac{dp_s}{dt} + \dot{s}[\vec{\tau} \times \vec{p}^C]_1\right) - \tau_1 y \left(\frac{dp_x}{dt} + \dot{s}[\vec{\tau} \times \vec{p}^C]_2\right)$$
$$+ \tau_1 x \left(\frac{dp_y}{dt} + \dot{s}[\vec{\tau} \times \vec{p}^C]_3\right) - v_s[\vec{\tau} \times \vec{p}^C]_1 - v_x[\vec{\tau} \times \vec{p}^C]_2 - v_y[\vec{\tau} \times \vec{p}^C]_3$$
$$= \alpha \left(\frac{dp_s}{dt} + \dot{s}[\vec{\tau} \times \vec{p}^C]_1\right) - \tau_1 y \left(\frac{dp_x}{dt} + \dot{s}[\vec{\tau} \times \vec{p}^C]_2\right)$$
$$+ \tau_1 x \left(\frac{dp_y}{dt} + \dot{s}[\vec{\tau} \times \vec{p}^C]_3\right),$$

where Eq. (5.19) is used from the second step to the third step, and

$$v_s[\vec{\tau} \times \vec{p}^C]_1 + v_x[\vec{\tau} \times \vec{p}^C]_2 + v_y[\vec{\tau} \times \vec{p}^C]_3 = \vec{v}^C \cdot (\vec{\tau} \times \vec{p}^C) = 0$$

is used in the last step. Therefore, the Lagrange equation for s simplifies to

$$\alpha \left(\frac{dp_s}{dt} + \dot{s}[\vec{\tau} \times \vec{p}^C]_1 \right) - \tau_1 y \left(\frac{dp_x}{dt} + \dot{s}[\vec{\tau} \times \vec{p}^C]_2 \right)$$
$$+ \tau_1 x \left(\frac{dp_y}{dt} + \dot{s}[\vec{\tau} \times \vec{p}^C]_3 \right)$$
$$= e \left[-\alpha \frac{dA_s}{dt} + \tau_1 y \frac{dA_x}{dt} - \tau_1 x \frac{dA_y}{dt} + A_s \frac{d}{dt}(\tau_3 x - \tau_2 y) \right.$$
$$\left. + A_x \frac{d}{dt}(\tau_1 y) + A_y \frac{d}{dt}(-\tau_1 x) - \frac{\partial}{\partial s}(\Phi - \vec{v}^C \cdot \vec{A}^C) \right].$$

The equations for x and y (Eqs. 5.25 and 5.26) can be used to simplify the previous equation. Doing this, we obtain

$$\alpha \left(\frac{dp_s}{dt} + \dot{s}[\vec{\tau} \times \vec{p}^C]_1 \right)$$
$$= e \left[-\alpha \frac{dA_s}{dt} - \left(\frac{\partial}{\partial s} + \tau_1 y \frac{\partial}{\partial x} - \tau_1 x \frac{\partial}{\partial y} \right) (\Phi - \vec{v}^C \cdot \vec{A}^C) \right.$$
$$\left. + A_s \frac{d}{dt}(\tau_3 x - \tau_2 y) + A_x \frac{d}{dt}(\tau_1 y) + A_y \frac{d}{dt}(-\tau_1 x) \right],$$

and with the requirement that x and y are small enough such that $\alpha = 1 - \tau_3 x + \tau_2 y > 0$, the equation can be written as

$$\frac{dp_s}{dt} + \dot{s}[\vec{\tau} \times \vec{p}^C]_1$$
$$= e \left[-\frac{dA_s}{dt} - \frac{1}{\alpha} \left(\frac{\partial}{\partial s} + \tau_1 y \frac{\partial}{\partial x} - \tau_1 x \frac{\partial}{\partial y} \right) (\Phi - \vec{v}^C \cdot \vec{A}^C) \right.$$
$$\left. + \frac{1}{\alpha} \left\{ A_s \frac{d}{dt}(\tau_3 x - \tau_2 y) + A_x \frac{d}{dt}(\tau_1 y) + A_y \frac{d}{dt}(-\tau_1 x) \right\} \right].$$

Thus, the set of three Lagrange equations can be summarized as follows: It apparently agrees with Newton's equations in curvilinear coordinates [see (Berz et al., 1999)].

$$\frac{d}{dt} \begin{pmatrix} p_s \\ p_x \\ p_y \end{pmatrix} + \dot{s} \cdot \begin{pmatrix} \tau_1 \\ \tau_2 \\ \tau_3 \end{pmatrix} \times \begin{pmatrix} p_s \\ p_x \\ p_y \end{pmatrix}$$

$$= -\frac{d}{dt} \begin{pmatrix} eA_s \\ eA_x \\ eA_y \end{pmatrix} - \frac{e}{\alpha} \cdot \begin{pmatrix} \frac{\partial}{\partial s} + \tau_1 y \frac{\partial}{\partial x} - \tau_1 x \frac{\partial}{\partial y} \\ \alpha \frac{\partial}{\partial x} \\ \alpha \frac{\partial}{\partial y} \end{pmatrix} (\Phi - \vec{v}^C \cdot \vec{A}^C)$$

$$+ \frac{e}{\alpha} \left\{ A_s \frac{d}{dt}(\tau_3 x - \tau_2 y) + A_x \frac{d}{dt}(\tau_1 y) + A_y \frac{d}{dt}(-\tau_1 x) \right\} \vec{e}_s. \tag{5.29}$$

5.1.3 The Hamiltonian and Hamilton's Equations in Curvilinear Coordinates

To obtain the Hamiltonian now is conceptually standard fare, although practically it is somewhat involved. We adopt the curvilinear coordinates $\{s, x, y\}$ as generalized coordinates, and we denote the corresponding generalized momentum by $\vec{P}^G = \left(P_s^G, P_x^G, P_y^G \right)$. The generalized momentum is obtained via the partials of L with respect to the generalized velocities; using Eqs. (5.24) and (5.27), we obtain

$$P_s^G = \frac{\partial L}{\partial \dot{s}} = (p_s + eA_s)\alpha - (p_x + eA_x)\tau_1 y + (p_y + eA_y)\tau_1 x$$

$$P_x^G = \frac{\partial L}{\partial \dot{x}} = p_x + eA_x$$

$$P_y^G = \frac{\partial L}{\partial \dot{y}} = p_y + eA_y. \tag{5.30}$$

It is worthwhile to express the mechanical momentum $\vec{p}_{Mech}^C$, namely $\vec{p}^C$, in terms of the generalized momentum $\vec{P}^G = \left(P_s^G, P_x^G, P_y^G \right)$. By combining the expressions in Eq. (5.30), we have

$$P_s^G = (p_s + eA_s)\alpha - P_x^G \tau_1 y + P_y^G \tau_1 x,$$

and so

$$p_s + eA_s = \frac{1}{\alpha} \left(P_s^G + P_x^G \tau_1 y - P_y^G \tau_1 x \right),$$

and altogether

$$\vec{p}_{Mech}^C = \vec{p}^C = \begin{pmatrix} \frac{1}{\alpha} \left(P_s^G + P_x^G \tau_1 y - P_y^G \tau_1 x \right) - eA_s \\ P_x^G - eA_x \\ P_y^G - eA_y \end{pmatrix}. \tag{5.31}$$

Squaring $\vec{p}^C = \gamma m \vec{v}^C = m\vec{v}^C/\sqrt{1-(\vec{v}^C)^2/c^2}$ and reorganizing yields

$$(\vec{v}^C)^2 = \frac{c^2(\vec{p}^C)^2}{(\vec{p}^C)^2 + m^2c^2},$$

and because $\vec{v}^C$ and $\vec{p}^C$ are parallel we even have

$$\vec{v}^C = \frac{c\vec{p}^C}{\sqrt{(\vec{p}^C)^2 + m^2c^2}}. \tag{5.32}$$

We also observe that

$$\frac{1}{\gamma} = \sqrt{1-(\vec{v}^C)^2/c^2} = \sqrt{1 - \frac{(\vec{p}^C)^2}{(\vec{p}^C)^2 + m^2c^2}} = \frac{mc}{\sqrt{(\vec{p}^C)^2 + m^2c^2}}. \tag{5.33}$$

The Hamiltonian in the curvilinear system H is defined from the Lagrangian L (Eq. 5.22) and the generalized momentum $\vec{P}^G$ (Eq. 5.30) via the Legendre transformation

$$H = \dot{s} P_s^G + \dot{x} P_x^G + \dot{y} P_y^G - L$$

$$= \dot{s} P_s^G + \dot{x} P_x^G + \dot{y} P_y^G + mc^2 \sqrt{1 - \frac{\vec{v}^{C2}}{c^2}} + e\Phi - e\vec{v}^C \cdot \vec{A}^C,$$

and the subsequent expression in terms of only s, x, y, P_s^G, P_x^G, P_y^G and t, if this is possible. Using Eqs. (5.31), (5.32), and (5.33), we have from Eq. (5.19) that

$$\dot{s} = \frac{v_s}{\alpha} = \frac{1}{m\gamma}\frac{1}{\alpha}\left\{\frac{1}{\alpha}\left(P_s^G + P_x^G \tau_1 y - P_y^G \tau_1 x\right) - eA_s\right\},$$

$$\dot{x} = v_x + \dot{s}\tau_1 y$$

$$= \frac{1}{m\gamma}\left[P_x^G - eA_x + \frac{\tau_1 y}{\alpha}\left\{\frac{1}{\alpha}\left(P_s^G + P_x^G \tau_1 y - P_y^G \tau_1 x\right) - eA_s\right\}\right]$$

$$= \frac{1}{m\gamma}\frac{1}{\alpha^2}\left\{\tau_1 y\, P_s^G + \left(\alpha^2 + \tau_1^2 y^2\right) P_x^G - \tau_1^2 xy\, P_y^G - e\tau_1 y\alpha\, A_s - e\alpha^2 A_x\right\}$$

$$\dot{y} = v_y - \dot{s}\tau_1 x$$

$$= \frac{1}{m\gamma}\frac{1}{\alpha^2}\left\{-\tau_1 x\, P_s^G - \tau_1^2 xy\, P_x^G + \left(\alpha^2 + \tau_1^2 x^2\right) P_y^G + e\tau_1 x\alpha\, A_s - e\alpha^2 A_y\right\},$$

where we used the abbreviation γ from Eq. (5.33), which is in terms of the generalized coordinates and the generalized momenta

$$\frac{1}{m\gamma} = \frac{c}{\sqrt{(P_s^G + P_x^G \tau_1 y - P_y^G \tau_1 x - \alpha e A_s)^2 \alpha^2 + (P_x^G - eA_x)^2 + (P_y^G - eA_y)^2 + m^2 c^2}}. \tag{5.34}$$

We also have

$$\vec{v}^C \cdot \vec{A}^C = \frac{1}{m\gamma} \vec{p}^C \cdot \vec{A}^C$$

$$= \frac{1}{m\gamma} \left[\left\{ \frac{1}{\alpha} \left(P_s^G + P_x^G \tau_1 y - P_y^G \tau_1 x \right) - eA_s \right\} A_s \right.$$
$$\left. + (P_x^G - eA_x) A_x + (P_y^G - eA_y) A_y \right],$$

and in particular it proved possible to invert the relationships between generalized velocities and generalized momenta. Hence, the Hamiltonian H can be expressed in curvilinear coordinates, and it is given by

$$H = \frac{1}{m\gamma} \left[\frac{1}{\alpha^2} \left(P_s^G + P_x^G \tau_1 y - P_y^G \tau_1 x \right) P_s^G + \frac{\tau_1 y}{\alpha^2} P_s^G P_x^G \right.$$
$$+ \left\{ 1 + \frac{(\tau_1 y)^2}{\alpha^2} \right\} \left(P_x^G \right)^2 - \frac{\tau_1 x}{\alpha^2} P_s^G P_y^G + \left\{ 1 + \frac{(\tau_1 x)^2}{\alpha^2} \right\} \left(P_y^G \right)^2$$
$$- \frac{2\tau_1^2 xy}{\alpha^2} P_x^G P_y^G - \frac{1}{\alpha} P_s^G e A_s - \frac{\tau_1 y}{\alpha} P_x^G e A_s - P_x^G e A_x + \frac{\tau_1 x}{\alpha} P_y^G e A_s$$
$$- P_y^G e A_y - \frac{1}{\alpha} \left(P_s^G + P_x^G \tau_1 y - P_y^G \tau_1 x \right) e A_s + e^2 A_s^2$$
$$\left. - (P_x^G - eA_x) e A_x - (P_y^G - eA_y) e A_y + m^2 c^2 \right] + e\Phi$$

$$= \frac{1}{m\gamma} \left[\frac{1}{\alpha^2} \left(P_s^G + P_x^G \tau_1 y - P_y^G \tau_1 x \right)^2 \right.$$
$$- 2 \frac{1}{\alpha} \left(P_s^G + P_x^G \tau_1 y - P_y^G \tau_1 x \right) e A_s + e^2 A_s^2$$
$$\left. + (P_x^G - eA_x)^2 + (P_y^G - eA_y)^2 + m^2 c^2 \right] + e\Phi$$

$$= \frac{1}{m\gamma} \left[\left\{ \frac{1}{\alpha} \left(P_s^G + P_x^G \tau_1 y - P_y^G \tau_1 x \right) - eA_s \right\}^2 \right.$$
$$\left. + (P_x^G - eA_x)^2 + (P_y^G - eA_y)^2 + m^2 c^2 \right] + e\Phi$$

$$= \frac{1}{m\gamma}(mc\gamma)^2 + e\Phi = mc^2\gamma + e\Phi.$$

Explicitly, the Hamiltonian in curvilinear coordinates is

$$H = c\sqrt{(P_s^G + P_x^G \tau_1 y - P_y^G \tau_1 x - \alpha e A_s)^2 \alpha^2 + (P_x^G - eA_x)^2 + (P_y^G - eA_y)^2 + m^2 c^2} \\ + e\Phi, \tag{5.35}$$

where $\alpha = 1 - \tau_3 x + \tau_2 y$. Thus we derive Hamilton's equations as follows.

$$\dot{s} = \frac{\partial H}{\partial P_s^G} = \frac{1}{m\gamma}\frac{1}{\alpha}\left\{\frac{1}{\alpha}\left(P_s^G + P_x^G \tau_1 y - P_y^G \tau_1 x\right) - eA_s\right\},$$

$$\dot{x} = \frac{\partial H}{\partial P_x^G} = \frac{1}{m\gamma}\left[\frac{\tau_1 y}{\alpha}\left\{\frac{1}{\alpha}\left(P_s^G + P_x^G \tau_1 y - P_y^G \tau_1 x\right) - eA_s\right\} \right. \\ \left. + P_x^G - eA_x\right],$$

$$\dot{y} = \frac{\partial H}{\partial P_y^G} = \frac{1}{m\gamma}\left[-\frac{\tau_1 x}{\alpha}\left\{\frac{1}{\alpha}\left(P_s^G + P_x^G \tau_1 y - P_y^G \tau_1 x\right) - eA_s\right\} \right. \\ \left. + P_y^G - eA_y\right],$$

$$\dot{P}_s^G = -\frac{\partial H}{\partial s} = \frac{1}{m\gamma}\left[-\left\{\frac{1}{\alpha}\left(P_s^G + P_x^G \tau_1 y - P_y^G \tau_1 x\right) - eA_s\right\} \right. \\ \cdot \left\{\frac{1}{\alpha^2}\left(\frac{d\tau_3}{ds}x - \frac{d\tau_2}{ds}y\right)\left(P_s^G + P_x^G \tau_1 y - P_y^G \tau_1 x\right) \right. \\ \left. + \frac{1}{\alpha}\left(P_x^G \frac{d\tau_1}{ds}y - P_y^G \frac{d\tau_1}{ds}x\right) - e\frac{\partial A_s}{\partial s}\right\} \\ \left. + e\left(P_x^G - eA_x\right)\frac{\partial A_x}{\partial s} + e\left(P_y^G - eA_y\right)\frac{\partial A_y}{\partial s}\right] - e\frac{\partial \Phi}{\partial s}, \tag{5.36}$$

$$\dot{P}_x^G = -\frac{\partial H}{\partial x} = \frac{1}{m\gamma}\left[-\left\{\frac{1}{\alpha}\left(P_s^G + P_x^G \tau_1 y - P_y^G \tau_1 x\right) - eA_s\right\} \right. \\ \cdot \left\{\frac{\tau_3}{\alpha^2}\left(P_s^G + P_x^G \tau_1 y - P_y^G \tau_1 x\right) - \frac{\tau_1}{\alpha}P_y^G - e\frac{\partial A_s}{\partial x}\right\} \\ \left. + e\left(P_x^G - eA_x\right)\frac{\partial A_x}{\partial x} + e\left(P_y^G - eA_y\right)\frac{\partial A_y}{\partial x}\right] - e\frac{\partial \Phi}{\partial x}, \tag{5.37}$$

THE PARTICLE OPTICAL EQUATIONS OF MOTION

$$\dot{P}_y^G = -\frac{\partial H}{\partial y} = \frac{1}{m\gamma}\left[-\left\{\frac{1}{\alpha}\left(P_s^G + P_x^G \tau_1 y - P_y^G \tau_1 x\right) - eA_s\right\}\right.$$

$$\cdot\left\{-\frac{\tau_2}{\alpha^2}\left(P_s^G + P_x^G \tau_1 y - P_y^G \tau_1 x\right) + \frac{\tau_1}{\alpha}P_x^G - e\frac{\partial A_s}{\partial y}\right\}$$

$$\left. + e\left(P_x^G - eA_x\right)\frac{\partial A_x}{\partial y} + e\left(P_y^G - eA_y\right)\frac{\partial A_y}{\partial y}\right] - e\frac{\partial \Phi}{\partial y}, \quad (5.38)$$

where the abbreviation (Eq. 5.34) is used.

To verify the derivations, we check that Hamilton's equations agree with previous results. It is shown easily that the first three equations agree with Eq. (5.19). The last three equations are shown to agree with Lagrange's equations (5.25), (5.26), and (5.28). We have from Eq. (5.37)

$$\dot{P}_x^G = -\frac{1}{\alpha}v_s\left\{\frac{\tau_3}{\alpha}\left(P_s^G + P_x^G \tau_1 y - P_y^G \tau_1 x\right) - \tau_1 P_y^G\right\}$$

$$+ ev_s\frac{\partial A_s}{\partial x} + ev_x\frac{\partial A_x}{\partial x} + ev_y\frac{\partial A_y}{\partial x} - e\frac{\partial \Phi}{\partial x}$$

$$= -\frac{\dot{s}\tau_3}{\alpha}P_s^G - \frac{\dot{s}\tau_1\tau_3 y}{\alpha}P_x^G + \dot{s}\tau_1\left(\frac{\tau_3 x}{\alpha} + 1\right)P_y^G$$

$$- e\left(\frac{\partial \Phi}{\partial x} - v_s\frac{\partial A_s}{\partial x} - v_x\frac{\partial A_x}{\partial x} - v_y\frac{\partial A_y}{\partial x}\right).$$

Expressing the equation in terms of the mechanical momentum $\vec{p}^C$ rather than the generalized momentum $\vec{P}^G$ according to Eq. (5.30) and using Eq. (5.23), we have

$$\dot{p}_x + e\frac{dA_x}{dt} = -\frac{\dot{s}\tau_3}{\alpha}\left\{(p_s + eA_s)\alpha - (p_x + eA_x)\tau_1 y + (p_y + eA_y)\tau_1 x\right\}$$

$$- \frac{\dot{s}\tau_1\tau_3 y}{\alpha}(p_x + eA_x) + \dot{s}\tau_1\left(\frac{\tau_3 x}{\alpha} + 1\right)(p_y + eA_y)$$

$$- e\frac{\partial}{\partial x}(\Phi - v_s A_s - v_x A_x - v_y A_y)$$

$$- e\left(A_s\frac{\partial v_s}{\partial x} + A_x\frac{\partial v_x}{\partial x} + A_y\frac{\partial v_y}{\partial x}\right)$$

$$= -\dot{s}(\tau_3 p_s - \tau_1 p_y) - e\frac{\partial}{\partial x}(\Phi - \vec{v}^C \cdot \vec{A}^C),$$

which is in agreement with the first Lagrange equation (Eq. 5.25). The Hamilton equation for y is similarly modified from (Eq. 5.38)

$$\dot{P}_y^G = \frac{\dot{s}\tau_2}{\alpha}P_s^G + \dot{s}\tau_1\left(\frac{\tau_2 y}{\alpha} - 1\right)P_x^G - \frac{\dot{s}\tau_1\tau_2 x}{\alpha}P_y^G$$

$$-e\left(\frac{\partial \Phi}{\partial y} - v_s \frac{\partial A_s}{\partial y} - v_x \frac{\partial A_x}{\partial y} - v_y \frac{\partial A_y}{\partial y}\right).$$

In terms of the mechanical momentum $\vec{p}^C$, we have

$$\dot{p}_y + e\frac{dA_y}{dt} = -\dot{s}(\tau_1 p_x - \tau_2 p_s) - e\frac{\partial}{\partial y}(\Phi - \vec{v}^C \cdot \vec{A}^C),$$

and it agrees with the second Lagrange equation (Eq. 5.26). Similarly, the Hamilton equation for s is modified from Eq. (5.36)

$$\dot{P}_s^G = \frac{\dot{s}}{\alpha}\left(-\frac{d\tau_3}{ds}x + \frac{d\tau_2}{ds}y\right)P_s^G + \left\{\frac{\dot{s}}{\alpha}\left(-\frac{d\tau_3}{ds}x + \frac{d\tau_2}{ds}y\right)\tau_1 y - \dot{s}\frac{d\tau_1}{ds}y\right\}P_x^G$$
$$+ \left\{-\frac{\dot{s}}{\alpha}\left(-\frac{d\tau_3}{ds}x + \frac{d\tau_2}{ds}y\right)\tau_1 x + \dot{s}\frac{d\tau_1}{ds}x\right\}P_y^G$$
$$- e\left(\frac{\partial \Phi}{\partial s} - v_s \frac{\partial A_s}{\partial s} - v_x \frac{\partial A_x}{\partial s} - v_y \frac{\partial A_y}{\partial s}\right).$$

In terms of the mechanical momentum $\vec{p}^C$, it takes the form

$$\frac{d}{dt}\left\{(p_s + eA_s)\alpha - (p_x + eA_x)\tau_1 y + (p_y + eA_y)\tau_1 x\right\}$$
$$= \frac{\dot{s}}{\alpha}\left(-\frac{d\tau_3}{ds}x + \frac{d\tau_2}{ds}y\right)\left\{(p_s + eA_s)\alpha - (p_x + eA_x)\tau_1 y + (p_y + eA_y)\tau_1 x\right\}$$
$$+ \left\{\frac{\dot{s}}{\alpha}\left(-\frac{d\tau_3}{ds}x + \frac{d\tau_2}{ds}y\right)\tau_1 y - \dot{s}\frac{d\tau_1}{ds}y\right\}(p_x + eA_x)$$
$$+ \left\{-\frac{\dot{s}}{\alpha}\left(-\frac{d\tau_3}{ds}x + \frac{d\tau_2}{ds}y\right)\tau_1 x + \dot{s}\frac{d\tau_1}{ds}x\right\}(p_y + eA_y)$$
$$- e\left(\frac{\partial \Phi}{\partial s} - v_s \frac{\partial A_s}{\partial s} - v_x \frac{\partial A_x}{\partial s} - v_y \frac{\partial A_y}{\partial s}\right),$$

and reorganization leads to

$$\frac{d}{dt}\left(p_s\alpha - p_x\tau_1 y + p_y\tau_1 x\right) + \dot{s}x\left[\frac{d\vec{\tau}}{ds} \times \vec{p}^C\right]_2 + \dot{s}y\left[\frac{d\vec{\tau}}{ds} \times \vec{p}^C\right]_3$$
$$= e\left[-\frac{d}{dt}\left(A_s\alpha - A_x\tau_1 y + A_y\tau_1 x\right) - \frac{\partial}{\partial s}(\Phi - \vec{v}^C \cdot \vec{A}^C)\right],$$

which agrees with the third Lagrange equation (Eq. 5.28).

5.1.4 Arc Length as an Independent Variable for the Hamiltonian

As the last step, we perform a change of the independent variable from the time t to the space coordinate s. For such an interchange, there is a surprisingly simple procedure which merely requires viewing t as a new position variable, $-H$ as the associated momentum, and $-P_s^G$ as the new Hamiltonian, and expressing the interchange in terms of the new variables, if this is possible. Then the equations are

$$\frac{dx}{ds} = \frac{\partial(-P_s^G)}{\partial P_x^G}, \quad \frac{dy}{ds} = \frac{\partial(-P_s^G)}{\partial P_y^G}, \quad \frac{dt}{ds} = \frac{\partial(-P_s^G)}{\partial(-H)},$$

$$\frac{dP_x^G}{ds} = -\frac{\partial(-P_s^G)}{\partial x}, \quad \frac{dP_y^G}{ds} = -\frac{\partial(-P_s^G)}{\partial y}, \quad \frac{d(-H)}{ds} = -\frac{\partial(-P_s^G)}{\partial t}.$$

To begin, let us try to express $-P_s^G$ in terms of $t, x, y, -H, P_x^G$, and P_y^G. From Eq. (5.35), we obtain that

$$\left\{\frac{1}{\alpha}\left(P_s^G + P_x^G \tau_1 y - P_y^G \tau_1 x\right) - eA_s\right\}^2$$
$$+ (P_x^G - eA_x)^2 + (P_y^G - eA_y)^2 + m^2 c^2$$
$$= \frac{1}{c^2}(H - e\Phi)^2.$$

Therefore,

$$\left(P_s^G + P_x^G \tau_1 y - P_y^G \tau_1 x - \alpha e A_s\right)^2$$
$$= \alpha^2 \left\{\frac{1}{c^2}(H - e\Phi)^2 - (P_x^G - eA_x)^2 - (P_y^G - eA_y)^2 - m^2 c^2\right\}.$$

Considering the case that $\vec{A} = 0$ and x and y are small, we demand p_s should be positive (and stay that way throughout); it must also be remembered that $\alpha > 0$, and hence the choice of sign is made such that

$$P_s^G = -P_x^G \tau_1 y + P_y^G \tau_1 x + \alpha e A_s$$
$$+ \alpha \sqrt{\frac{1}{c^2}(H - e\Phi)^2 - (P_x^G - eA_x)^2 - (P_y^G - eA_y)^2 - m^2 c^2}.$$

Thus, $-P_s^G$ and hence the new Hamiltonian H^s is obtained as

$$H^s = -P_s^G = P_x^G \tau_1 y - P_y^G \tau_1 x - \alpha e A_s$$
$$- \alpha \sqrt{\frac{1}{c^2}(H - e\Phi)^2 - (P_x^G - eA_x)^2 - (P_y^G - eA_y)^2 - m^2 c^2}.$$

MAPS: CALCULATION

Here, for later convenience, note that

$$\sqrt{\frac{1}{c^2}(H - e\Phi)^2 - (P_x^G - eA_x)^2 - (P_y^G - eA_y)^2 - m^2c^2}$$
$$= \frac{1}{\alpha}\left(P_s^G + P_x^G \tau_1 y - P_y^G \tau_1 x\right) - eA_s = p_s. \tag{5.39}$$

Then, the equations of motion are

$$\frac{dx}{ds} = \frac{\partial(-P_s^G)}{\partial P_x^G} = \tau_1 y$$
$$+ \frac{\alpha(P_x^G - eA_x)}{\sqrt{\frac{1}{c^2}(H - e\Phi)^2 - (P_x^G - eA_x)^2 - (P_y^G - eA_y)^2 - m^2c^2}}, \tag{5.40}$$

$$\frac{dy}{ds} = \frac{\partial(-P_s^G)}{\partial P_y^G} = -\tau_1 x$$
$$+ \frac{\alpha(P_y^G - eA_y)}{\sqrt{\frac{1}{c^2}(H - e\Phi)^2 - (P_x^G - eA_x)^2 - (P_y^G - eA_y)^2 - m^2c^2}}, \tag{5.41}$$

$$\frac{dt}{ds} = \frac{\partial(-P_s^G)}{\partial(-H)}$$
$$= \frac{\alpha\frac{1}{c^2}(H - e\Phi)}{\sqrt{\frac{1}{c^2}(H - e\Phi)^2 - (P_x^G - eA_x)^2 - (P_y^G - eA_y)^2 - m^2c^2}}, \tag{5.42}$$

$$\frac{dP_x^G}{ds} = -\frac{\partial(-P_s^G)}{\partial x} = P_y^G \tau_1 - e\tau_3 A_s + \alpha e\frac{\partial A_s}{\partial x} \tag{5.43}$$
$$- \tau_3 \sqrt{\frac{1}{c^2}(H - e\Phi)^2 - (P_x^G - eA_x)^2 - (P_y^G - eA_y)^2 - m^2c^2}$$
$$- \alpha e \frac{\frac{1}{c^2}(H - e\Phi)\frac{\partial \Phi}{\partial x} - (P_x^G - eA_x)\frac{\partial A_x}{\partial x} - (P_y^G - eA_y)\frac{\partial A_y}{\partial x}}{\sqrt{\frac{1}{c^2}(H - e\Phi)^2 - (P_x^G - eA_x)^2 - (P_y^G - eA_y)^2 - m^2c^2}},$$

$$\frac{dP_y^G}{ds} = -\frac{\partial(-P_s^G)}{\partial y} = -P_x^G \tau_1 + e\tau_2 A_s + \alpha e\frac{\partial A_s}{\partial y} \tag{5.44}$$

THE PARTICLE OPTICAL EQUATIONS OF MOTION

$$+ \tau_2 \sqrt{\frac{1}{c^2}(H - e\Phi)^2 - (P_x^G - eA_x)^2 - (P_y^G - eA_y)^2 - m^2c^2}$$

$$- \alpha e \frac{\frac{1}{c^2}(H - e\Phi)\frac{\partial \Phi}{\partial y} - (P_x^G - eA_x)\frac{\partial A_x}{\partial y} - (P_y^G - eA_y)\frac{\partial A_y}{\partial y}}{\sqrt{\frac{1}{c^2}(H - e\Phi)^2 - (P_x^G - eA_x)^2 - (P_y^G - eA_y)^2 - m^2c^2}},$$

$$\frac{d(-H)}{ds} = -\frac{\partial(-P_s^G)}{\partial t} = e\alpha \left[\frac{\partial A_s}{\partial t} \right. \tag{5.45}$$

$$\left. - \frac{\frac{1}{c^2}(H - e\Phi)\frac{\partial \Phi}{\partial t} - (P_x^G - eA_x)\frac{\partial A_x}{\partial t} - (P_y^G - eA_y)\frac{\partial A_y}{\partial t}}{\sqrt{\frac{1}{c^2}(H - e\Phi)^2 - (P_x^G - eA_x)^2 - (P_y^G - eA_y)^2 - m^2c^2}} \right].$$

For the sake of convenience and checking purposes, we replace P_x^G, P_y^G, and H by $\vec{p}^C$ using Eqs. (5.30) and (5.35), with the help of Eqs. (5.32) and (5.39). Then we have from Eqs. (5.40), (5.41) and (5.42)

$$\frac{dx}{ds} = \tau_1 y + \alpha \frac{p_x}{p_s} \tag{5.46}$$

$$\frac{dy}{ds} = -\tau_1 x + \alpha \frac{p_y}{p_s} \tag{5.47}$$

$$\frac{dt}{ds} = \alpha \frac{1}{p_s} \frac{\sqrt{(\vec{p}^C)^2 + m^2 c^2}}{c}.$$

We have from Eq. (5.43)

$$\frac{dp_x}{ds} + e\frac{dA_x}{ds} = (p_y + eA_y)\tau_1 - e\tau_3 A_s + \alpha e \frac{\partial A_s}{\partial x} - \tau_3 p_s$$

$$- \alpha \frac{e}{p_s}\left\{ \frac{\sqrt{(\vec{p}^C)^2 + m^2 c^2}}{c}\frac{\partial \Phi}{\partial x} - p_x\frac{\partial A_x}{\partial x} - p_y\frac{\partial A_y}{\partial x} \right\},$$

and organizing the expression using Eqs. (5.23), (5.19), and (5.32) we find

$$\frac{dp_x}{ds} + \left[\vec{\tau} \times \vec{p}^C\right]_x = e\left[-\frac{dA_x}{ds} - \frac{1}{\dot{s}}\frac{\partial}{\partial x}(\Phi - \vec{v}^C \cdot \vec{A}^C)\right]. \tag{5.48}$$

In a similar way, we obtain from Eq. (5.44)

$$\frac{dp_y}{ds} + \left[\vec{\tau} \times \vec{p}^C\right]_y = e\left[-\frac{dA_y}{ds} - \frac{1}{\dot{s}}\frac{\partial}{\partial y}(\Phi - \vec{v}^C \cdot \vec{A}^C)\right], \tag{5.49}$$

and from Eq. (5.45)

$$\frac{dH}{ds} = \frac{1}{\dot{s}} \frac{\partial}{\partial t} \left[e(\Phi - \vec{v}^C \cdot \vec{A}^C) \right].$$

This concludes the derivations of dynamics in curvilinear coordinates. In particular, we have succeeded in deriving the equations of motion of a particle moving in an electromagnetic field in curvilinear coordinates, with the arc length s as the independent variable. Moreover, we know that these equations of motion are Hamiltonian in nature, which has important consequences for theoretical studies.

5.1.5 Curvilinear Coordinates for Planar Motion

As an application of the concepts just derived, we consider a particularly important special case, namely, the situation in which the reference curve stays in the $x_1 x_2$ plane. This so-called two-dimensional (2-D) curvilinear system occurs frequently in practice, in particular if the reference curve is an actual orbit and the fields governing the motion have a symmetry around the horizontal plane. The basis vectors in this 2-D curvilinear system can be expressed by the Cartesian basis vectors via

$$\vec{e}_y = \vec{e}_3$$
$$\vec{e}_s = \cos\theta \vec{e}_1 - \sin\theta \vec{e}_2$$
$$\vec{e}_x = \sin\theta \vec{e}_1 + \cos\theta \vec{e}_2,$$

where θ depends on the arc length s, and its derivative is denoted by h, i.e.,

$$h = h(s) = \frac{d\theta(s)}{ds}.$$

From Eq. (5.9), all the elements of the matrix $\hat{O}$ are determined as

$$\hat{O} = \begin{pmatrix} \cos\theta & \sin\theta & 0 \\ -\sin\theta & \cos\theta & 0 \\ 0 & 0 & 1 \end{pmatrix}.$$

Therefore, the antisymmetric matrix $\hat{T}$ of Eq. (5.11) has the form

$$\hat{T} = \hat{O}^t \cdot \frac{d\hat{O}}{ds} = \begin{pmatrix} \cos\theta & -\sin\theta & 0 \\ \sin\theta & \cos\theta & 0 \\ 0 & 0 & 1 \end{pmatrix} \cdot \begin{pmatrix} -\sin\theta \cdot h & \cos\theta \cdot h & 0 \\ -\cos\theta \cdot h & -\sin\theta \cdot h & 0 \\ 0 & 0 & 0 \end{pmatrix}$$

$$= \begin{pmatrix} 0 & h & 0 \\ -h & 0 & 0 \\ 0 & 0 & 0 \end{pmatrix}.$$

THE PARTICLE OPTICAL EQUATIONS OF MOTION

Thus, the elements of $\hat{T}$ and hence $\vec{\tau}$ are given as

$$\vec{\tau} = \begin{pmatrix} \tau_1 \\ \tau_2 \\ \tau_3 \end{pmatrix} = \begin{pmatrix} 0 \\ 0 \\ -h \end{pmatrix};$$

finally, we have

$$\alpha = 1 - \tau_3 x + \tau_2 y = 1 + hx.$$

The velocity expressed in this system is, from Eq. (5.19),

$$\vec{v}^C = \begin{pmatrix} v_s \\ v_x \\ v_y \end{pmatrix} = \begin{pmatrix} \dot{s}(1+hx) \\ \dot{x} \\ \dot{y} \end{pmatrix}.$$

Thus, the equations of motion expressed in this system are

$$\frac{d}{dt}\begin{pmatrix} p_s \\ p_x \\ p_y \end{pmatrix} + \dot{s}\begin{pmatrix} 0 \\ 0 \\ -h \end{pmatrix} \times \begin{pmatrix} p_s \\ p_x \\ p_y \end{pmatrix} = \begin{pmatrix} \frac{dp_s}{dt} + \dot{s}hp_x \\ \frac{dp_x}{dt} - \dot{s}hp_s \\ \frac{dp_y}{dt} \end{pmatrix}$$

$$= e\left[-\frac{d}{dt}\begin{pmatrix} A_s \\ A_x \\ A_y \end{pmatrix} - \begin{pmatrix} \frac{1}{1+hx}\frac{\partial}{\partial s} \\ \frac{\partial}{\partial x} \\ \frac{\partial}{\partial y} \end{pmatrix}(\Phi - \vec{v}^C \cdot \vec{A}^C) \right.$$

$$\left. + \begin{pmatrix} A_s d/dt(-hx) \\ 1+hx \\ 0 \\ 0 \end{pmatrix} \right].$$

Furthermore, the equations of motion after space–time interchange in this system are, from Eqs. (5.48) and (5.49),

$$\frac{dp_x}{ds} - hp_s = e\left[-\frac{dA_x}{ds} - \frac{1}{\dot{s}}\frac{\partial}{\partial x}(\Phi - \vec{v}^C \cdot \vec{A}^C) \right]$$

$$\frac{dp_y}{ds} = e\left[-\frac{dA_y}{ds} - \frac{1}{\dot{s}}\frac{\partial}{\partial y}(\Phi - \vec{v}^C \cdot \vec{A}^C) \right]$$

$$= e\left[\frac{1}{\dot{s}}E_y + (1+hx)B_x - \frac{dx}{ds}B_s\right].$$

Here, $E_{x,y,s}$ and $B_{x,y,s}$ are the electric and magnetic field components in the x, y, and s directions, respectively.

It is customary to express the equations of motion in terms of the normalized coordinates $a = p_x/p_0$ and $b = p_y/p_0$, where p_0 is a reference momentum. Expressed in these coordinates, the complete equations of motion take the form

$$\frac{dx}{ds} = (1+hx)\frac{a}{\sqrt{(p/p_0)^2 - a^2 - b^2}} \tag{5.50}$$

$$\frac{dy}{ds} = (1+hx)\frac{b}{\sqrt{(p/p_0)^2 - a^2 - b^2}} \tag{5.51}$$

$$\frac{da}{ds} = h\cdot\sqrt{(p/p_0)^2 - a^2 - b^2} + \frac{e}{p_0}\left[\frac{1}{\dot{s}}E_x + \frac{dy}{ds}B_s - (1+hx)B_y\right] \tag{5.52}$$

$$\frac{db}{ds} = \frac{e}{p_0}\left[\frac{1}{\dot{s}}E_y - \frac{dx}{ds}B_s + (1+hx)B_x\right] \tag{5.53}$$

5.2 Equations of Motion for Spin

The equation for the classical spin vector $\vec{S}$ of a particle in the electromagnetic field is generally assumed to have the form of the Thomas–BMT equation (Thomas, 1927; Michel et al., 1959):

$$\frac{d\vec{S}}{dt} = \vec{W}\times\vec{S}, \tag{5.54}$$

where the vector $\vec{W}$ is given by

$$\vec{W} = -\frac{e}{m\gamma}\left\{(1+\gamma G)\vec{B} - \frac{G}{(1+\gamma)}\frac{(\vec{p}\cdot\vec{B})}{mc^2}\vec{p} - \left(G + \frac{1}{1+\gamma}\right)\frac{1}{mc^2}\vec{p}\times\vec{E}\right\}. \tag{5.55}$$

Here, $G = (g-2)/2$ quantifies the anomalous spin g factor, $\vec{p}$ is the kinetic momentum of the particle, and the time t is the independent variable.

A careful analysis reveals that the above formula perhaps poses more questions than it answers, because its detailed derivation hinges on several assumptions. First, the equation should be relativistically covariant; it should reduce to the proper nonrelativistic equation in the rest frame of the particle; and it should be linear in the field. While these assumptions appear plausible, caution may already be in order because the orbital motion of the particle in the field is also accelerated, perhaps opening the door for the requirement to treatment in general relativity. But many further assumptions are necessary, including that spin is described by a four-vector (and not connected to a tensor like the magnetic field it represents), about the maximal orders of dependencies on four-velocities, and the preservation of various products of four-vectors and tensors. For some details, refer to (Thomas, 1927; Michel *et al.*, 1959; Parrott, 1987; Rohrlich, 1990). Some of the complexities of the arguments are illuminated when comparing the results stated in the first and second edition of (Rohrlich, 1990). While Eq. (5.55) seems to agree well with experimental evidence, its deep theoretical foundations at this point appear rather unclear and worthy of further study.

If we express the motion in curvilinear coordinates (s, x, y) with arc length s as an independent variable as in Eq. (5.2), the spin motion equation (Eq. 5.55) takes the form

$$\frac{d\vec{S}}{ds} = \vec{W}^* \times \vec{S}, \tag{5.56}$$

where

$$W_s^* = (1 - \tau_3 x + \tau_2 y) \frac{m\gamma}{p_s} W_s;$$

$$W_x^* = -\tau_2 + (1 - \tau_3 x + \tau_2 y) \frac{m\gamma}{p_s} W_x,$$

$$W_y^* = -\tau_3 + (1 - \tau_3 x + \tau_2 y) \frac{m\gamma}{p_s} W_y; \tag{5.57}$$

here (W_s^*, W_x^*, W_y^*) are the components of the vector $\vec{W}^*$ expressed in the new variables (s, x, y), p_s is the longitudinal kinetic momentum, τ_3 and τ_2 are the curvatures in the x and y directions introduced in Eq. (5.12), where τ_1 is assumed to be zero.

For further discussion, it is useful to introduce the matrix $\hat{W}(\vec{z})$, which is made from the vector $\vec{W}^*(\vec{z})$ via

$$\hat{W}(\vec{z}) = \begin{pmatrix} 0 & -W_y^* & W_x^* \\ W_y^* & 0 & -W_s^* \\ -W_x^* & W_s^* & 0 \end{pmatrix}. \tag{5.58}$$

Note that the matrix $\hat{W}$ is **antisymmetric** and, using $\hat{W}$, the spin equations of motion can be written as

$$\frac{d\vec{S}}{ds} = \hat{W} \cdot \vec{S}. \tag{5.59}$$

The particular form of the equation of motion (Eq. 5.56) entails that the solution always has a particular form. Because the equations of motion are linear in the spin, the final transformation of the spin variables can be described in terms of a **matrix** that depends only on the **orbital quantities**. The orbital quantities themselves are unaffected by the spin motion, such that altogether the map has the form

$$\begin{cases} \vec{z}_f = \mathcal{M}(\vec{z}_i, s), \\ \vec{S}_f = \hat{A}(\vec{z}_i, s) \cdot \vec{S}_i. \end{cases} \tag{5.60}$$

The special cross product form of the spin motion imposes a restriction on the matrix $\hat{A}$ in that the matrix is **orthogonal** with determinant 1, i.e., $\hat{A}(\vec{z}) \in SO(3)$, and $\hat{A}(\vec{z})$ satisfies $\hat{A}^t(\vec{z}) \cdot \hat{A}(\vec{z}) = \hat{I}$ and $\det(\hat{A}(\vec{z})) = 1$. We will demonstrate this using similar arguments as presented in Section 1.4.7 in the proof that flows of Hamiltonian systems are symplectic. Let us consider an ODE $d/dt\, \vec{r} = \hat{W}(t) \cdot \vec{r}$, where the n-dimensional matrix $\hat{W}$ is antisymmetric. Since the equation is linear, so is the flow, and we have

$$\vec{r}(t) = \hat{A}(t) \cdot \vec{r}_0, \tag{5.61}$$

where $\hat{A}$ is the Jacobian of the flow. Define

$$\hat{P}(t) = \hat{A}^t(t) \cdot \hat{A}(t). \tag{5.62}$$

Initially, $\hat{P}$ equals the identity, and so $\hat{P}(t_0) = \hat{I}$. The equations of motion entail that the motion of the matrix $\hat{A}$ can be described by

$$\frac{d}{dt}\hat{A}(t) = \hat{W} \cdot \hat{A}(t), \tag{5.63}$$

which in turn entails that

$$\frac{d}{dt}\hat{P}(t) = \frac{d}{dt}(\hat{A}^t(t) \cdot \hat{A}(t)) = \hat{A}^t \cdot \hat{W}^t \cdot \hat{A} + \hat{A}^t \cdot \hat{W} \cdot \hat{A}$$
$$= \hat{A}^t \cdot (-\hat{W}) \cdot \hat{A} + \hat{A}^t \cdot \hat{W} \cdot \hat{A} = 0. \tag{5.64}$$

This entails that one solution of the ODE for $\hat{P}(t)$ that satisfies the necessary initial condition $\hat{P}(t_0) = \hat{I}$ is

$$\hat{P}(t) = \hat{I} \text{ for all } t. \tag{5.65}$$

However, because of the uniqueness theorem for differential equations, this is also the **only solution**, and we have proved what we wanted to show. We now consider the determinant of the matrix. Because $1 = \det(\hat{P}(t)) = \det(\hat{A}^t(t) \cdot \hat{A}(t)) =$

$\det(\hat{A}(t))^2$, we have $\det(\hat{M}) = \pm 1$. However, since $\hat{A}$ and hence $\det(\hat{A})$ depend continuously on time and $\det(\hat{A}) = 1$ for $t = t_0$, we must have that

$$\det(\hat{A}) = +1. \tag{5.66}$$

Therefore, $\hat{A} \in SO(n)$, as we wanted to show.

The practical computation of the spin-orbit map can be achieved in a variety of ways. Conceptually, the **simplest way** is to interpret it as a motion in the **nine variables** consisting of orbit and spin described by the orbit equations as well as the spin equations (Eq. 5.56). In this case, the DA method allows the computation of the spin–orbit map in two conventional ways (see Section 5.4), namely, via a propagation operator for the case of the s-independent fields, such as main fields, and via integration of the equations of motion with DA, as described in Chaper 2, Section 2.3.4. However, in this simplest method, the number of independent variables increases from six to nine, which particularly in higher orders entails a substantial increase in computational and storage requirements. This severely limits the ability to perform analysis and computation of spin motion to high orders.

Instead, it is more advantageous to keep only **six independent variables** but to augment the differential equations for the orbit motion by the equations of the orbit-dependent spin matrix which have the form

$$\hat{A}'(\vec{z}) = \hat{W}(\vec{z}) \cdot \hat{A}(\vec{z}). \tag{5.67}$$

If desired, the $SO(3)$ structure of $\hat{A}$ can be used to reduce the number of additional differential equations at the cost of slight subsequent computational expense. The most straightforward simplification results from the fact that since orthogonal matrices have orthogonal columns and their determinant is unity and hence the orientation of a Dreibein is preserved, it follows that the third column of the matrix $\hat{A}(\vec{z}) = (\vec{A}_1(\vec{z}), \vec{A}_2(\vec{z}), \vec{A}_3(\vec{z}))$ can be uniquely calculated via

$$\vec{A}_3(\vec{z}) = \vec{A}_1(\vec{z}) \times \vec{A}_2(\vec{z}). \tag{5.68}$$

Thus, in this scenario, only six additional differential equations without new independent variables are needed for the description of the dynamics. For the case of integrative solution of the equations of motion, which is necessary in the case of s-dependent elements, these equations can be integrated in DA with any numerical integrator.

However, for the case of main fields, the explicit avoidance of the spin variables in the previously described manner is not possible since, for reasons of computational expense, it is desirable to phrase the problem in terms of a **propagator operator**

$$\begin{pmatrix} \vec{z}_f \\ \vec{S}_f \end{pmatrix} = \exp(\Delta s \cdot L_{\vec{F}}) \begin{pmatrix} \vec{z} \\ \vec{S} \end{pmatrix}. \tag{5.69}$$

Here, $L_{\vec{F}} = \vec{F} \cdot \vec{\nabla}$ is the nine-dimensional vector field belonging to the spin–orbit motion. In this case, the differential vector field $L_{\vec{F}}$ describes the entire motion including that of the spin, i.e., $d/ds(\vec{z}, \vec{S}) = \vec{F}(\vec{z}, \vec{S}) = (\vec{f}(\vec{z}), \hat{W}^* \times \vec{S})$. In particular, the operator $L_{\vec{F}}$ contains differentiation with respect to the spin variables, which requires their presence. Therefore, the original propagator is not directly applicable for the case in which the spin variables are dropped and has to be rephrased for the new choice of variables. For this purpose, we define two spaces of functions $g(\vec{z}, \vec{S})$ on spin–orbit phase space as follows:

Z: Space of functions depending only on $\vec{z}$
S: Space of linear forms in $\vec{S}$ with coefficients in Z

Then, we have for $g \in Z$,

$$L_{\vec{F}} g = \left(\vec{f}^t \cdot \vec{\nabla}_{\vec{z}} + (\hat{W} \cdot \vec{S})^t \cdot \vec{\nabla}_{\vec{S}} \right) g = \vec{f}^t \cdot \vec{\nabla}_{\vec{z}} g = L_{\vec{f}} g, \quad (5.70)$$

and in particular, the action of $L_{\vec{F}}$ can be computed without using the spin variables; furthermore, since $\vec{f}$ depends only on $\vec{z}$, we have $L_{\vec{F}} g \in Z$. Similarly, we have for $g = |a_1, a_2, a_3\rangle = \sum_{j=1}^{3} a_j \cdot S_j \in S$,

$$L_{\vec{F}} | |a_1, a_2, a_3\rangle = (\vec{f}^t \cdot \vec{\nabla}_{\vec{z}} + (\hat{W} \cdot \vec{S})^t \cdot \vec{\nabla}_{\vec{S}}) \left(\sum_{j=1}^{3} a_j \cdot S_j \right)$$

$$= \sum_{j=1}^{3} (\vec{f}^t \cdot \vec{\nabla}_{\vec{z}}) a_j \cdot S_j + \sum_{j,k=1}^{3} S_j W_{kj} a_k$$

$$= \Big| L_{\vec{f}} a_1 + \sum_{k=1}^{3} W_{k1} a_k, \ L_{\vec{f}} a_2 + \sum_{k=1}^{3} W_{k2} a_k,$$

$$L_{\vec{f}} a_3 + \sum_{k=1}^{3} W_{k3} a_k \Big\rangle, \quad (5.71)$$

and in particular, the action of $L_{\vec{F}}$ can be computed without using the spin variables; furthermore, $L_{\vec{F}} | a_1, a_2, a_3\rangle \in S$. Thus, Z and S are **invariant subspaces** of the operator $L_{\vec{F}}$. Furthermore, the action of the nine-dimensional differential operator $L_{\vec{F}}$ on S is uniquely described by Eq. (5.71), which expresses it in terms of the six-dimensional differential operator $L_{\vec{f}}$. This allows the computation of the action of the **original propagator** $\exp(\Delta s \cdot L_{\vec{F}})$ on the identity in R^9, the result of which actually describes the total nine-dimensional map. For the top six lines of the identity, note that the components are in Z, and hence the repeated application of $L_{\vec{F}}$ will stay in Z; for the bottom three lines, those of the identity map are in S, and hence the repeated application of $L_{\vec{F}}$ will stay in S, allowing the utilization of the invariant subspaces. Since elements in either space are characterized by just six-dimensional functions, $\exp(\Delta s \cdot L_{\vec{F}})$ can be computed in a merely six-dimensional differential algebra.

To conclude, we note that for the problem of **composition** of maps involving spin, in the formalism outlined previously we observe that when composing the maps $(\mathcal{M}_{1,2}, \hat{A}_{1,2})$ and $(\mathcal{M}_{2,3}, \hat{A}_{2,3})$, it is necessary to evaluate $\hat{A}_{2,3}$ at the intermediate position $\mathcal{M}_{1,2}$ before multiplying the spin matrices. Altogether, we have

$$\mathcal{M}_{1,3} = \mathcal{M}_{2,3} \circ \mathcal{M}_{1,2}$$
$$\hat{A}_{1,3}(\vec{z}) = \hat{A}_{2,3}(\mathcal{M}_{1,2}(\vec{z})) \cdot \hat{A}_{1,2}(\vec{z}). \tag{5.72}$$

5.3 Maps Determined by Algebraic Relations

The most straightforward methods for the determination of maps in the neighborhood of the reference trajectory through Taylor expansion are applicable where the motion under consideration is determined merely by algebraic equations. These cases include lens-based light optics, the motion in magnetic dipoles, and an approximation frequently used for high-energy accelerators based on drifts and kicks.

5.3.1 Lens Optics

Lens optics represents the special case of light optics in which the system under consideration consists of an ensemble of individual glass objects g_i with index of refraction n_i, separated by drifts of length l_i. It thus represents the practically and historically most important case of the general optical problem, which is described in terms of an index of refraction $n(x, y, z)$ that is dependent on position and that affects local propagation velocity via $v = c/n(x, y, z)$. The motion is described by **Fermat's principle**, which states that the light ray between two points follows the fastest path. The general problem is described in terms of Lagrange equations with the **Lagrangian**

$$L = n(x, y, z) \cdot \sqrt{1 + \left(\frac{dx}{dz}\right)^2 + \left(\frac{dy}{dz}\right)^2}. \tag{5.73}$$

As a consequence of Fermat's principle, light rays follow straight paths in regions where the index of refraction is constant. On the other hand, at transitions between regions in which the index of refraction is constant, the direction of travel of the light ray changes as dictated by **Snell's Law** of refraction.

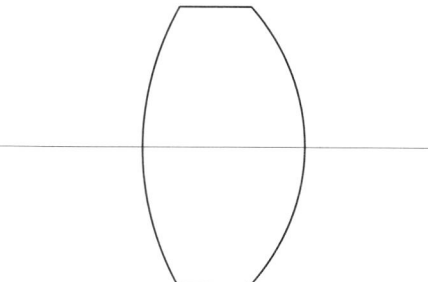

FIGURE 5.4. A glass lens with spherical surfaces.

The most common glass element is the lens shown in Fig. 5.4. In the most general case, the left and right surfaces of the glass lens are given by functions

$$S_1(x, y) \text{ and } S_2(x, y), \tag{5.74}$$

describing the shape of the surface relative to the so-called entrance and exit planes. Frequently, the surfaces S_1 and S_2 are spherical.

The motion of a ray is completely determined by the **ray-tracing** scheme, in which the ray is propagated until it hits a transition surface, then refracted following Snell's law, then propagated to the next surface, and so on. In the case of spherical lenses, the intersection with the next surface can usually be calculated in closed form. In the case of a general surface $S(x, y)$, the ray is usually propagated to the plane tangent to S at $(x, y) = (0, 0)$. Denoting its position in this plane by (x_r, y_r), its distance to the lens is given approximately by $S(x_r, y_r)$, and the particle is propagated along a straight line by this distance. Its new positions (x_r, y_r) are determined, and the method is iterated until convergence to sufficient accuracy is obtained.

To determine the map of such a glass optical system, one merely has to recognize that the ray-tracing scheme provides a **functional dependence** of final conditions on initial conditions and parameters, and to evaluate the entire scheme within DA utilizing Eq. (2.96).

As an example of the approach, we calculate nonlinearities of the map of the Kitt Peak telescope (Wynne, 1973) up to order 13. To illustrate the effects, we plot the focal point positions of an ensemble of parallel rays of an angle of 20 minutes of arc, striking the main mirror at different positions. The results are shown in Fig. 5.5. Apparently, the telescope is designed so that fifth-order terms help compensate third-order terms. Going from fifth to seventh order then shows an increase in spot size, apparently because seventh-order terms have not been corrected. Going from seventh to thirteenth order does not result in additional effects, suggesting that for the rays selected, the device is well described by order seven.

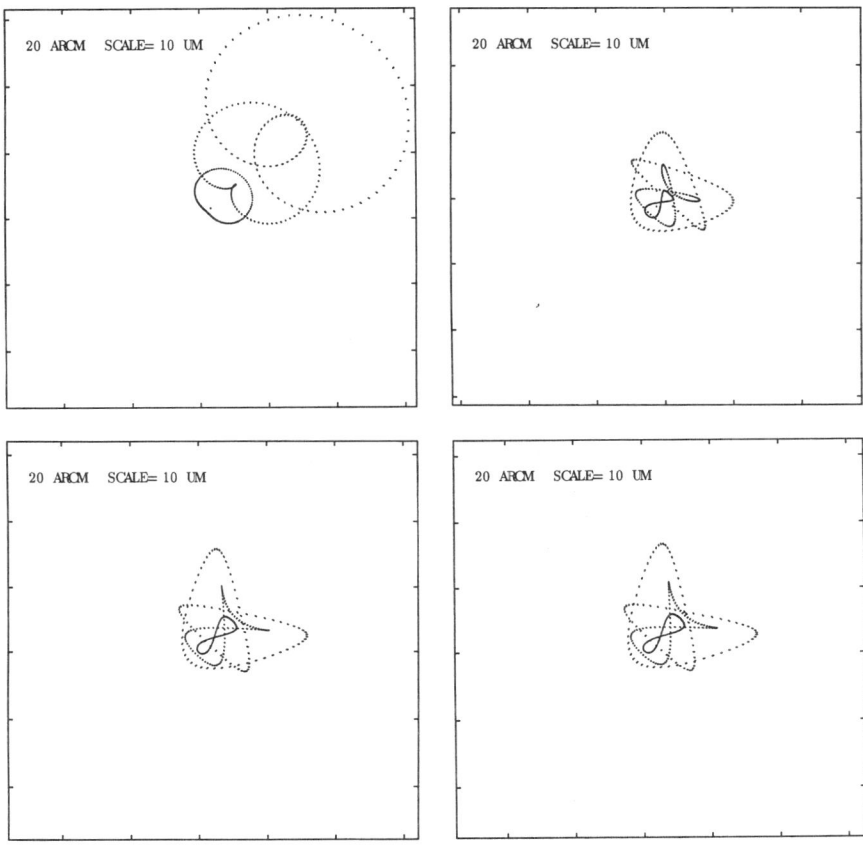

FIGURE 5.5. The effects of nonlinear aberrations on a parallel bundle of rays in the Kitt Peak telescope. Shown are orders 3, 5, 7, and 13.

5.3.2 The Dipole

Another case where the map of a system is uniquely determined through mere geometric relationships is the motion in a dipole magnet with a uniform magnetic field. The general shape of such a magnet, the reference trajectory, and a particle trajectory are shown in Fig. 5.6.

Analytic computation is possible, because charged particles experience a force in the plane perpendicular to the field direction, which produces circular motion in the projection on that plane, which we call the x–s plane. In the direction parallel to the homogeneous field, no force is produced and the particles perform a force-free drift in that direction.

We are looking for the map which relates canonical particle optical coordinates $\vec{z}_i$ in front of the magnet to the final coordinates $\vec{z}_f$ behind the magnet. The initial

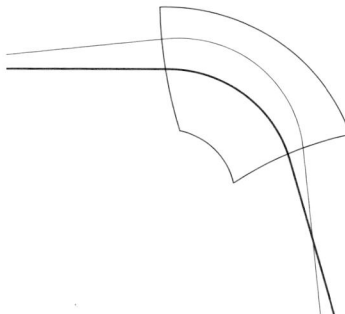

FIGURE 5.6. An example dipole for which the transfer map can be computed analytically, when fringe fields are neglected.

plane is denoted by s_i and the final plane by s_f. The entrance curve of the magnet relative to the entrance plane is given by the function $f_i(x)$ and the exit curve is described by the function $f_f(x)$. The projected motion through the magnet can therefore be given by successively computing three separate parts of the motion:

1. motion on a line from s_i to the curve f_i
2. motion on a circle from f_i to f_f
3. motion on a line from f_f to s_f.

Since the analytic solution is based on the geometry of the motion, we first have to transform from canonical coordinate $\vec{z}_i$ to geometric coordinates $\vec{x}_i$ containing positions and slopes. After that, the trajectory can be determined analytically by straightforward geometry. Finally, the geometric coordinates $\vec{x}_f$ have to be transformed back to the canonical notation $\vec{z}_f$.

As an example for the analytical calculation of high-order maps of bending magnets, we consider the aberrations of the so-called Browne–Buechner magnet. It consists of a single dipole magnet with a deflection angle of 90°. The entrance and exit pole faces are not tilted, but rather curved with a radius equal to the deflection radius. Therefore, the resulting dipole can be made from a simple circular magnet. The layout of a Browne–Buechner magnet is shown in Fig. 5.7.

The magnet is usually preceded and followed by equal drifts to obtain horizontal imaging, i.e., to make the final coordinates independent of initial directions in linear approximation. According to Barber's rule, this requires the drifts to have a length equal to the deflection radius R. It has been shown by basic yet elaborate geometrical arguments that this geometrical layout entails that all second-order geometric aberrations in the x–s projection vanish, which is very helpful for improving the resolution of the device (see Chapter 6, Section 6.3.1).

DA methods can now be used to obtain the high-order map of the system. To this end, we perform the various steps of the geometric calculations to determine the position of the particle as a function of the initial (x, a) coordinates. Similar

MAPS DETERMINED BY ALGEBRAIC RELATIONS

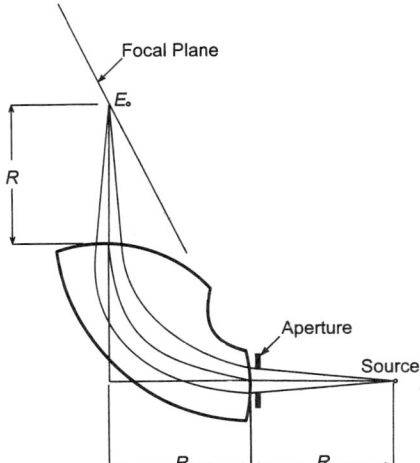

FIGURE 5.7. The layout of a Browne–Buechner magnet, showing the circular geometry of the pole faces, and the target and focal drifts.

to the previous section, they are merely evaluated with DA following Eq. (2.96) to obtain as many high-order derivatives as desired.

Table I shows the resulting aberrations for both position and angle in the focal plane to fifth order. The calculation has been performed for a deflection radius of 1 m; since this deflection radius is the only free parameter, and obviously

TABLE I
ABERRATIONS OF UP TO ORDER FIVE OF THE BROWNE–BUECHNER MAGNET. ALL SECOND-ORDER ABERRATIONS VANISH.

$(x_f \| x_i^{i_x} a_i^{i_a})$	$(a_f \| x_i^{i_x} a_i^{i_a})$	i_x	i_a
−1.000000000000000	−1.000000000000000	1	0
−0.500000000000000	−1.000000000000000	3	0
−2.000000000000000	−0.500000000000000	2	1
−3.000000000000000	−1.000000000000000	1	2
−2.000000000000000	−1.000000000000000	0	3
−0.375000000000000	0.000000000000000	5	0
−2.750000000000000	−0.125000000000000	4	1
−8.000000000000000	−1.000000000000000	3	2
−12.500000000000000	−2.500000000000000	2	3
−10.750000000000000	−3.000000000000000	1	4
−4.500000000000000	−1.750000000000000	0	5

the whole geometry scales with it, this device is representative of all Browne–Buechner magnets.

It is possible in principle to compute analytical formulas for the aberrations of higher orders by differentiating the analytical formulas relating final coordinates to initial coordinates. However, due to the complexity of the formulas, this quickly leads to cumbersome expressions. The use of DA, on the other hand, allows the calculation of higher-order aberrations as effortlessly as did the plain numeric evaluation of formulas with numbers, except for an increase of computation time and storage requirements.

To illustrate this point, we computed all aberrations of the Browne–Buechner magnet up to order 39. The results are shown in Table II, where for reasons of space, we restrict ourselves to the opening aberrations of the form (x, a^n).

All aberrations of even order vanish. Furthermore, the aberrations have a tendency to be quite benign, and there is much interesting asymptotic behavior, which is also exemplified in Table II.

TABLE II

ABERRATIONS OF UP TO ORDER 39 OF THE BROWNE–BUECHNER MAGNET. FOR REASONS OF SPACE, WE LIMIT OURSELVES TO THE SO-CALLED OPENING ABERRATIONS (x, a^n).

$(x_f \mid a_i^n)$	n
-1.000000000000000	1
$-.5000000000000000$	3
$-.3750000000000000$	5
$-.3125000000000000$	7
$-.2734375000000000$	9
$-.2460937500000000$	11
$-.2255859375000000$	13
$-.2094726562500000$	15
$-.1963806152343750$	17
$-.1854705810546870$	19
$-.1761970520019532$	21
$-.1681880950927736$	23
$-.1611802577972413$	25
$-.1549810171127320$	27
$-.1494459807872773$	29
$-.1444644480943681$	31
$-.1399499340914191$	33
$-.1358337595593185$	35
$-.1320605995715597$	37
$-.1285853206354660$	39

5.3.3 Drifts and Kicks

The method of kicks and drifts is an approximation for the motion through particle optical elements that is often employed in high energy accelerators, and often yields satisfactory results. To this end, the motion through the element under consideration that is governed by the ODE $d(x, y, d, a, b, t)/dt = \vec{f}(x, y, d, a, b, t)$ is assumed to begin with a drift to the middle of the element of length L, followed by a kick changing only momenta and given by

$$\Delta a = L \cdot f_a(x, y, d, a, b, t)$$
$$\Delta b = L \cdot f_a(x, y, d, a, b, t), \tag{5.75}$$

and finally conclude with another drift to the end of the element. The accuracy of this approximation depends on the actual amount of curvature of the orbit; if each element effects the motion by only a very small amount, then the force on the particle is reasonably approximated by the force experienced by a particle merely following a straight line. By evaluating the force in the middle of the trajectory, any variation over the course of the straight line is averaged out to some degree.

Apparently, the simple kick model described by Eq. (5.75) leads to a relationship of final coordinates in terms of initial coordinates by a sequence of elementary operations and functions, and, via Eq. (2.96), allows the computation of maps of the approximated motion to any order of interest.

5.4 MAPS DETERMINED BY DIFFERENTIAL EQUATIONS

For maps determined by differential equations, the DA methods introduced in Chapter 2 allow the determination of high-order nonlinearities in a convenient and rather straightforward way. DA methods can be used to construct algorithms that allow the computation of maps to arbitrary orders, including parameter dependence and arbitrary fields. In this section we outline various approaches to this goal, based both on conventional ODE solvers and on new algorithms intimately connected to the differential algebraic framework.

5.4.1 Differentiating ODE Solvers

Intuitively, the most direct method for obtaining Taylor expansions for the flow of an ODE is to recognize that a numerical **ODE solver describes a functional dependency** between initial conditions and final conditions. Thus, by replacing all arithmetic operations in it by the corresponding ones in the differential algebra $_nD_v$, we readily obtain the derivatives of order n in the v initial conditions.

While this method is practically straightforward and very robust, several comments and cautions are in order. First, if the step sizes in the original integration scheme are chosen to be sufficient to obtain the final conditions with sufficient accuracy, this does not necessarily guarantee any **accuracy of the derivatives** obtained by this approach. In fact, in most practical cases we observe that the accuracy of the derivatives decreases rapidly with order. One straightforward and robust way to circumvent this problem in practice is to utilize an integration scheme with automatic step size control, where the norm utilized in the accuracy control is replaced by the corresponding norm in Eq. (2.68). In this way, the automatic step size control assures that all derivatives in the components of the DA vector are determined with suitable accuracy. While this leads to a straightforward algorithm, it often entails significantly smaller step sizes than in the original algorithm, but is unavoidable in order to obtain accurate derivatives.

Another practical difficulty arises when parts of the functional dependencies in the integration scheme are **not differentiable**. This situation, that often arises when direct use of measured field data is made, can sometimes lead to very inaccurate or even wrong values for higher derivatives, unless suitable precautions are taken, as in the representation of the field by the smooth model in Eq. (3.20).

Altogether, the method is robust and straightforward to implement, but has potential pitfalls. Furthermore, its computational expense is proportional to the expense of the original integrator times the computational expense for the elementary DA operations, and for high orders and many variables can become significant. It has been implemented in the code COSY INFINITY (Berz, 1997a; Berz *et al.*, 1998) as well as a number of other recent codes (Yan, 1993; Yan and Yan, 1990; van Zeijts and Neri, 1993; Michelotti, 1990; Davis *et al.*, 1993).

5.4.2 DA Solvers for Differential Equations

Besides differentiation through the numerical integrator, the various DA-based ODE solvers discussed in Section 2.3.4 often represent significant advantages. The algorithm based on **fixed-point iteration** (Eq. 2.115), a natural approach within the DA framework, can outperform the integrator discussed in the previous section in speed, especially for ODEs for which the right-hand sides consist of a large number of arithmetic operations. Similar to the previously discussed integrators, it offers straightforward step size control via the use of the DA norm (Eq. 2.68).

For autonomous systems, which occur, for example, in the main field of particle optical elements, the use of the propagator (Eq. 2.120)

$$\vec{z}_f = \sum_{i=1}^{\infty} \frac{t^i \cdot L_{\vec{f}}^i}{i!} \mathcal{I}$$

is particularly suitable. In this case, the right-hand side $\vec{f}$ has to be evaluated only once per step, and each new order in the power series requires only the application of the derivation operation and is hence only a small multiple of the cost of evaluation of $\vec{f}$. Therefore, this method is most efficient at high orders and

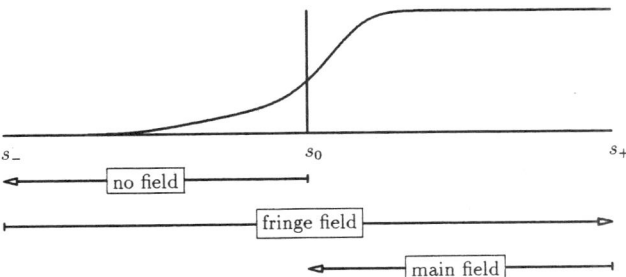

FIGURE 5.8. The fringe–field region of an optical element with effective edge at s_0. For computational purposes, effects of the s dependent field are usually concentrated in a map of length zero sandwiched between a pure drift map up to s_0 and a main-field map beginning at s_0.

large time steps. In the COSY INFINITY code, step sizes are fixed and orders adjusted dynamically, typically falling in a range between 25 and 30. Altogether, the DA-based ODE solvers are the main integration tools in the COSY INFINITY code (Berz, 1997a; Berz *et al.*, 1998) and represent the method of choice where applicable.

5.4.3 Fast Perturbative Approximations

In the following, we develop fast approximate methods for the computation of maps for which the use of the autonomous solver is not applicable, and which would otherwise require more time-consuming approaches. The tools are perturbative in that they are applicable for various systems near a reference system to be specified. In particular, the method allows efficient treatment of **fringe field effects**.

Because of the simplifications available for autonomous systems, the effect of a particle optical device is represented traditionally by a field-free drift and a main-field region. If effects of the s dependent parts of an element, so called fringe-field effects, are considered, they are represented by a fringe-field map, which is sandwiched between the drift and the main-field map at the position s_0. Hence, the fringe-field map consists of a negative drift to the region where the field vanishes, the map through the varying field, and the application of an inverse main-field map back to s_0 as shown in Fig. 5.8. So the fringe-field map represents the necessary corrections to the simple step function field model described by

$$\vec{B}_{mf}(x, y, s) = \begin{cases} \vec{B}(x, y, s_m) & \text{for } s \text{ in the main field} \\ \vec{0} & \text{for } s \text{ outside the main field} \end{cases}, \quad (5.76)$$

where s_m describes the center of the main field. The abrupt change of the field in this model of course violates Laplace's equation and cannot therefore represent a physical system. To describe a realistic system, the main-field map must be composed with fringe-field maps, which describe the connection of the main-field to the field-free region outside the element. As outlined in the previous section, the main-field map of optical elements can be computed more easily than that of the fringe fields; the situation is similar in older low-order codes utilizing analytically derived formulas (Matsuo and Matsuda, 1976; Wollnik, 1965; Brown, 1979a; Dragt *et al.*, 1985; Berz *et al.*, 1987).

As mentioned previously, the fringe-field map $\vec{M}_{ff}$ is defined as a correction to be inserted at the edge of the element, which can be formalized as follows. Let s_0 denote the effective field boundary at the entrance of the element, s_- a position so far before the optical device that the field can be neglected, and s_+ a position so far inside the element that $\vec{B}(x, y, s)$ changes very little with s. Let the map $\vec{M}_{mf, s_0 \to s_+}$ describe particle motion through the main field given in Eq. (5.76) from the effective field boundary to s_+. The fringe-field map is constructed in such a way that a drift $\vec{D}_{s_- \to s_0}$ from s_- to the effective field boundary composed first with $\vec{M}_{ff}$ and then with $\vec{M}_{mf, s_0 \to s_+}$ yields the transfer map $\vec{M}_{s_- \to s_+}$ from s_- to s_+:

$$\vec{M}_{s_- \to s_+} = \vec{M}_{mf, s_0 \to s_+} \circ \vec{M}_{ff} \circ \vec{D}_{s_- \to s_0} . \tag{5.77}$$

Hence the **fringe-field map** has the form

$$\vec{M}_{ff} = \vec{M}_{mf, s_0 \to s_+}^{-1} \circ \vec{M}_{s_- \to s_+} \circ \vec{D}_{s_- \to s_0}^{-1} . \tag{5.78}$$

Computing fringe-field maps requires the computation of the map $\vec{M}_{s_- \to s_+}$ of a system where $\vec{B}(x, y, s)$ and, hence, the differential equation depends on s. While the resulting computational effort is substantially greater, ignoring the fringe-field effects unfortunately leads to substantial errors in the nonlinearities of the element.

The importance of fringe fields becomes apparent when realizing that many nonlinear properties of an electric or magnetic field are due to the dependence of the field on s, as seen in Eq. (3.9). Many nonlinear contributions arise only because of the non-vanishing s derivatives. However, the functions involved and also their derivatives can be quite complicated; one frequently used model is the so-called Enge function, which has the form

$$E(s) = \frac{1}{1 + \exp(b_{k,0} + b_{k,1}(s/a) + b_{k,2}(s/a)^2 + \dots)} , \tag{5.79}$$

in which a is the aperture of the device under consideration, and $b_{k,l}$ are real coefficients modeling the details of the field's fall-off.

A careful treatment of fringe-field effects is imperative for a detailed study of nonlinear effects, and the advantages of very fast DA evaluation of propagators can be used most dramatically with efficient fringe-field approximations.

Any approximation for such fringe-field effects should satisfy various **requirements**. It should:

1. lead to order n symplectic maps
2. represent the s dependent element well for a wide range of apertures
3. be usable for arbitrary orders.

The simplest approximation, already described, is SCOFF, where s dependent fields are simply ignored. As illustrated previously, this method strongly violates the accuracy requirement and point 2. The impulse approximation (Helm, 1963) used in the code TRANSPORT (Brown, 1979a) violates points 2 and 3, and the method of fringe-field integrals (Hartmann, 1990; Wollnik, 1965) used in the computer code GIOS (Wollnik et al., 1988) violates points 1 and 3.

The general problem is to determine the transfer map of interest for a beam with reference particle of energy E, mass m, and charge q in a magnetic or electric particle optical device. The device is characterized by a size parameter A, a reference field strength F, and, possibly, additional parameters $\vec{\delta}$. Thus, the task is to find the transfer map

$$\vec{M}^{E,m,q,A,F}(\vec{z},\vec{\delta}),\qquad(5.80)$$

that will be achieved by relating it through a sequence of transformations to a suitably chosen previously computed and stored reference map. The **SYSCA** method (Hoffstätter, 1994; Hoffstätter and Berz, 1996) outlined in the following relies on two different scaling mechanisms, one based on geometric observations, and the other relying on the linear relation between rigidity of a particle and the field.

The first scaling method used in the determination of the map of a general element is based on a simple **geometric** observation. Assume a certain space-dependent field and an orbit through it are given. If the geometry of the field is scaled up by a factor α and, simultaneously, the field strength is scaled down by the factor α, then the original orbit can be scaled up by the factor α along with the geometry of the device. The reason for this phenomenon is that the momentary radius of curvature of the orbit, which is geometric in nature, is inversely proportional to the field, which follows directly from the Lorentz equation. If a trajectory described by $x(t)$ and $p(t)$ satisfies

$$\frac{d\vec{p}}{dt}=q\left(\frac{d\vec{x}}{dt}\times\vec{B}(\vec{x})+\vec{E}(\vec{x})\right),\qquad(5.81)$$

then a scaled trajectory $\vec{X}(t)=\alpha\vec{x}(t/\alpha)$ and $\vec{P}=\vec{p}(t/\alpha)$ satisfies

$$\frac{d\vec{P}}{dt}=q\left(\frac{d\vec{X}}{dt}\times\frac{1}{\alpha}\vec{B}(\vec{X}/\alpha)+\frac{1}{\alpha}\vec{E}(\vec{X}/\alpha)\right).\qquad(5.82)$$

A limitation to this approach is that the shape of the field is not only determined by the geometry generating it, but also by possible saturation effects that depend

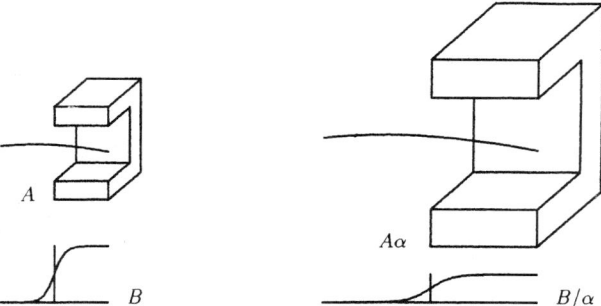

FIGURE 5.9. If the size of an element is scaled by a factor α and simultaneously the strength of the field is scaled by a factor $1/\alpha$, then coordinates of particle trajectories scale with the factor α.

on the strength of the field. This limits the allowed factor α to a size where changes of saturation can be ignored.

It is also important to note that once the ratio of length to aperture of a device exceeds a certain minimum, the exact shape of the fringe field is almost unaffected by the actual length of the device. Therefore, for the treatment of fringe fields, in good approximation the aperture of the element can be used as size parameter A.

For the actual use of the method, it is important to observe that only geometric quantities associated with the particle, like positions, slopes, and lengths of trajectories, scale with the geometry, while the canonical momenta do not scale properly. Thus before scaling it is necessary to transform the map to purely **geometric coordinates** like those used in the TRANSPORT code (Brown, 1979a). The transformation between these two sets of coordinates depends on the energy E_0 and the mass m_0 of the reference particle, and is denoted by $\vec{T}(E_0, m_0)$. The coordinates are denoted by $\vec{z}_{E_0} = (x, a, y, b, \delta_E, \tau)$ and $\vec{z}_{p_0} = (x, x', y, y', l, \delta_p)$ and the transformation is given by

$$x' = \frac{a}{\sqrt{(\frac{p}{p_0})^2 - a^2 - b^2}}, \quad y' = \frac{b}{\sqrt{(\frac{p}{p_0})^2 - a^2 - b^2}},$$

$$l = l_i - \frac{v}{v_0}\frac{2+\eta_0}{1+\eta_0}(\tau - \tau_i) + \left(\frac{v}{v_0} - 1\right)s,$$

$$\delta_p = \frac{\sqrt{(E_0(1+\delta_E))^2 + 2E_0 mc^2(1+\delta_E)}}{p_0 c} - 1, \qquad (5.83)$$

which has the inverse

$$a = \left(\frac{p}{p_0}\right)\frac{x'}{\sqrt{(1+x'^2+y'^2)}}, \quad b = \left(\frac{p}{p_0}\right)\frac{y'}{\sqrt{1+x'^2+y'^2}},$$

$$\tau = \tau_i - \frac{v_0}{v}\frac{1+\eta_0}{2+\eta_0}\left(l - l_i - \left(\frac{v}{v_0}-1\right)s\right),$$

$$\delta_E = \frac{\sqrt{(p_0 c(1+\delta_p))^2 + (mc^2)^2} - mc^2}{E_0} - 1. \tag{5.84}$$

The change of coordinates can be performed conveniently to arbitrary order in DA following Eq. (4.5). It is worthwhile to point out that in order to perform the transformation, the knowledge of the total arc length s of the system under consideration as well as the mass and energy of the reference particle is required; thus, these quantities have to be stored along with the reference map. In geometric coordinates, the map $\vec{S}_\alpha$ performing the scaling is characterized by

$$\begin{array}{lll} x_2 = x_1 \alpha, & y_2 = y_1 \alpha, & l_2 = l_1 \alpha, \\ x_2' = x_1', & y_2' = y_1', & \delta_{p2} = \delta_{p1}; \end{array} \tag{5.85}$$

if there are any parameters $\vec{\delta}$, their scaling behavior must also be considered.

The geometric scaling thus allows expression of a map that is associated with the size scale A, in terms of a stored reference map whose size scale is $A^* = \alpha \cdot A$. This is accomplished by the transformations

$$\vec{M}^{E,m,q,F,A}(\vec{z}_E,\vec{\delta}) = \vec{T}^{-1} \circ \vec{S}_\alpha^{-1} \circ \vec{T} \circ \vec{M}^{E,q,m,A^*,F/\alpha}(\vec{z}_E,\vec{\delta}) \circ \vec{T}^{-1} \circ \vec{S}_\alpha \circ \vec{T}. \tag{5.86}$$

Since the stored reference map has to be evaluated at F/α, it is clear that the reference map has to be known as a function of the field strength.

The next step is to **transform the properties of the reference particle** to those stored in the reference file. To this end, the rigidity χ of the reference particle of interest is computed and compared to that of the stored map χ^*; since electric and magnetic rigidities depend on the quantities E, q, m in different ways, this requires that electric and magnetic fields not be present simultaneously. In the following, we discuss the case of the magnetic field in detail and only briefly mention the electric case. Let $\beta = \chi^*/\chi$ be the ratio of the rigidity associated with the stored map and the map under consideration. Because of $v\vec{p}' = q\vec{v} \times \vec{B}$, a simultaneous scaling of magnetic rigidity and magnetic field has no influence on the orbit, we have $\vec{M}^{E,m,q,A^*,F/\alpha}(\vec{z}_p,\vec{\delta}) = \vec{M}^{E^*,m^*,q^*,A^*,F\cdot\beta/\alpha}(\vec{z}_p,\vec{\delta})$. The change of a trajectory induced by a relative energy deviation δ_E depends on the energy E of the reference particle. This cannot be true, however, for δ_p. Due to the scaling law for magnetic rigidity, a relative momentum deviation δ_p creates the same changes in a trajectory, no matter which reference momentum p is used. Thus, the full transformation to the reference map is obtained as

$$\vec{M}^{E,m,q,A,F}(\vec{z}_E,\vec{\delta}) = \vec{T}^{-1}(E,m) \circ \vec{S}_\alpha^{-1} \circ \tag{5.87}$$

$$\{\vec{T}(E^*,m^*) \circ \vec{M}^{E^*,q^*,m^*,A^*}(\vec{z}_E,\vec{\delta},\vec{F}) \circ \vec{T}^{-1}(E^*,m^*)\}|_{\vec{F}=F\cdot\beta/\alpha} \circ \vec{S}_\alpha \circ \vec{T}(E,m).$$

Besides the dependence of the original map on the parameters $\vec{\delta}$, its dependence on the field F has to be known. While the exact dependence of the map on the field is usually difficult to obtain, the DA method conveniently allows determination of the expansion in terms of the field strength.

For an electric field of strength F the trajectory does not change if the quantity Fq/v_p does not change, due to $v\vec{p}' = q\vec{E}$. Therefore, TRANSPORT coordinates $\vec{z}_p$ with δ_p are not appropriate but δ_{vp} must be used, which denotes the relative deviation of vp from the corresponding value for the reference particle.

Even though this expansion can be obtained to rather high order, it still represents a source of errors, and hence a study of the influence of these errors is warranted. The first question is that of the accuracy of the expansion; for this purpose we analyze the range of values the quantity

$$\bar{F} = F \cdot \frac{\beta}{\alpha} \tag{5.88}$$

can assume; this establishes how far away from a chosen expansion point the extrapolation will have to be made. However, as it turns out, the quantity $\bar{F}$ has a rather simple geometric meaning. Note that both in the process of size scaling and in the process of rigidity scaling, the total deflection of a particle transversing the field at the aperture is not changed. Thus, the quantity $\bar{F}$ plays the role of a universal strength parameter.

The other important consideration in the process of approximating the field dependence in terms of an expansion is the question of **symplecticity**. Even if the errors produced by inserting into the truncated polynomial are minor, without additional considerations they violate the symplectic symmetry of the map.

This problem can be avoided by storing a symplectic representation of the map. For the nonlinear part of the map, it appears advantageous to choose a one-operator **flow representation** (Eq. 4.79) using the pseudo–Hamiltonian, which can be determined from the map

$$\vec{M}(\vec{z}_E, \vec{\delta}, F) = L(\vec{\delta}, F) e^{:P(\vec{z}_E, \vec{\delta}, F):}. \tag{5.89}$$

On the other hand, for the linear part this representation just contains the linear matrix in ordinary nonsymplectic form. To preserve the symplecticity of the linear part, it is advantageous to represent it in terms of a **generating function** (Eq. 4.81), which again can be conveniently calculated from the map.

There are a variety of ways to extend the method of symplectic scaling. First, note that the required knowledge of the dependence of the map on a relative field change δ_F can be indirectly obtained from the conventional map itself. To this end, we can express the chromatic dependence of the map, not in terms of the conventional energy deviation δ_E, but in terms of the momentum deviation δ_p, and substitute

$$\vec{M}^{E,m,q,A,F}(\vec{z}_E, \delta_F) = \vec{T}(E, m) \circ \vec{M}^{E,m,q,A,F}(x, x', y, y', l, \delta_p/(1+\delta_F)). \tag{5.90}$$

This observation is particularly useful for the practical use of the method since it does not require any special setup to determine the stored map.

Similarly, the two scaling properties can be used to compute the map's dependence on a relative change of aperture δ_A, mass δ_m, and charge δ_q. Therefore, no special parameters $\vec{\delta}$ have to be introduced to compute these dependencies if they are desired, and we write the most general parameter dependent map as $\vec{M}^{E,m,q,A,F}(\vec{z}_E, \vec{\delta}, \delta_m, \delta_q, \delta_A, \delta_F)$.

Another possible extension is the use of the parameters $\vec{\delta}$ for the description of other degrees of freedom of the field under consideration. These could include details about the fall off, or about the geometric form of the effective field boundary. The last topic is of great practical importance and can also be treated conveniently by rotating the fringe-field map of a wedge dipole appropriately, as discussed in the following.

Finally, for practical considerations it is sufficient to store only maps of entrance fringe fields. Given the entrance fringe field map $\vec{L} \circ (\exp(: P :)\vec{z})$, the exit fringe-field map is described by the reversed map in (Eq. 4.8). In addition, if a representation approximates maps of 2ν poles with fields close to B, then a rotation by an angle of $180/\nu°$ allows approximations close to the field $-B$.

To avoid the computation of a reference file for different possible **edge angles and curvatures**, a further approximation is used. The effect of the edge shape is approximated by first applying the fringe-field map of a straight edge dipole, and then taking curvatures up to second order into account analytically. Higher-order curvatures are evaluated with nonlinear kicks.

The following table demonstrates the applicability of this approximation. We used a dipole of radius 2 m, a bend angle of 30°, and an aperture of one inch. The Taylor coefficients (a, xa) and (x, xaa) were computed with SYSCA (left) and

angle	(a, xa)m	
5°	$-0.1360314E - 02$	$-0.1368305E - 02$
10°	$-0.7231982E - 03$	$-0.7253955E - 03$
15°	$-0.1033785E - 06$	$-0.2891228E - 08$
20°	$-0.1718367E - 02$	$-0.1721275E - 02$
25°	$-0.9628906E - 02$	$-0.9643736E - 02$
30°	$-0.2977263E - 01$	$-0.2981580E - 01$
	(x, xaa)	
5°	$-0.6524050E - 01$	$-0.6520405E - 01$
10°	0.5269389E-03	0.5795750E-03
15°	0.7179696E-01	0.7179678E-01
20°	0.1463239	0.1461203
25°	0.2187553	0.2180152
30°	0.2775432	0.2755090

with accurate numerical integration in DA (right) for different edge angles. In all examples the entrance and the exit edge angle are equal.

To approximate the effects of **superimposed multipoles**, we compute fringe-field maps of the separate multipoles via SYSCA. These fringe-field maps are then composed to a single map, to approximate the fringe-field map of the instrument with superimposed multipoles. While this approximation may appear crude, it is often quite accurate, as shown in the following table. SYSCA (left) and DA integration (right) was used to compute coefficients of the fringe-field map of a magnetic quadrupole that is superimposed with a magnetic hexapole. The instrument has an aperture of one inch and was 0.5 m long. The pole-tip field of the quadrupole was $1T$ and the pole-tip field B_H of the hexapole is given in the table.

B_H/T	(a, xa)m		(x, aaa)/m	
0.0	0.00000	0.00000	0.0109946	0.0109946
0.1	−2.54411	−2.54412	0.0400742	0.0400755
0.2	−5.08823	−5.08824	0.1273131	0.1273182
0.3	−7.63235	−7.63236	0.2727113	0.2727228
0.4	−10.17647	−10.17649	0.4762688	0.4762891
0.5	−12.72058	−12.72061	0.7379855	0.7380173

For a variety of examples showing the validity and accuracy of the SYSCA approach for various specific systems, refer to (Hoffstätter, 1994; Hoffstätter and Berz, 1996).

Chapter 6

Imaging Systems

6.1 INTRODUCTION

Imaging systems are devices used for the purpose of **measuring position, momentum, energy**, or **mass** of charged particles. **Microscopes** provide detailed information of initial positions by magnifying these into more easily measurable final conditions. **Momentum spectrometers** are used for the determination of the momentum distribution of nuclear reaction products. Most momentum spectrometers are magnetic because the energies that need to be analyzed are too high to allow sufficient deflection by electric fields. In addition, magnets have two more advantages: They automatically preserve the momentum, and they can be built big enough to achieve large acceptance.

Mass spectrometers are used mainly for the analysis of masses of molecules, and they can be operated at much lower energies. They have a long history, and their applications pervade many disciplines from physics and chemistry to biology, environmental sciences, etc. (Berz *et al.*, 1999; Watson, 1985; White and Wood, 1986; Fenselau, 1994; Gijbels *et al.*, 1988; Glish *et al.*, 1988; Cotter, 1994). Mass spectrometers are also more diverse; the **major types** include sector field, quadrupole, accelerator, energy loss, time-of-flight, Fourier transform ion cyclotron resonance, and ion trap. A detailed overview of the various types of imaging systems can be found in (Berz *et al.*, 1999).

For all the different types of imaging systems, the goal is to achieve **high resolution** and in many cases **large acceptance** at the same time. As resolution improves, the need for better understanding and correction of high-order aberrations increases. In this chapter, the linear theory of various types of systems will be outlined, followed by the study of aberrations and their correction.

In order to use the **final position** as a measure of initial position, momentum, energy, or mass of the particle, it is necessary that the final position be independent of other quantities, and particularly that the device be focusing such that

$$(x|a) = 0. \tag{6.1}$$

In microscopes, it is desirable to have (x, x) as large as possible; on the other hand, for spectrometers it is useful to have (x, x) as small as possible and the dependence on the spectroscopic quantity of interest δ, the so-called **dispersion**

$$(x, \delta) \qquad (6.2)$$

as large as possible. Finally, in a mass spectrometer in which particles of different energies are present, the dependence on energy (x, d) should vanish. In the map picture, the linear behavior is thus given by the transfer matrix of the horizontal motion:

$$M = \begin{pmatrix} (x|x) & 0 & (x|\delta) \\ (a|x) & (a|a) & (a|\delta) \\ 0 & 0 & 1 \end{pmatrix}. \qquad (6.3)$$

Let $2D_i$ be the width of the source. From Eq. (6.3), it is clear that the particles to be detected focus around a spot at $(x|\delta)\delta$ with a width of $|2(x|x)D_i|$. Hence, the distance between the centers of particles of different energies must be larger than the width, i.e.,

$$|(x|\delta)\delta| > |2(x|x)D_i|. \qquad (6.4)$$

This sets an upper bound for $1/\delta$, which we call the linear resolving power (or linear resolution),

$$R_l = \left(\frac{1}{\delta}\right)_{\max} = \left|\frac{(x|\delta)}{2(x|x)D_i}\right|. \qquad (6.5)$$

In order to increase the resolution, it is necessary to increase $|(x|\delta)|$ and/or decrease $|D_i|$.

As an example, we study the first broad-range momentum spectrometer, the **Browne–Buechner** spectrometer (Browne and Buechner, 1956). It contains only a homogeneous dipole with 90° bending and circular pole boundaries. The reason why this layout is chosen will be shown in Section 6.2. The layout is depicted in Fig. 6.2, and it is applicable for particles with energies up to 25 MeV/u. As one can see from Barber's rule in Fig 6.1 (or simple matrix algebra), the system is x focusing.

In general, it is not hard to obtain the first-order map by hand, and it is very easy to obtain it using a computer code; the result is shown in Table I. With the typical assumption that the half width D_i is 0.25 mm, the resulting linear energy resolution is

$$R_l = \frac{1}{\delta_{\min}} = \left|\frac{(x|\delta)}{2(x|x)D_i}\right| \approx 1000. \qquad (6.6)$$

Since all electric and magnetic devices produce nonlinear terms in the map called **aberrations**, their impact on the resolution has to be studied whenever necessary. The nonlinear effects are very important in the case of the momentum

INTRODUCTION

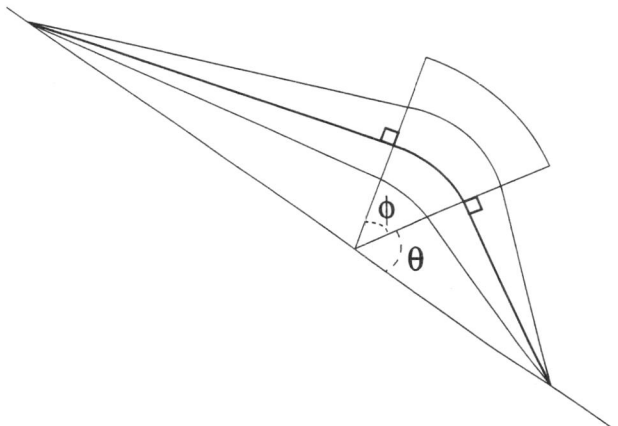

FIGURE 6.1. Barber's Rule

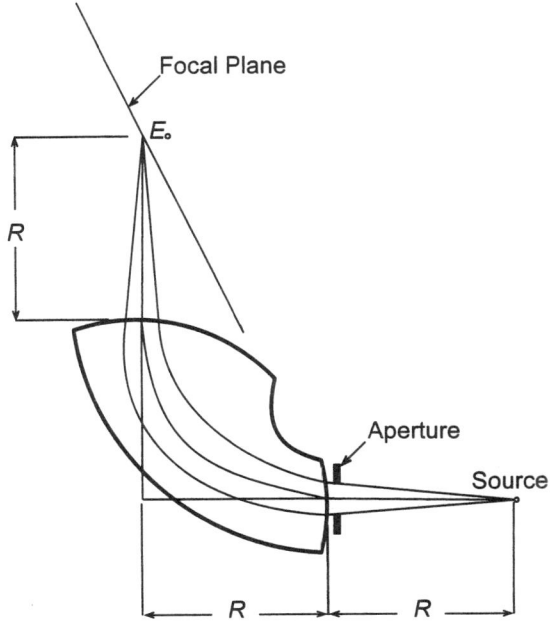

FIGURE 6.2. The Browne-Buechner spectrograph

TABLE I

THE FIRST ORDER MAP OF THE BROWNE-BUECHNER SPECTROGRAPH. THE COLUMNS REPRESENT x_f, a_f, y_f AND b_f, RESPECTIVELY, AND THE INTEGERS ON THE RIGHT CORRESPOND TO THE RESPECTIVE EXPONENTS IN THE INITIAL VARIABLES.

```
-1.000000      -1.950458       0.0000000E+00 0.0000000E+00 100000
0.0000000E+00  -1.000000       0.0000000E+00 0.0000000E+00 010000
0.0000000E+00  0.0000000E+00   1.000000      0.0000000E+00 001000
0.0000000E+00  0.0000000E+00   1.830747      1.000000      000100
0.0000000E+00  0.0000000E+00   0.0000000E+00 0.0000000E+00 000010
0.5194406      0.5065736       0.0000000E+00 0.0000000E+00 000001
```

spectrometers due to their large angular and momentum acceptances. Considering the aberrations, the final width will be a new value Δx_{ab} instead of $|(x|x)D_i|$, which has as an upper bound

$$\Delta x_{ab} = (2|(x|x)D_i| + |(x|x^2)|D_i^2 + |(x|xa)D_i A_i| + \cdots) \quad (6.7)$$

where A_i is the half-width of the spread in the quantity a. Therefore, the actual resolution R_{ab} is

$$R_{ab} = \frac{|(x|\delta)|}{\Delta x_{ab}}. \quad (6.8)$$

A detailed study of aberrations and methods of their corrections is presented in Section 6.2.

A parameter often used as a comprehensive quality indicator for a spectrometer is the so-called **Q value**:

$$Q = \frac{\Omega \ln(p_{\max}/p_{\min})}{\ln 2}, \quad (6.9)$$

where Ω is the nominal solid angle from which a reasonable resolution is expected. The Q value shows both the **geometric** and **momentum acceptance** of a spectrometer. For example, the Ω and $p_{\max}/p_{\min}$ for the Browne–Buechner spectrometer are 0.4 msr and 1.5, respectively. Large Ω translates into high intensity, which is important for nuclear studies and other situations in which there are a small number of available particles. Large momentum acceptance can reduce the number of exposures to cover a certain momentum range.

6.2 ABERRATIONS AND THEIR CORRECTION

For many imaging systems, **nonlinear effects** are a concern in design. For example, in the case of the **Browne–Buechner** spectrometer, the linear energy resolution obtained was approximately 1000. When aberrations are considered, the resolution calculated to the eighth order drops sharply to 61, which is far below the resolution actually desired, and which shows the importance of aberrations. Thus, it is imperative to study them carefully and to correct the prominent ones.

Since the entrance slit D_i is usually small and the solid angle large, only the angle and dispersion aberrations are important. Both map calculations and geometric considerations show that all the terms $(x|x^m a^n)$ $(m+n$ even) vanish. Since $(x|b^2)$ is small (Table II), the only second-order term that has a strong impact is $(x|a\delta)$, which can be as large as 8 mm when $a = 40$ mrad and $\delta = 20\%$. In fact, this is the most important factor that causes the decrease of the resolution. This becomes apparent when the resolution considering $(x|a\delta)$ is calculated:

$$R_{ab} = \frac{(x|\delta)}{(x|a\delta)a_i\delta} = 63. \tag{6.10}$$

Fortunately, $(x|a\delta)$ is easy to correct because it only causes a tilt of the focal plane, as we show now. Suppose a particle of energy $E_0(1+\delta)$ starts from the origin with slope a_0 and goes through an angle-focusing system. The final position and angle to first order are, respectively,

$$x_1 = (x|\delta)\delta$$
$$a_1 = (a|a)a_0 + (a|\delta)\delta$$

Taking into account $(x|a\delta)$, the result becomes

$$\tilde{x}_1 = (x|\delta)\delta + (x|a\delta)a_0\delta,$$
$$\tilde{a}_1 = (a|a)a_0 + (a|\delta)\delta.$$

Consequently, the system is not focusing anymore. Now consider a second particle of the same energy starting from the same point but with a different angle $a_0 + \Delta a_0$. The differences in final position and angle between the two particles are

$$\Delta x_1 = (x|a\delta)\Delta a_0 \delta,$$
$$\Delta a_1 = (a|a)\Delta a_0.$$

The fact that $\Delta x_1 / \Delta a_1$ is independent of Δa_0 indicates that particles of energy $E_0(1+\delta)$ are focusing at

$$\Delta z = -\frac{\Delta x_1}{\Delta a_1} = -\frac{(x|a\delta)\delta}{(a|a)},$$

which is proportional to δ. Therefore, the tilting angle is

$$\tan \psi = \frac{\Delta z}{x_1} = -\frac{(x|a\delta)}{(a|a)(x|\delta)}, \tag{6.11}$$

where ψ is the angle between the normal to the focal plane and the z-axis.

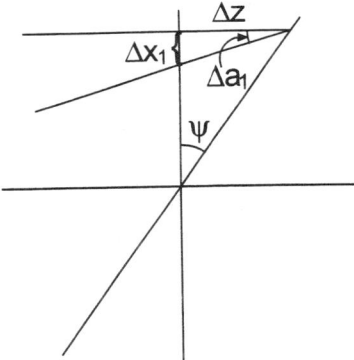

FIGURE 6.3. The Effect of the Aberration $(x|a\delta)$

Furthermore, the correction of $(x|a\delta)$ increases the resolution in certain circumstances. When Δx_{ab} in Eq. (6.7) is smaller than the detector resolution Δx_d, Δx_d becomes the limitation of the momentum resolution and is independent of ψ. Since the distance between two peaks increases by a factor of $1/\cos\psi$ while Δx_d remains unchanged, the resolution is

$$R_{ab} = \frac{(x|\delta)}{\Delta x_d \cos\psi}, \qquad (6.12)$$

which is greater than the linear resolution.

Rigorous computation of the actual resolution requires that the aberrations on the tilted focal plane be calculated. For the Browne–Buechner spectrometer, eighth-order maps of both straight and tilted focal planes are computed. Table II shows the aberrations on the straight focal plane, where $(x|a\delta)$ is clearly the major contributor. Table III contains the aberrations on the tilted focal plane, where $(x|a\delta)$ vanishes and others are either reduced or unchanged. The resolution after the cancellation of $(x|a\delta)$ recovers back to 780 (or 1560 in momentum), which is quite close to the linear resolution.

When a higher resolution is required, more aberrations have to be corrected. Usually $(x|a^2)$ and $(x|b^2)$ are corrected first. Then $(x|a^3)$, $(x|a^2b)$, and $(x|ab^2)$ are tackled. If necessary, fourth-order terms such as $(x|a^4)$ also have to be minimized. At least in one instance, eighth-order terms such as $(x|a^5b^3)$ are also corrected. This is done in the QDQ spectrometer at the Indiana University Cyclotron Facility (IUCF), where the pole faces of the last quadrupole are shaped to produce quadrupole, sextupole, octupole and decapole field components.

There are many different approaches to hardware correction, which usually involve the introduction of magnetic multipoles or edge effects of the same order as the terms to be corrected, and careful optimization. Because of the wide variety of approaches that are practically feasible in various situations, we refrain from an attempt to characterize them here, but refer to (Berz et al., 1999) and other literature.

TABLE II

HORIZONTAL (x) ABERRATIONS OF THE BROWNE–BUECHNER SPECTROMETER AT THE STRAIGHT FOCAL PLANE EXCEEDING 10μM. THE PARAMETERS USED ARE $x_{max} = 0.23$ MM, $a_{max} = 40$ MRAD, $y_{max} = 1$ MM, $b_{max} = 10$ MRAD AND $\delta_{max} = 20\%$.

I	COEFFICIENT	ORDER	EXPONENTS					
1	-.2300000000000000E-03	1	1	0	0	0	0	0
2	0.1038881145421565	1	0	0	0	0	0	1
3	0.4660477149345817E-04	2	1	0	0	0	0	1
4	0.8311049163372523E-02	2	0	1	0	0	0	1
5	-.5127000000000000E-04	2	0	0	0	2	0	0
6	-.1025577239401090E-01	2	0	0	0	0	0	2
7	-.6562560000000000E-04	3	0	3	0	0	0	0
8	0.3324419665349009E-03	3	0	2	0	0	0	1
9	-.1873001321765092E-02	3	0	1	0	0	0	2
10	0.1038881145421565E-04	3	0	0	0	2	0	1
11	0.1544932687715805E-02	3	0	0	0	0	0	3
12	0.4654187531488613E-04	4	0	3	0	0	0	1
13	-.1338622665642800E-03	4	0	2	0	0	0	2
14	0.4380445064894873E-03	4	0	1	0	0	0	3
15	-.2577160791614789E-03	4	0	0	0	0	0	4
16	0.4622130002260186E-04	5	0	2	0	0	0	3
17	-.9970354551245065E-04	5	0	1	0	0	0	4
18	0.4511716453676560E-04	5	0	0	0	0	0	5
19	0.2219821460952441E-04	6	0	1	0	0	0	5

6.3 RECONSTRUCTIVE CORRECTION OF ABERRATIONS

In many applications the number of particles to be studied is so small that it becomes of prime importance to collect and analyze as many of them as possible. In this situation, it is necessary to maximize the angular acceptance as well as the momentum acceptance, which is often referred to as the "momentum bite." While such large acceptances increase the count rate of particles that can be analyzed, at the same time they almost inevitably introduce large aberrations. Higher-order aberrations that scale with high powers of the phase space variables are usually substantially enhanced. One important category of such spectrometers are those used for the study of nuclear processes that can be found in many nuclear physics laboratories.

Hence, such devices pose **particularly challenging correction problems**; in many cases they are so severe that it is nearly impossible to correct them by hardware methods as discussed in ((Berz et al., 1999)), and an entirely different approach becomes necessary. Furthermore, in some cases is it also desirable to not only determine the initial value of δ, but knowledge of the initial values of a and b is also required.

The questions of correction of aberrations and the determination of the initial angles can be analyzed in an elegant way using the method of trajectory recon-

TABLE III
Aberrations of the Browne–Buechner Spectrometer at the Tilted Focal Plane That Are 10 μm or Larger. The Parameters Used Are $x_{max} = 0.23$ mm, $a_{max} = 40$ mrad, $y_{max} = 1$ mm, b_{max} 10 mrad and $\delta_{max} = 20\%$.

I	COEFFICIENT	ORDER EXPONENTS						
1	-.2300000000000000E-03	1	1	0	0	0	0	0
2	0.1038881145421565	1	0	0	0	0	0	1
3	-.9320954298691634E-04	2	1	0	0	0	0	1
4	-.5127000000000000E-04	2	0	0	0	2	0	0
5	0.1079501821087351E-01	2	0	0	0	0	0	2
6	-.6562560000000000E-04	3	0	3	0	0	0	0
7	0.3324419665349009E-03	3	0	2	0	0	0	1
8	-.2105079060488440E-03	3	0	1	0	0	0	2
9	-.1038881145421566E-04	3	0	0	0	2	0	1
10	0.1654199766297046E-02	3	0	0	0	0	0	3
11	-.1329767866139604E-04	4	0	3	0	0	0	1
12	0.5138469075870278E-04	4	0	2	0	0	0	2
13	-.2242022047923136E-04	4	0	1	0	0	0	3
14	0.1745846601755523E-03	4	0	0	0	0	0	4
15	0.1305877080464639E-04	5	0	2	0	0	0	3
16	0.2771149616039620E-04	5	0	0	0	0	0	5

struction. For the purposes of simplicity in notation, in the following we base the discussion on the case of spectrographs where the primary quantity of interest is the initial energy d; but the arguments also apply for imaging devices by merely exchanging the energy by the initial position x.

6.3.1 Trajectory Reconstruction

In a conventional spectrometer, a particle passes through a target of negligible thickness, passes through the spectrometer, and finally strikes a detector where its position is recorded. This position is then used to infer the particle's energy, a process which, as discussed in the previous section, is limited by the aberrations as well as the resolution of the detector.

Besides its final position, usually nothing is known about the actual orbit the particle followed within the spectrometer, and hence it is not possible to determine to what extent its final position was influenced by any aberrations in the system. On the other hand, if the orbit of the particle is known in detail, one can compute the effects of the aberrations on the position of the particle and possibly correct for their influence. This principle is applied in many detectors used in particle physics, in which usually the actual orbits of the particles are recorded and irregularities in the fields that lead to aberrations in the conventional sense are quite acceptable as long as they are known.

While in conventional nuclear physics spectrometers it is not possible to record the actual orbit of the particle throughout the system, by using a second set of detectors, it is possible to simultaneously determine **position and angle** at which it emerges from the spectrometer up to the resolution of the detectors. In addition to these data, it is known that at the target position the particle passed through the target slit, enabling the determination of its initial position in the deflecting plane up to the width of the slit.

There are currently five parameters available to describe the orbit of the particle, namely, the vector

$$(x_i, x_f, a_f, y_f, b_f). \tag{6.13}$$

If the details of the fields inside the device are known, these parameters actually allow the determination of the entire trajectory of the particle and with it the particle's energy as well as the initial angles. An intuitive way to understand this fact is based on **backwards integration** of the motion of the particle through the spectrometer, which requires the knowledge of positions and momenta at the exit as well as the particle's energy. While the first four of these parameters are known from the measurement, the last one is not. However, since all orbits integrated backwards with the same initial position and angle strike the target plane at a different position depending on their energy, the knowledge of the initial position can be used to implicitly determine the particle's energy. In practice this approach requires iterative "shooting" in which the particle's energy is varied in the integration until it is found to hit the required target position. While practically possible, in this form the method is apparently rather cumbersome and requires substantial calculation to determine the energy and reaction angles of just one particle.

There is an elegant way to **directly** infer this information from the knowledge of a **high-order transfer map** of the spectrometer (Berz et al., 1993). Such a map relates initial conditions at the reaction plane to final conditions at the detector via

$$\begin{pmatrix} x_f \\ a_f \\ y_f \\ b_f \\ d_f \end{pmatrix} = \mathcal{M} \begin{pmatrix} x_r \\ a_r \\ y_i \\ b_r \\ d_r \end{pmatrix}, \tag{6.14}$$

where $d_f = d_r$ because no change of energy occurred in the element proper. Since all particles reaching the detector have passed through the target slit, it is known that x_r is always very small. We approximate it by setting $x_r = 0$ and insert this in the equation. In addition, the equation which states that $d_r = d_r$ is dropped. We obtain the approximation

$$\begin{pmatrix} x_f \\ a_f \\ y_f \\ b_f \end{pmatrix} = \mathcal{S} \begin{pmatrix} a_r \\ y_i \\ b_r \\ d_r \end{pmatrix}, \tag{6.15}$$

the accuracy of which is limited by the slit width. This nonlinear map relates the quantities (x_f, a_f, y_f, b_f), which can be measured in the two planes, to the quantities of interest, a_r, b_r, and d_r, which characterize the nuclear reaction being studied. In addition, they depend on the initial position y_i which has no effect on the reaction processes and therefore in most cases is of little interest.

The map $\mathcal{S}$ is not a regular transfer map, and in particular its linear part does not have to be a priori nonsingular. Specifically, in a particle spectrometer with point-to-point imaging in the dispersive plane, the linear part S of the map $\mathcal{S}$ has the form

$$\begin{pmatrix} x_f \\ a_f \\ y_f \\ b_f \end{pmatrix} = \begin{pmatrix} 0 & 0 & 0 & * \\ * & 0 & 0 & * \\ 0 & * & * & 0 \\ 0 & * & * & 0 \end{pmatrix} \cdot \begin{pmatrix} a_r \\ y_i \\ b_r \\ d_r \end{pmatrix} \qquad (6.16)$$

where the asterisks denote nonzero entries. The term (x, a) vanishes because the spectrometer is imaging, and all the other terms vanish because of midplane symmetry. The element $(x|d)$ is usually maximized in spectrometer design, and $(a|a)$ cannot vanish in an imaging system because of symplecticity (Eq. 1.4.5). In fact, to reduce the effect of the finite target, $(x|x)$ is often minimized within the constraints, and therefore $(a|a) = 1/(x|x)$ is maximized.

Because of symplecticity, we have $(y|y)(b|b) - (y|b)(b|y) = 1$, and so we obtain for the total determinant of S

$$|S| = -(x|d) \cdot (a|a) = -\frac{(x|d)}{(x|x)}. \qquad (6.17)$$

The size of the determinant is a measure for the physical quality of the spectrometer; indeed, the linear resolution

$$R_l = \frac{(x|d)}{(x|x) \cdot x_t} = -\frac{|S|}{x_t} \qquad (6.18)$$

equals $|S|$ divided by the thickness x_t of the target.

In particular, this entails that the linear matrix S is invertible, and according to Section (4.1.1), this means that the whole nonlinear map $\mathcal{S}$ is invertible to arbitrary order. Thus, it is possible to compute the initial quantities of interest as a function of the **measurable final quantities** to arbitrary order via the map

$$\begin{pmatrix} a_r \\ y_i \\ b_r \\ d_r \end{pmatrix} = \mathcal{S}^{-1} \begin{pmatrix} x_f \\ a_f \\ y_f \\ b_f \end{pmatrix}; \qquad (6.19)$$

compared to the numerical shooting method, the determination of the relevant quantities is now direct and can be performed with a few polynomial operations instead of extensive numerical integration.

A closer inspection of the algorithm to invert maps shows that the result is multiplied by the inverse of the linear matrix S. Since the determinant of this inverse is the inverse of the original determinant and is thus small for a well-designed spectrometer, this entails that the originally potentially large terms in the nonlinear part of the original map are suppressed accordingly. Therefore, for a given accuracy, this may reduce the number of terms required to represent the reconstruction map. Thus, in the case of trajectory reconstruction, the original investment in the quality of the spectrometer and its linear resolution directly influences the quality of the reconstruction.

6.3.2 Reconstruction in Energy Loss Mode

The method outlined in the previous section allows a precise determination of the energy of a particle traversing the spectrometer. However, this energy is composed of the initial energy before reaction with the target and the change of energy experienced in the reaction. In the study and analysis of nuclear processes, however, **only the change in energy is relevant**, along with the angular distribution of the products. Therefore, unless the energy spread of the beam particles is smaller than the resolution of the device, which is rarely the case, the resolution of the energy measurement would be limited by the energy resolution of the beam.

There is a remedy to this fundamental problem—the so-called **energy loss mode** or **dispersion matching technique** (Blosser *et al.*, 1971). In this method, the dispersive spectrometer is preceded by another dispersive device, the so-called beam-matching section. The dispersive properties of the matching section are chosen in such a way that the resulting system consisting of the matching section and the spectrometer proper is achromatic to first order, i.e.,

$$(x|d) = 0 \quad \text{and} \quad (a|d) = 0. \tag{6.20}$$

The solid line in Fig. 6.4 schematically shows the trajectory of an off-energy particle as a function of position going through the matching section and the spectrometer.

Besides the energy spread, which typically may be approximately 0.1%, it is assumed that the phase space going through the matching section is small compared to the phase space after the target. This entails that any one energy in the target plane between the matching section and the spectrometer corresponds to a relatively small spot in the x direction, the center of which depends on the energy. Similarly, any one energy corresponds to a relatively small spot in the a direction, and the center of the spot again depends on the energy.

The matching system is now adjusted to **minimize the size of the spot for a given energy** component of the beam, the so-called **incoherent spot size**. The centers x_c and a_c of the incoherent spot in the x and a directions can be described in terms of the beam energy spread d_i as

$$x_c = k_x \cdot d_i, \quad a_c = k_a \cdot d_i, \tag{6.21}$$

where the matching condition connects the factors k_x and k_a to the matrix elements of the main part of the spectrometer such that the **overall system is achromatic**. Elementary matrix algebra shows that this requires

$$k_x = (x|a)(a|d) - (a|a)(x|d)$$
$$k_a = (a|x)(x|d) - (x|x)(a|d). \qquad (6.22)$$

To achieve the desired linearity of x_c and a_c and to limit the size of the incoherent spot, some aberration correction may be required. However, since the phase space of the original beam is small compared to that after the reaction, this task is substantially simpler than it would be for the main part of the spectrometer.

The dispersion matching condition can now be used to remove most of the influence of the initial energy spread in the beam. It will be possible to determine the reaction energy d_r proper in a reconstructive way similar to that described in the previous section. As a first step, we perform an approximation similar to the one discussed in the previous section by assuming that all beam particles are actually located at the energy-dependent center of the incoherent spot x_c and neglect the finite width of the spot. In a similar way, we assume that all beam particle slopes a are actually equal to the center slope a_c. In the vertical direction, we assume that the beam slopes b are all zero, but we do allow for a nonzero spot size in y direction. Similar to the nondispersive case, it is this incoherent spot size which determines the first-order resolution of the system and hence is a limit for the resolution achievable in higher orders.

The reaction that is to be studied now takes place at the target plane, illustrated as position T in Fig. 6.4. In the reaction, the slopes a and b of the particle as well as its energy d_i are being changed and replaced by the quantities a_r, b_r, and d_r, respectively, which are characteristic of the reaction. The range of possible angular changes is much larger than the spread in these values for the beam. Therefore, after the reaction, we have the following approximations:

$$x = k_x \cdot d_i$$
$$a = k_a \cdot d_i + a_r$$
$$y = y_i$$
$$b = b_r$$
$$d = d_i + d_r \qquad (6.23)$$

These five quantities now have to be propagated through the main part of the spectrometer. In terms of transfer maps, this requires inserting the linear transfer map represented by Eq. (6.23) into the nonlinear transfer map describing the spectrometer. The result is a map relating the **measurable final quantities** after the system x_f, a_f, y_f, and b_f to the quantities d_i, d_r, a_r, y_i, and b_r.

RECONSTRUCTIVE CORRECTION OF ABERRATIONS

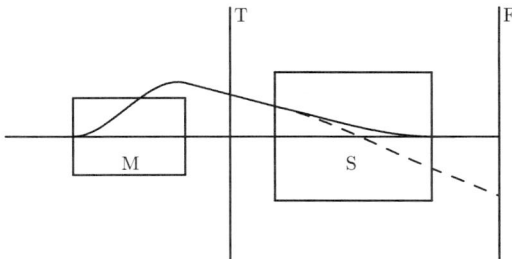

FIGURE 6.4. The principle of dispersion matching. The linear dispersion properties of the matching section (M; the beam analysis system) are chosen such that M followed by the spectrometer (S) becomes linearly achromatic and hence off-energy particles follow the solid line. Particles which experience a change in energy due to interactions in the target (T) hit the focal plane (F) at a different location (dashed trajectory).

Inserting Eq. (6.23) into the spectrometer map reveals that as a consequence of the dispersion matching, the final quantities no longer depend linearly on the unknown initial energy d_i. Higher order dependences on d_i usually still exist, but since they contain products of small quantities they are of reduced importance. In ignoring these higher order contributions of d_i, a final approximation is performed. We are left with a map of the following form:

$$\begin{pmatrix} x_f \\ a_f \\ y_f \\ b_f \end{pmatrix} = \mathcal{S} \begin{pmatrix} a_r \\ y_i \\ b_r \\ d_r \end{pmatrix}. \tag{6.24}$$

While the map $\mathcal{S}$ is different from that in the previous section, the linear part of this map has the same structure as before, the determinant has the same form, and we can apply the same techniques and reasoning to find the reconstruction map

$$\begin{pmatrix} a_r \\ y \\ b_r \\ d_r \end{pmatrix} = \mathcal{S}^{-1} \begin{pmatrix} x_f \\ a_f \\ y_f \\ b_f \end{pmatrix}. \tag{6.25}$$

6.3.3 Examples and Applications

The method for trajectory reconstruction and reconstructive correction of aberrations has been implemented in the code COSY INFINITY (Berz, 1993a). In this section we will apply the method to the study of two simple hypothetical spectrometers; here, we provide examples of a realistic device used in nuclear physics.

First, we study the **Browne–Buechner** spectrometer that was discussed previously. The spot size was assumed to be 0.46 mm. In this situation, the linear

TABLE IV

THE RESOLUTION OF THE SIMPLE BROWNE–BUECHNER SPECTROMETER FOR VARIOUS CORRECTION ORDERS. THE RESOLUTIONS ARE BASED ON A DETECTOR RESOLUTION OF .4MM FWHM AND .5 MRAD FWHM AND A RECTANGULAR PHASE SPACE OF $(\pm.3\text{MM}) \cdot (\pm 200\text{MR})$ (X), $(\pm 1\text{MM}) \cdot (\pm 100\text{MRAD})$ (Y) AND $\pm 5\%$ (D).

	Energy Resolution $\Delta E/E$
Linear Resolution	1,129
Uncorrected Nonlinear	65
Order 2 Reconstruction	291
Order 3 Reconstruction	866
Order 4 Reconstruction	941
Order 5 Reconstruction	945
Order 8 Reconstruction	945

energy resolution is about 1100, while the aberrations decrease it to only about 60. As can be seen in Table IV, reconstructive correction succeeds in eliminating most of the influence of these aberrations and almost recovers the linear resolution.

The second example spectrometer is a **very crude 90° dipole** preceded and followed by a drift equal to its bending radius. Such a system is known to be imaging in the dispersive plane, but it also possesses very substantial aberrations that limit its practical use to low resolutions or small phase spaces. However, because of this poor performance, the system provides an interesting test for the trajectory reconstruction method.

In order to obtain high linear resolutions, the deflection radius is chosen as 10 m, and the spot size is chosen as 0.1 mm. In order to produce substantial aberrations, the phase space was chosen as $0.2 \cdot 300$ mm mrad horizontally and $20 \cdot 200$ mm mrad vertically at an energy spread of 10%. While the linear resolution of the device is 50,000, the size of the aberrations limits the uncorrected nonlinear resolution to only about 20.

Reconstructive correction was applied to the system utilizing various orders of the reconstruction map. At reconstruction orders of approximately seven, the linear resolution, which represents the upper bound of any correction method, is almost reached. Details are listed in Table V. In this example, the reconstructive correction boosts the resolution by a factor of about 2500, illustrating the versatility of the technique for very crude systems.

It is worthwhile to point out that in extreme cases such as the one discussed here, aside from the unavoidable limitations due to detector resolution and incoherent spot size, the major limitation is the exact knowledge of the aberrations of the system. Using the DA methods, these can be computed with sufficient accuracy once details of the fields of the elements are known; in particular, this requires a detailed knowledge of the fringe fields. On the other hand, the actual design of the device is not critical, and even substantial aberrations can be corrected with the method. This suggests that spectrometers geared toward high-

TABLE V

CRUDE 10 M RADIUS 90° SINGLE DIPOLE SPECTROMETER ACCEPTING A PHASE SPACE OF 0.2 * 300 MM MRAD (X), 20 * 200 MM MRAD (Y) AND ±10% (D). THE DETECTOR RESOLUTION IS ZERO. THE LINEAR RESOLUTION OF 50,000 IS ALMOST FULLY RECOVERED DESPITE ABERRATIONS THAT LIMIT THE UNCORRECTED RESOLUTION TO ABOUT 20.

	Energy Resolution $\Delta E/E$	Angle Resolution $1/rad$
Linear Resolution	50,000	—
Uncorrected Nonlinear	20	200
Order 2 Reconstruction	800	3,900
Order 3 Reconstruction	5,400	23,600
Order 4 Reconstruction	23,600	63,500
Order 5 Reconstruction	43,000	96,500
Order 6 Reconstruction	47,000	99,000
Order 7 Reconstruction	48,000	99,000
Order 8 Reconstruction	48,000	99,000

resolution reconstruction should be designed with different criteria than those for conventional devices. Because the main aspect is the **knowledge** of the fields, **regardless of specific shape,** one should concentrate on devices for which the fields in space are known very precisely in advance.

In the following, we apply the method to the **S800** (Nolen *et al.*, 1989), a superconducting magnetic spectrograph at the National Superconducting Cyclotron Laboratory at Michigan State University. The spectrometer and its associated analysis line allow the study of heavy ion reactions with momenta of up to 1.2 GeV/c. It is designed for an energy resolution of one part in 10,000 with a large solid angle of 20 msr and an energy acceptance of 10%.

The S800 consists of two superconducting quadrupoles followed by two superconducting dipoles deflecting by 75° with y-focusing edge angles. Table VI lists the parameters of the system, and Fig. 6.5 shows the layout of the device. It should be noted that the S800 disperses in the vertical direction, while the scattering angle is measured in the transverse plane. Thus, the first-order focal plane is two-dimensional, with energy mostly measured in one dimension and the scattering angle measured in the other.

The aberrations of the S800 have been calculated using the code COSY INFINITY. Care is taken to include a detailed treatment of the actual fields of the system. To utilize measured field data, the methods developed in Chapter 3 can be used. Furthermore, in a simplified model of the device, the falloff of the dipole field as well as the quadrupole strengths can be described by Enge-type functions (Eq. 5.79). It was found that a transfer map of order seven represents the actual spectrometer with an accuracy sufficient for the desired resolution.

The linear resolution of the S800 spectrograph is limited by the incoherent spot size of 0.46 mm and the detector resolution, which is conservatively assumed to be 0.4 mm and 0.3 mrad FWHM in both planes of the detectors (Morris, 1989).

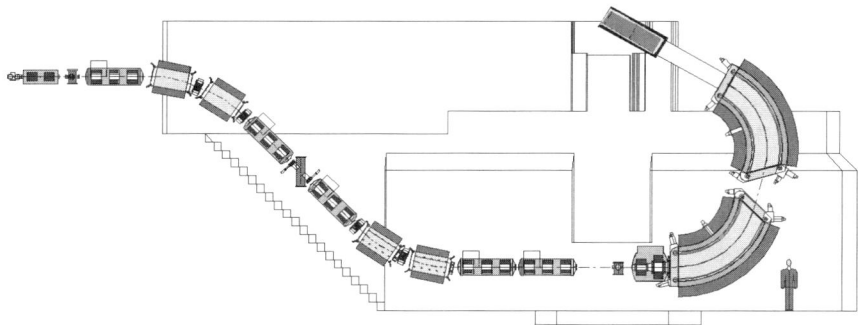

FIGURE 6.5. The layout of the S800 spectrograph.

These two effects limit the energy resolution to about 10,000. Since no hardware correction of higher order aberrations has been incorporated in the design, the map of the spectrograph exhibits large aberrations, which if left uncorrected would decrease the resulting resolution to only little more than 100. In particular, the slope-dependent aberrations are contributing substantially, with (x, aa) being the largest.

Using the map of the S800, the trajectory reconstruction was performed to various orders and the resolutions were computed following the methods outlined in Section 6.3. Table VII shows the resulting resolutions as a function of the reconstruction order.

The resolution with trajectory reconstruction correction improves with order; with third-order reconstruction, about 65% of the linear resolution is recovered, and with fifth-order correction, this increases to about 96% of the linear resolution and thus satisfies the design specifications.

Besides the computation of the corrected energy, the two-plane reconstruction technique also allows the computation of the angle under which the particle en-

TABLE VI
THE S800 SPECTROGRAPH.

Drift	l = 60 cm				
Quad	l = 40 cm,	$G_{max} = 21 \frac{T}{m}$,	r = 0.1 m		
Drift	l = 20 cm				
Quad	l = 40 cm,	$G_{max} = 6.8 \frac{T}{m}$,	r = 0.2 m		
Drift	l = 50 cm				
Dipole	ρ = 2.6667 m,	$B_{max} = 1.5 T$,	$\phi = 75°$,	$\epsilon_1 = 0°$,	$\epsilon_2 = 30°$
Drift	l = 140 cm				
Dipole	ρ = 2.6667 m,	$B_{max} = 1.5 T$,	$\phi = 75°$,	$\epsilon_1 = 30°$,	$\epsilon_2 = 0°$
Drift	l = 257.5 cm				

TABLE VII

THE RESOLUTION OF THE S800 FOR VARIOUS CORRECTION ORDERS. THE RESOLUTIONS ARE BASED ON A DETECTOR RESOLUTION OF 0.4 MM AND 0.3 MRAD FWHM AND A RECTANGULAR PHASE SPACE OF $(\pm.23$ MM)$\cdot(\pm 60$ MR) (X), $(\pm 10$ MM) $\cdot$ $(\pm 90$ MRAD) (Y) AND $\pm 5\%$ (D)

Correction Order	Energy Resolution ($\Delta E/E$)	Angle resolution (1/rad)
Linear resolution	9,400	—
Uncorrected nonlinear	100	25
Order 2 reconstruction	1,500	150
Order 3 reconstruction	6,300	600
Order 4 reconstruction	8,700	700
Order 5 reconstruction	9,100	1300
Order 8 reconstruction	9,100	1300

tered the spectrometer, which is dominated by the scattering angle. The prediction accuracy for this initial angle is approximately 1,300 at the fifth order, which corresponds to an angle resolution of about 0.05° (0.75 mr).

It is worthwhile to study the influence of possible improvements in the detector resolution on the resolution of the spectrometer. For this purpose, the focal plane position resolution was assumed to be 0.2 mm. Since the resolution limit due to the spot size now corresponds to the detector resolution limit, the incoherent spot size was also reduced by a factor of two. Table VIII shows the resulting resolutions as a function of the order of the reconstruction. With an eighth-order reconstruction, about 90% of the now twice as large resolution is recovered. Note that contrary to the previous operation mode, a third-order reconstruction gives less than half of the linear resolution.

TABLE VIII

THE RESOLUTION OF THE S800 FOR VARIOUS CORRECTION ORDERS FOR A DETECTOR RESOLUTION OF 0.2 MM AND 0.3 MRAD FWHM AND A REDUCED PHASE SPACE OF $(\pm.115$ MM) $\cdot$ $(\pm 60$ MRAD) (X), $(\pm 10$ MM) $\cdot$ $(\pm 90$ MRAD) (Y) AND $\pm 5\%$ (D)

Correction Order	Energy Resolution ($\Delta E/E$)	Angle Resolution (1/rad)
Linear resolution	19,000	—
Uncorrected nonlinear	100	25
Order 2 reconstruction	1500	150
Order 3 reconstruction	7900	650
Order 4 reconstruction	14,600	750
Order 5 reconstruction	16,800	1500
Order 8 reconstruction	17,000	1800

TABLE IX
S800 INCREASED IN SIZE BY FACTOR FIVE.[a]

Correction Order	Energy Resolution ($\Delta E/E$)	Angle Resolution (1/rad)
Linear resolution	95,000	—
Uncorrected nonlinear	100	25
Order 2 reconstruction	1,500	300
Order 3 reconstruction	9,700	750
Order 4 reconstruction	32,000	22,000
Order 5 reconstruction	85,000	28,000
Order 6 reconstruction	91,000	29,000
Order 7 reconstruction	93,000	29,000
Order 8 reconstruction	93,000	29,000

[a] The phase space is the same as for the S800 in Table 6.9, the position resolution of the detectors is 0.2 mm, and the angular resolution is assumed to be 0.15 mrad.

From the previous tables it is apparent that for the case of the S800, the trajectory reconstruction using the method presented here allows a correction of the device to nearly the linear limit and design specifications; with better detectors and smaller incoherent spot size, even increased resolutions are possible. It seems worthwhile to study to what extent the resolution could be improved by improvements in these areas.

Since gains in detector resolution by large factors do not appear likely, it is advantageous to increase the sheer size of the device. In the following example, the S800 was scaled up by a factor of five in all its linear dimensions. The phase space is assumed to be like that in the high-resolution mode of the S800, and so is the position detector resolution. This implies that the linear resolution is now five times larger than that in Table VIII and is nearly 100,000. The angular detector resolution is assumed to have increased by another factor of two. As Table IX shows, the linear resolution is eventually almost fully recovered. Because of the much higher accuracy demand, this requires seventh-order correction. At third order, only the resolution of the original S800 is achieved.

6.4 ABERRATION CORRECTION VIA REPETITIVE SYMMETRY

From the very early days of optics it was known that certain longitudinal symmetries can be used beneficially to cancel some, and possibly even all, aberrations of a system. In this section we discuss the use of such symmetries by analyzing which **aberrations can be removed by mere exploitation of symmetry**. We also discuss the case of complete aberration correction, the case of the so-called **achromats**. The choice of this name is historical and perhaps somewhat unfortunate, because it seems to imply independence of chromatic effects, whereas indeed not only chromatic terms but also all aberrations are removed. In the fol-

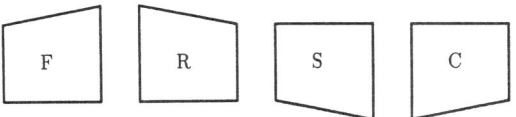

FIGURE 6.6. The four elementary cells: Forward, Reverse, Switched, and Combined

lowing we develop a theory of correction of particular aberrations, or even all of them, that is in principle applicable to arbitrary order.

In order to exploit the longitudinal symmetry, we introduce four kinds of cells (shown in Fig. 6.6): a **forward** cell (F), a **reversed** cell (R), a **switched** cell in which the direction of bending is switched (S), and the **switched-reversed** cell in which reversion and switching is combined (C). With these considerations, we are able to study systems with an arbitrary number of cells and obtain solutions that are independent of the arrangements inside a cell.

Because of their freedom from redundancy, **Lie transformations** are often adopted to represent symplectic maps. Instead of order-by-order factorization, we use a factorization formed by a linear matrix and a single Lie operator, describing the linear and nonlinear parts, respectively. In the following, the basic ideas and important results of the achromat theory will be presented, many of which without proof; for details, refer to (Wan and Berz, 1996); (Wan, 1995).

To begin our discussion, we consider a sequence of n identical F cells and study under what condition such a system will be an achromat. Let L_x be the x matrix of one cell, which has the form

$$L_x = \begin{pmatrix} (x|x) & (x|a) & (x|d) \\ (a|x) & (a|a) & (a|d) \\ 0 & 0 & 1 \end{pmatrix} = \begin{pmatrix} M & \vec{\omega} \\ 0 & 1 \end{pmatrix}. \tag{6.26}$$

After n iterations of the forward cell, the total x matrix T_x is

$$T_x = L_x^n = \begin{pmatrix} M^n & (M^{n-1} + M^{n-2} + \cdots + \hat{I})\vec{\omega} \\ 0 & 1 \end{pmatrix}$$

$$= \begin{pmatrix} M^n & (M^n - \hat{I})(M - \hat{I})^{-1}\vec{\omega} \\ 0 & 1 \end{pmatrix}. \tag{6.27}$$

Equation (6.27) shows that when $M^n = \hat{I}$, i.e., the tunes (phases of the eigenvalues of M over 2π) are multiples of $1/n$, the dispersion vanishes and the geometric part becomes unity. Together with the requirement that $L_y = \hat{I}$, a first-order achromat is reached when the tunes of the whole system are integers.

Although the concept of first-order achromats had been widely used in various beam optical systems and accelerators for a long time, it was only in the 1970s that a theory developed by K. Brown enabled the design of realistic second-order achromats in a systematic and elegant way (Brown, 1979b). The theory is based on the following observations. First, any system of n identical F cells ($n > 1$) with the overall first-order matrix equaling to unity in both transverse planes gives a first-order achromat, as previously shown. Second, when n is not equal to three, it cancels all second-order geometric aberrations. Finally, of all second-order chromatic aberrations, only two are independent. Therefore, they can be corrected by two families of sextupoles each responsible for one in each transverse plane. These findings make it possible to design a four-cell second-order achromat with only one dipole, two quadrupoles, and two sextupoles per cell. Detailed studies on this theory will be discussed in Section 6.4.1.

Because of its simplicity, the second-order achromat concept has been applied to the design of various beam optical systems such as the time-of-flight mass spectrometers, both single-pass in the spectrometer **TOFI** (Wouters *et al.*, 1987) and multipass as in the **ESR** ring (Wollnik, 1987; Wollnik, 1987), the Arcs and the Final Focus System of **SLC**; (Fischer *et al.*, 1987; Murray *et al.*, 1987; Schwarzschild, 1994), and the MIT South Hall Ring (**SHR**) (Flanz, 1989).

Since it is difficult to generalize the second-order achromat theory to higher orders, the first third-order achromat theory was developed along a different line of reasoning by Alex Dragt based on **normal form methods** (Dragt, 1987) and the observation that an achromat in normal form coordinates is also an achromat in regular coordinates. Since as discussed in Chapter 7, Section 7.3, in normal form coordinates the motion is particularly simple and straightforward, the amount of necessary correction is considerably reduced. Specifically, a system of n identical cells is a third-order achromat if the following conditions are met: (i) The tunes of a cell T_x and T_y are not full-, half-, third-, or quarter-integer resonant, but nT_x and nT_y are integers; and (ii) The two first-order and two second-order chromaticities and three anharmonicities are zero.

Two third-order achromats have been designed. The first design, by Dragt, contains 30 cells with $T_x = 1/5$ and $T_y = 1/6$. Each contains 10 bends, two quadrupoles, two sextupoles, and five octupoles. The whole system forms a 180° bending arc. The second design was done by Neri (Neri *et al.*, 1991). It is a seven-cell system with only 1 bend per cell and the total bend is also 180°. The tunes of a cell are $T_x = 1/7$ and $T_y = 2/7$, which seems to violate the theory because of the third-order resonance $2T_x - T_y = 0$. However, the achromaticity can still be achieved because the respective driving terms (see Chapter 7, Section 7.3) are canceled by midplane symmetry, which greatly reduces the number of cells.

6.4.1 Second-Order Achromats

Brown's second-order achromat theory is built on the repetitive first-order achromat described previously. It is based on two fundamental observations. The **first observation** is that if a system contains N identical cells ($N > 1$ and $N \neq 3$), all second-order geometric aberrations vanish when both transverse planes have the same noninteger tunes for a cell and the phase advance of the system is a multiple of 2π.

For the proof, we adopt Brown's original notation in which linear matrix is represented by R and second-order matrix is T such that

$$x_{i,1} = \sum_j R_{ij} x_{j,0} + \sum_{j,k} T_{ijk} x_{j,0} x_{k,0}, \tag{6.28}$$

where $x_{i,0}$ and $x_{i,1}$ are initial and final coordinates, respectively. Perturbative methods (see Brown, 1979b; Berz et al., 1999) allow us to express T_{ijk} as an integral involving R_{ij} in the form

$$T_{ijk} = \int_0^L K_p(s)(R_{ij}(s))^n (R_{ik}(s))^m ds \text{ with } (n+m) = 3, \tag{6.29}$$

where $K_p(s)$ is the multipole strength at s. For geometric aberrations, R_{ij} should come from the geometric part of R only, which is written as

$$\begin{pmatrix} \cos\psi(s) + \alpha(s)\sin\psi(s) & \beta(s)\sin\psi(s) \\ -\gamma(s)\sin\psi(s) & \cos\psi(s) - \alpha(s)\sin\psi(s) \end{pmatrix}.$$

As a result, T_{ijk} can be written as

$$T_{ijk} = \int_0^L F_p(s) \sin^n(\psi(s)) \cos^m(\psi(s)) ds.$$

Since for $m+n = 3$, the terms $\sin^n(\psi(s)) \cos^m(\psi(s))$ written in terms of complex exponentials contain only $e^{\pm i\psi(s)}$ and $e^{\pm 3i\psi(s)}$, the conditions for all second-order geometric aberrations to vanish are

$$\int_0^L F_p e^{\pm i\psi} ds = 0 \quad \text{and} \quad \int_0^L F_p e^{\pm 3i\psi} ds = 0.$$

Because the system consists of individual elements centered at s_0 with length Δs, the integral conditions become the following sums

$$\sum_{k=1}^N \tilde{F}_k e^{\pm i\psi_k} = 0 \quad \text{and} \quad \sum_{k=1}^N \tilde{F}'_k e^{\pm 3i\psi_k} = 0, \tag{6.30}$$

where

$$\tilde{F}_k = \int_{\frac{k}{N}L+s_0-\Delta s}^{\frac{k}{N}L+s_0+\Delta s} F_p(s) e^{\pm i(\psi(s)-\psi_k(s_0))} ds$$

$$= \int_{-\Delta s}^{\Delta s} F_p\left(\frac{k}{N}L + s_0 + \tilde{s}\right) e^{\pm i(\psi(\frac{k}{N}L+s_0+\tilde{s})-\psi_k(s_0))} d\tilde{s}$$

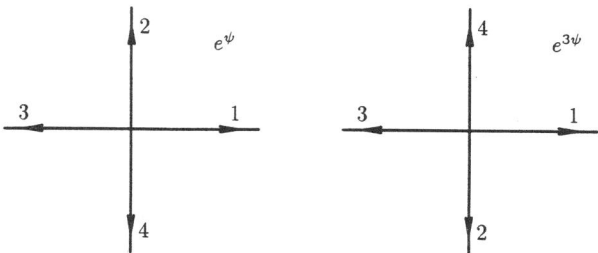

FIGURE 6.7. Complex plane diagram for second-order geometric aberrations of a four-cell repetitive system with phase advances 2π.

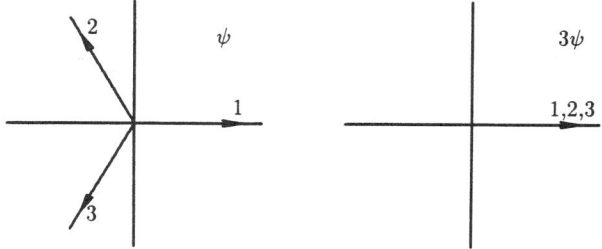

FIGURE 6.8. Complex plane diagram for second-order geometric aberrations of a three-cell repetitive system with phase advances 2π.

$$\tilde{F}_k = \int_{-\Delta s}^{\Delta s} F_p(\tilde{s}) e^{\pm i(\Delta \psi(\tilde{s}))} d\tilde{s}, \tag{6.31}$$

$$\tilde{F}'_k = \int_{-\Delta s}^{\Delta s} F_p(\tilde{s}) e^{\pm 3i(\Delta \psi(\tilde{s}))} d\tilde{s}. \tag{6.32}$$

Here, repetition of the system is used to obtain Eqs. (6.31) and (6.32). Since $\tilde{F}_k$ and $\tilde{F}'_k$ are independent of k, Eq. (6.30) is further reduced to

$$\sum_{k=1}^{N} e^{\pm i \psi_k} = 0 \quad \text{and} \quad \sum_{k=1}^{N} e^{\pm 3i \psi_k} = 0.$$

In conclusion, **all second-order aberrations vanish** when $N \neq 3$, $N\psi_{x,y} = 2m_{x,y}\pi$ and $m_{x,y} \neq 2mN$ ($m = 1, 2, \ldots$) (see Figs. 6.7 and 6.8).

The **second observation** deals with the correction of second-order chromatic aberrations left in a system satisfying the previous requirements: For a system that has no geometric aberrations based on the previous reasoning and $N > 3$, a second-order achromat is achieved when two families of sextupole components are adjusted so as to make one chromatic aberration vanish in each transverse plane. In another words, only **two chromatic aberrations** are independent. The

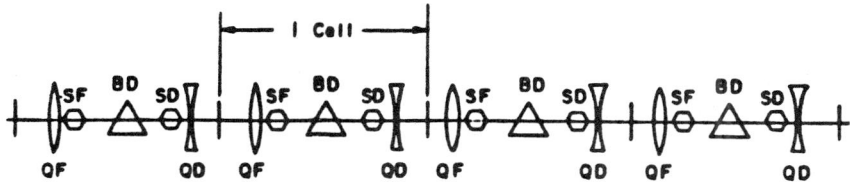

FIGURE 6.9. Brown's four-cell second-order achromat. Quadrupoles are used to tune the system to phase advance 2π in both transfer planes and two families sextupoles SF and SD are used to correct chromatic second-order aberrations in x and y planes, respectively. To make the sextupoles weak, they are placed such that β_x is larger at SF and β_y is larger at SD.

proof of this theorem can be found in (Carey, 1992). Another proof will be given in Section 6.4.3 as part of the third-order achromat theory. A typical four-cell second order achromat is shown in Fig. 6.9.

6.4.2 Map Representations

For further discussion, we will first discuss various representations of symplectic maps. We consider a phase space consisting of $2n$ variables $(q_1, \ldots, q_n, p_1, \ldots, p_n)$. Since we are not considering systems subject to synchrotron radiation and acceleration, the transverse motion is described by a symplectic map. It is well known that for any symplectic map $\mathcal{M}$ of arbitrary order m, there exists a matrix $\hat{L}$ and a polynomial H of order from 3 up to $m+1$, such that

$$\mathcal{M} =_m (\hat{L}\mathcal{I}) \circ (\exp(: H :)\mathcal{I}). \tag{6.33}$$

Here $\mathcal{I} = (q_1, \ldots, q_n, p_1, \ldots, p_n)$ is the identity function, and "$\circ$" denotes the composition of functions. Hence, the inverse is

$$\mathcal{M}^{-1} =_m (\exp(- : H :)\mathcal{I}) \circ (\hat{L}^{-1}\mathcal{I}). \tag{6.34}$$

Now let us define a "standard" and a "sub-standard" form of symplectic maps, the advantages of which will become evident later.

For a symplectic map $\mathcal{M}_s$, the **standard form** is defined as

$$\mathcal{M}_s = \exp(: H :)(M_L \mathcal{I}),$$

where H is the pseudo-Hamiltonian with orders three and higher, and M_L is the linear matrix. A representation of the form

$$\mathcal{M}_s = \prod_i \exp(: H_i :)(M_L \mathcal{I})$$

is called a **substandard form**.

Apparently, use of the Baker–Campbell–Hausdorff **(BCH) formula** in principle allows the transformation of a substandard form into a standard form. In practice, this may often be complicated to higher orders because of the quickly increasing complexity of the required algebraic manipulations.

As a direct result of a well-known equality,

$$g(\exp(: H :)\mathcal{I}) = \exp(: H :)g(\mathcal{I}), \tag{6.35}$$

provided that $\exp(: H :)\mathcal{I}$ and $\exp(: H :)g(\mathcal{I})$ both converge, the map $\mathcal{M}$, which we call the forward map $\mathcal{M}^F$, can be written in the standard form,

$$\mathcal{M}^F = \exp(: H :)(\hat{L}\mathcal{I}). \tag{6.36}$$

To obtain the maps of the R, S, and C cells, we need the following: If a map $\mathcal{M}$ is **symplectic**, i.e., $\text{Jac}(\mathcal{M}) \cdot \hat{J} \cdot \text{Jac}(\mathcal{M})^t = \hat{J}$, we have

$$(\exp(: H :)g(\mathcal{I})) \circ (\mathcal{M}(\mathcal{I})) = \exp(: H(\mathcal{M}) :)g(\mathcal{M}). \tag{6.37}$$

If a map $\mathcal{M}$ is **antisymplectic**, i.e., $\text{Jac}(\mathcal{M}) \cdot \hat{J} \cdot \text{Jac}(\mathcal{M})^t = -\hat{J}$, we have

$$(\exp(: H :)g(\mathcal{I})) \circ (\mathcal{M}(\mathcal{I})) = \exp(- : H(\mathcal{M}) :)g(\mathcal{M}). \tag{6.38}$$

Here $\text{Jac}(\mathcal{M})$ denotes the Jacobian matrix of the map $\mathcal{M}$, and $\hat{J}$ is the antisymmetric $2n \times 2n$ matrix from Eq. (1.86). For the proof, we observe that in case $\mathcal{M}$ is a symplectic map, according to (1.191), the Poisson bracket is an invariant under the action of the map, i.e.,

$$[H(\mathcal{M}), g(\mathcal{M})] = ([H, g])(\mathcal{M}). \tag{6.39}$$

From this, we readily obtain

$$\exp(: H(\mathcal{M}) :)g(\mathcal{M})$$
$$= g(\mathcal{M}) + [H(\mathcal{M}), g(\mathcal{M})] + \frac{1}{2}[H(\mathcal{M}), [H(\mathcal{M}), g(\mathcal{M})]] + \cdots$$
$$= g(\mathcal{M}) + ([H, g])(\mathcal{M}) + \frac{1}{2}[H(\mathcal{M}), ([H, g])(\mathcal{M})] + \cdots$$
$$= g(\mathcal{M}) + ([H, g])(\mathcal{M}) + \frac{1}{2}([H, [H, g]])(\mathcal{M}) + \cdots$$
$$= (\exp(: H :)g(\mathcal{I})) \circ (\mathcal{M}(\mathcal{I})).$$

If $\mathcal{M}$ is antisymplectic, the proof is very similar except that the Poisson bracket changes sign under the transformation, i.e.,

$$[H(\mathcal{M}), g(\mathcal{M})] = -([H, g])(\mathcal{M}). \tag{6.40}$$

Then we have

$$\exp(- : H(\mathcal{M}) :)g(\mathcal{M})$$
$$= g(\mathcal{M}) - [H(\mathcal{M}), g(\mathcal{M})] + \frac{1}{2}[H(\mathcal{M}), [H(\mathcal{M}), g(\mathcal{M})]] + \cdots$$
$$= g(\mathcal{M}) + ([H, g])(\mathcal{M}) + \frac{1}{2}[H(\mathcal{M}), -([H, g])(\mathcal{M})] + \cdots$$
$$= g(\mathcal{M}) + ([H, g])(\mathcal{M}) + \frac{1}{2}([H, [H, g]])(\mathcal{M}) + \cdots$$
$$= (\exp(: H :)g(\mathcal{I})) \circ (\mathcal{M}(\mathcal{I})).$$

A **reversed cell** is one in which the order of the elements is reversed relative to that of a forward cell; according to Eq. (4.8), it has the form

$$\mathcal{M}^R = (\hat{R}\mathcal{I}) \circ \mathcal{M}^{-1} \circ (\hat{R}^{-1}\mathcal{I}), \tag{6.41}$$

where matrix $\hat{R}$ in Eq. (4.8) is antisymplectic. As can be seen readily, we obtain the standard form of $\mathcal{M}^R$:

$$\mathcal{M}^R = (\hat{R}\mathcal{I}) \circ \mathcal{M}^{-1} \circ (\hat{R}^{-1}\mathcal{I})$$
$$= (\hat{R}\mathcal{I}) \circ (\exp(: -H :)\mathcal{I}) \circ (\hat{L}^{-1}\mathcal{I}) \circ (\hat{R}^{-1}\mathcal{I})$$
$$= (\hat{R}\mathcal{I}) \circ (\exp(: -H :)\mathcal{I}) \circ (\hat{L}^{-1}\hat{R}^{-1}\mathcal{I})$$
$$= (\hat{R}\mathcal{I}) \circ (\exp(: H(\hat{L}^{-1}\hat{R}^{-1}\mathcal{I}) :)(\hat{L}^{-1}\hat{R}^{-1}\mathcal{I}))$$
$$= \exp(: H(\hat{L}^{-1}\hat{R}^{-1}\mathcal{I}) :)((\hat{R}\mathcal{I}) \circ (\hat{L}^{-1}\hat{R}^{-1}\mathcal{I}))$$
$$= \exp(: H(\hat{L}^{-1}\hat{R}^{-1}\mathcal{I}) :)(\hat{R}\hat{L}^{-1}\hat{R}^{-1}\mathcal{I}). \tag{6.42}$$

A **switched cell** is a forward cell rotated 180° around the z-axis, i.e.,

$$\mathcal{M}^S = (\hat{S}\mathcal{I}) \circ \mathcal{M} \circ (\hat{S}^{-1}\mathcal{I}), \tag{6.43}$$

where

$$\hat{S}\mathcal{I} = \begin{pmatrix} -1 & 0 & 0 & 0 & 0 & 0 \\ 0 & -1 & 0 & 0 & 0 & 0 \\ 0 & 0 & 1 & 0 & 0 & 0 \\ 0 & 0 & 0 & -1 & 0 & 0 \\ 0 & 0 & 0 & 0 & -1 & 0 \\ 0 & 0 & 0 & 0 & 0 & 1 \end{pmatrix} \begin{pmatrix} x \\ y \\ d \\ a \\ b \\ t \end{pmatrix}. \tag{6.44}$$

It is easy to see that the matrix $\hat{S}$ is symplectic, and so

$$\begin{aligned}
\mathcal{M}^S &= (\hat{S}\mathcal{I}) \circ \mathcal{M} \circ (\hat{S}^{-1}\mathcal{I}) \\
&= (\hat{S}\mathcal{I}) \circ (\hat{L}\mathcal{I}) \circ (\exp(:H:)\mathcal{I}) \circ (\hat{S}^{-1}\mathcal{I}) \\
&= (\hat{S}\hat{L}\mathcal{I}) \circ (\exp(:H(\hat{S}^{-1}\mathcal{I}):)(\hat{S}^{-1}\mathcal{I})) \\
&= \exp(:H(\hat{S}^{-1}\mathcal{I}):)(\hat{S}\hat{L}\hat{S}^{-1}\mathcal{I}).
\end{aligned} \tag{6.45}$$

The cell that is simultaneously switched and reversed, the **switched–reversed cell**, has a map given by

$$\mathcal{M}^C = (\hat{S}\mathcal{I}) \circ (\hat{R}\mathcal{I}) \circ \mathcal{M}^{-1} \circ (\hat{R}^{-1}\mathcal{I}) \circ (\hat{S}^{-1}\mathcal{I}). \tag{6.46}$$

Noting that the matrix $\hat{S}\hat{R}$ is antisymplectic, like that of the reversed cell, $\mathcal{M}^C$ can be simplified to the standard form, where

$$\mathcal{M}^C = \exp(:H(\hat{L}^{-1}\hat{R}^{-1}\hat{S}^{-1}\mathcal{I}):)(\hat{S}\hat{R}\hat{L}^{-1}\hat{R}^{-1}\hat{S}^{-1}\mathcal{I}). \tag{6.47}$$

To summarize, we list the maps of all four kinds of cells:

$$\mathcal{M}^F = \exp(:H:)(\hat{L}\mathcal{I}), \tag{6.48}$$

$$\mathcal{M}^R = \exp(:H(\hat{L}^{-1}\hat{R}^{-1}\mathcal{I}):)(\hat{R}\hat{L}^{-1}\hat{R}^{-1}\mathcal{I}), \tag{6.49}$$

$$\mathcal{M}^S = \exp(:H(\hat{S}^{-1}\mathcal{I}):)(\hat{S}\hat{L}\hat{S}^{-1}\mathcal{I}), \tag{6.50}$$

$$\mathcal{M}^C = \exp(:H(\hat{L}^{-1}\hat{R}^{-1}\hat{S}^{-1}\mathcal{I}):)(\hat{S}\hat{R}\hat{L}^{-1}\hat{R}^{-1}\hat{S}^{-1}\mathcal{I}). \tag{6.51}$$

With the maps of the different kinds of cells, we are ready to construct the map of any multicell system. As examples, the total maps of some four-cell systems are presented here. We first introduce the following notation for the map of a multicell system. Let C_i be the ith cell in a k cell system, i.e., C_i can be F, $\hat{R}$, S, or C. Then we denote the **map of the total system** by $\mathcal{M}^{C_1 C_2 \cdots C_k}$.

For example, $\mathcal{M}^{FRSC}$ represents the map of a four-cell system consisting of a forward cell followed by the reversed cell, the switched cell, and the combined cell. In the rest of this section, we will need the substandard form and eventually the standard form for a variety of four-cell combinations. As a proof of principle, we show the derivation for $\mathcal{M}^{FRSC}$, where Eqs. (6.35, 6.37, 6.38) and the associativity of $\circ$ are repeatedly used:

$$\begin{aligned}
\mathcal{M}^{FRSC} &= \mathcal{M}^C \circ \mathcal{M}^S \circ \mathcal{M}^R \circ \mathcal{M}^F \\
&= [\exp(:H(\hat{L}^{-1}\hat{R}^{-1}\hat{S}^{-1}\mathcal{I}):)(\hat{S}\hat{R}\hat{L}^{-1}\hat{R}^{-1}\hat{S}^{-1}\mathcal{I})] \\
&\quad \circ [\exp(:H(\hat{S}^{-1}\mathcal{I}):)(\hat{S}\hat{L}\hat{S}^{-1}\mathcal{I})] \\
&\quad \circ [\exp(:H(\hat{L}^{-1}\hat{R}^{-1}\mathcal{I}):)(\hat{R}\hat{L}^{-1}\hat{R}^{-1}\mathcal{I})] \circ [\exp(:H(\mathcal{I}):)(\hat{L}\mathcal{I})]
\end{aligned}$$

ABERRATION CORRECTION VIA REPETITIVE SYMMETRY   237

$$\begin{aligned}
&= \exp(:H(\mathcal{I}):)\{[\exp(:H(\hat{L}^{-1}\hat{R}^{-1}\hat{S}^{-1}\mathcal{I}):)(\hat{S}\hat{R}\hat{L}^{-1}\hat{R}^{-1}\hat{S}^{-1}\mathcal{I})]\\
&\quad \circ [\exp(:H(\hat{S}^{-1}\mathcal{I}):)(\hat{S}\hat{L}\hat{S}^{-1}\mathcal{I})]\\
&\quad \circ [\exp(:H(\hat{L}^{-1}\hat{R}^{-1}\mathcal{I}):)(\hat{R}\hat{L}^{-1}\hat{R}^{-1}\mathcal{I})] \circ (\hat{L}\mathcal{I})\}\\
&= \exp(:H(\mathcal{I}):)\{[\exp(:H(\hat{L}^{-1}\hat{R}^{-1}\hat{S}^{-1}\mathcal{I}):)(\hat{S}\hat{R}\hat{L}^{-1}\hat{R}^{-1}\hat{S}^{-1}\mathcal{I})]\\
&\quad \circ [\exp(:H(\hat{S}^{-1}\mathcal{I}):)(\hat{S}\hat{L}\hat{S}^{-1}\mathcal{I})]\\
&\quad \circ [\exp(:H(\hat{L}^{-1}\hat{R}^{-1}\cdot\hat{L}\mathcal{I}):)(\hat{R}\hat{L}^{-1}\hat{R}^{-1}\cdot\hat{L}\mathcal{I})]\}\\
&= \exp(:H(\mathcal{I}):)\exp(:H(\hat{L}^{-1}\hat{R}^{-1}\cdot\hat{L}\mathcal{I}):)\\
&\quad \{[\exp(:H(\hat{L}^{-1}\hat{R}^{-1}\hat{S}^{-1}\mathcal{I}):)(\hat{S}\hat{R}\hat{L}^{-1}\hat{R}^{-1}\hat{S}^{-1}\mathcal{I})]\\
&\quad \circ [\exp(:H(\hat{S}^{-1}\mathcal{I}):)(\hat{S}\hat{L}\hat{S}^{-1}\mathcal{I})] \circ (\hat{R}\hat{L}^{-1}\hat{R}^{-1}\cdot\hat{L}\mathcal{I})\}\\
&= \exp(:H(\mathcal{I}):)\exp(:H(\hat{L}^{-1}\hat{R}^{-1}\cdot\hat{L}\mathcal{I}):)\\
&\quad \{[\exp(:H(\hat{L}^{-1}\hat{R}^{-1}\hat{S}^{-1}\mathcal{I}):)(S\hat{R}\hat{L}^{-1}\hat{R}^{-1}S^{-1}\mathcal{I})]\\
&\quad \circ [\exp(:H(\hat{S}^{-1}\cdot\hat{R}\hat{L}^{-1}\hat{R}^{-1}\cdot\hat{L}\mathcal{I}):)(\hat{S}\hat{L}\hat{S}^{-1}\cdot\hat{R}\hat{L}^{-1}\hat{R}^{-1}\cdot\hat{L}\mathcal{I})]\}\\
&= \exp(:H(\mathcal{I}):)\exp(:H(\hat{L}^{-1}\hat{R}^{-1}\cdot\hat{L}\mathcal{I}):)\\
&\quad \exp(:H(\hat{S}^{-1}\cdot\hat{R}\hat{L}^{-1}\hat{R}^{-1}\cdot\hat{L}\mathcal{I}):)\{[\exp(:H(\hat{L}^{-1}\hat{R}^{-1}\hat{S}^{-1}\mathcal{I}):)\\
&\quad (\hat{S}\hat{R}\hat{L}^{-1}\hat{R}^{-1}\hat{S}^{-1}\mathcal{I})] \circ (\hat{S}\hat{L}\hat{S}^{-1}\cdot\hat{R}\hat{L}^{-1}\hat{R}^{-1}\cdot\hat{L}\mathcal{I})\}\\
&= \exp(:H(\mathcal{I}):)\exp(:H(\hat{L}^{-1}\hat{R}^{-1}\cdot\hat{L}\mathcal{I}):)\\
&\quad \exp(:H(\hat{S}^{-1}\cdot\hat{R}\hat{L}^{-1}\hat{R}^{-1}\cdot\hat{L}\mathcal{I}):)\\
&\quad \exp(:H(\hat{L}^{-1}\hat{R}^{-1}\cdot\hat{L}\hat{S}^{-1}\cdot\hat{R}\hat{L}^{-1}\hat{R}^{-1}\cdot\hat{L}\mathcal{I}):)\\
&\quad (\hat{S}\hat{R}\hat{L}^{-1}\hat{R}^{-1}\cdot\hat{L}\hat{S}^{-1}\cdot\hat{R}\hat{L}^{-1}\hat{R}^{-1}\cdot\hat{L}\mathcal{I}).
\end{aligned}$$

Similarly, $\mathcal{M}^{FRFR}$, $\mathcal{M}^{FCSR}$, and $\mathcal{M}^{FCFC}$ are obtained. Altogether, we have

$$\begin{aligned}
\mathcal{M}^{FRSC} &= \exp(:H(\mathcal{I}):)\exp(:H(\hat{L}^{-1}\hat{R}^{-1}\cdot\hat{L}\mathcal{I}):)\\
&\quad \exp(:H(\hat{S}^{-1}\cdot\hat{R}\hat{L}^{-1}\hat{R}^{-1}\cdot\hat{L}\mathcal{I}):)\\
&\quad \exp(:H(\hat{L}^{-1}\hat{R}^{-1}\cdot\hat{L}\hat{S}^{-1}\cdot\hat{R}\hat{L}^{-1}\hat{R}^{-1}\cdot\hat{L}\mathcal{I}):)\\
&\quad (\hat{S}\hat{R}\hat{L}^{-1}\hat{R}^{-1}\cdot\hat{L}\hat{S}^{-1}\cdot\hat{R}\hat{L}^{-1}\hat{R}^{-1}\cdot\hat{L}\mathcal{I}), &(6.52)\\
\mathcal{M}^{FRFR} &= \exp(:H(\mathcal{I}):)\exp(:H(\hat{L}^{-1}\hat{R}^{-1}\cdot\hat{L}\mathcal{I}):)\\
&\quad \exp(:H(\hat{R}\hat{L}^{-1}\hat{R}^{-1}\cdot\hat{L}\mathcal{I}):)\\
&\quad \exp(:H(\hat{L}^{-1}\hat{R}^{-1}\cdot\hat{L}\cdot\hat{R}\hat{L}^{-1}\hat{R}^{-1}\cdot\hat{L}\mathcal{I}):)\\
&\quad (\hat{R}\hat{L}^{-1}\hat{R}^{-1}\cdot\hat{L}\cdot\hat{R}\hat{L}^{-1}\hat{R}^{-1}\cdot\hat{L}\mathcal{I}), &(6.53)
\end{aligned}$$

238 IMAGING SYSTEMS

$$\begin{aligned}\mathcal{M}^{FCSR} = &\exp(: H(\mathcal{I}) :) \exp(: H(\hat{L}^{-1}\hat{R}^{-1}\hat{S}^{-1} \cdot \hat{L}\mathcal{I}) :) \\ &\exp(: H(\hat{R}\hat{L}^{-1}\hat{R}^{-1}\hat{S}^{-1} \cdot \hat{L}\mathcal{I}) :) \\ &\exp(: H(\hat{L}^{-1}\hat{R}^{-1} \cdot \hat{S}\hat{L} \cdot \hat{R}\hat{L}^{-1}\hat{R}^{-1}\hat{S}^{-1} \cdot \hat{L}\mathcal{I}) :) \\ &(\hat{R}\hat{L}^{-1}\hat{R}^{-1} \cdot \hat{S}\hat{L} \cdot \hat{R}\hat{L}^{-1}\hat{R}^{-1}\hat{S}^{-1} \cdot \hat{L}\mathcal{I}),\end{aligned} \quad (6.54)$$

$$\begin{aligned}\mathcal{M}^{FCFC} = &\exp(: H(\mathcal{I}) :) \exp(: H(\hat{L}^{-1}\hat{R}^{-1}\hat{S}^{-1} \cdot \hat{L}\mathcal{I}) :) \\ &\exp(: H(\hat{S}\hat{R}\hat{L}^{-1}\hat{R}^{-1}\hat{S}^{-1} \cdot \hat{L}\mathcal{I}) :) \\ &\exp(: H(\hat{L}^{-1}\hat{R}^{-1}\hat{S}^{-1} \cdot \hat{L} \cdot \hat{S}\hat{R}\hat{L}^{-1}\hat{R}^{-1}\hat{S}^{-1} \cdot \hat{L}\mathcal{I}) :) \\ &(\hat{S}\hat{R}\hat{L}^{-1}\hat{R}^{-1}\hat{S}^{-1} \cdot \hat{L} \cdot \hat{S}\hat{R}\hat{L}^{-1}\hat{R}^{-1}\hat{S}^{-1} \cdot \hat{L}\mathcal{I}).\end{aligned} \quad (6.55)$$

This substandard form is important in that it is essential to proving many of the core theorems discussed later. Also, using the BCH formula, the substandard form can be brought to standard form.

6.4.3 Major Correction Theorems

In order to proceed to the discussion of removal of nonlinearities via longitudinal symmetries, we have to first study general multicell systems and then four-cell systems using the maps obtained previously. The goal is to find systems that require the smallest number of conditions to be achromats to a given order. A few definitions must first be provided, which are the key to the development of the whole theory.

In addition to symplecticity, **midplane symmetry** is usually preserved by accelerators and beamlines. Therefore, the transfer map of the forward cell of such a system can be represented by the standard form with the following pseudo-Hamiltonian:

$$H = \sum_{i_x i_a i_y i_b i_d} C_{i_x i_a i_y i_b i_d} x^{i_x} y^{i_y} d^{i_d} a^{i_a} b^{i_b}, \quad (6.56)$$

where $i_x + i_y + i_d + i_a + i_b \geq 3$ and $i_y + i_b$ is even because of midplane symmetry (see Eqs. (4.10) to (4.15)). To simplify further notation, we define the polynomials $A(H)$, $B(H)$, $C(H)$, and $D(H)$ as those parts of H that satisfy

$$A(H) = \sum_{i_x i_a i_y i_b i_d} C_{i_x i_y i_d i_a i_b} x^{i_x} y^{i_y} d^{i_d} a^{i_a} b^{i_b} \quad (i_x + i_a \text{ odd}, \quad i_a + i_b \text{ even})$$

$$B(H) = \sum_{i_x i_a i_y i_b i_d} C_{i_x i_y i_d i_a i_b} x^{i_x} y^{i_y} d^{i_d} a^{i_a} b^{i_b} \quad (i_x + i_a \text{ odd}, \quad i_a + i_b \text{ odd})$$

$$C(H) = \sum_{i_x i_a i_y i_b i_d} C_{i_x i_y i_d i_a i_b} x^{i_x} y^{i_y} d^{i_d} a^{i_a} b^{i_b} \quad (i_x + i_a \text{ even}, \quad i_a + i_b \text{ odd})$$

$$D(H) = \sum_{i_x i_a i_y i_b i_d} C_{i_x i_y i_d i_a i_b} x^{i_x} y^{i_y} d^{i_d} a^{i_a} b^{i_b} \quad (i_x + i_a \text{ even}, \quad i_a + i_b \text{ even}).$$

We also define

$$H^F = H(\mathcal{I}) = H(x, y, d, a, b)$$
$$H^R = H(\hat{R}\mathcal{I}) = H(x, y, d, -a, -b)$$
$$H^S = H(\hat{S}\mathcal{I}) = H(-x, y, d, -a, b)$$
$$H^C = H(\hat{S}\hat{R}\mathcal{I}) = H(-x, y, d, a, -b).$$

It is easy to show that

$$H^F = A(H) + B(H) + C(H) + D(H)$$
$$H^R = A(H) - B(H) - C(H) + D(H)$$
$$H^S = -A(H) - B(H) + C(H) + D(H)$$
$$H^C = -A(H) + B(H) - C(H) + D(H).$$

We are now ready to present some important results. Some of the proofs will not be given because they are beyond the scope of this book; for details, see (Wan and Berz, 1996; Wan, 1995).

First, we consider a general system of k cells consisting of the four kinds of cells whose transfer maps, as already discussed, can be written in the substandard form. They have the general form

$$\mathcal{M} = \exp(: H(\mathcal{I}) :) \exp(: H(M^{(1)}\mathcal{I}) :) \cdots \exp(: H(M^{(k-1)}\mathcal{I}) :)(M_T \mathcal{I}), \tag{6.57}$$

where M_T is the linear matrix of the system and $M^{(i)}$, $i = 1, 2, \ldots, k-1$ are midplane symmetric matrices obtained from combinations of the linear matrices of previous cells and $\hat{R}$, $\hat{R}^{-1}$, $\hat{S}$, $\hat{S}^{-1}$, and C, C^{-1}. Therefore, $\det(M^{(i)}) = 1$.

According to the BCH formula, $\mathcal{M}$ can be written as a single Lie operator acting on a linear map,

$$\mathcal{M} = \exp(: H_3(\mathcal{I}) + H_3(M^{(1)}\mathcal{I}) + \cdots + H_3(M^{(k-1)}\mathcal{I}) + \cdots :)(M_T \mathcal{I}). \tag{6.58}$$

Since the omitted part contains only monomials of order four and higher, all third-order terms are held in $H_3(\mathcal{I}) + H_3(M^{(1)}\mathcal{I}) + \cdots + H_3(M^{(k-1)}\mathcal{I})$. Therefore, they can only be canceled through symmetric arrangements and extra multipole elements. But $D(H)$ does not change sign under $\hat{R}$, $\hat{S}$ or C, which suggests that

240   IMAGING SYSTEMS

TABLE X
OPTIMAL SYSTEMS

System	Linear conditions
F R S C	$(a\|d) = 0$, $(x\|a) = (a\|x) = 0$
F R F R	$(a\|d) = 0$, $(x\|x) = (a\|a) = 0$
F C S R	$(x\|d) = 0$, $(x\|a) = (a\|x) = 0$
F C F C	$(x\|d) = 0$, $(x\|x) = (a\|a) = 0$

$D(H)$ cannot be canceled by symmetry. In fact, a large part of it does have this property, which is stated in Theorem 6.4.3. For the following, it is useful to define $\mathbf{D}^+(\mathbf{H}(\mathbf{I}))$ as the terms in $D(H(\mathcal{I}))$ with all exponents of x, a, y and b even, and has the form $\sum C_{2n_x,2n_y,i_d,2n_a,2n_b} x^{2n_x} y^{2n_y} d^{i_d} a^{2n_a} b^{2n_b}$. $\mathbf{D}^-(\mathbf{H}(\mathbf{I}))$ is defined as the terms in $D(H(\mathcal{I}))$ with all exponents of x, y, a, and b being odd, and it has the form $\sum C_{2n_x+1,2n_y+1,i_d,2n_a+1,2n_b+1} x^{2n_x+1} y^{2n_y+1} d^{i_d} a^{2n_a+1} b^{2n_b+1}$. n_x, n_y, n_a, and n_b are nonnegative integers.

We next list a collection of important **theorems** relating to k cell systems.
For a given k cell system, it is **impossible to cancel** any term from

$$D^+(H(\mathcal{I})) + D^+(H(M^{(1)}\mathcal{I})) + \cdots + D^+(H(M^{(k-1)}\mathcal{I})) \qquad (6.59)$$

solely by the symmetric arrangements of the cells and the choices of special linear matrices under the assumption that no relations are required among the Lie coefficients $C_{i_x i_a i_y i_b i_d}$.

Given a general k cell system, k has to be **at least four** to cancel $A_3(H(\mathcal{I})) + A_3(H(M^{(1)}\mathcal{I})) + \cdots + A_3(H(M^{(k-1)}\mathcal{I}))$, $B_3(H(\mathcal{I})) + B_3(H(M^{(1)}\mathcal{I})) + \cdots + B_3(H(M^{(k-1)}\mathcal{I}))$ and $C_3(H(\mathcal{I})) + C_3(H(M^{(1)}\mathcal{I})) + \cdots + C_3(H(M^{(k-1)}\mathcal{I}))$.

Given a four-cell system, it **cancels** $A_3(H(\mathcal{I})) + A_3(H(M^{(1)}\mathcal{I})) + A_3(H(M^{(2)}\mathcal{I})) + A_3(H(M^{(3)}\mathcal{I})$, $B_3(H(\mathcal{I})) + B_3(H(M^{(1)}\mathcal{I})) + B_3(H(M^{(2)}\mathcal{I})) + B_3(H(M^{(3)}\mathcal{I}))$, and $C_3(H(\mathcal{I})) + C_3(H(M^{(1)}\mathcal{I})) + C_3(H(M^{(2)}\mathcal{I})) + C_3(H(M^{(3)}\mathcal{I}))$ if and only if $H(M^{(1)}\mathcal{I})$, $H(M^{(2)}\mathcal{I})$, and $H(M^{(3)}\mathcal{I})$ are equal to H^R, H^S and H^C, respectively, or a permutation thereof.

Table X lists the systems that reach the optimum previously stated.

For the optimal systems, **fourth-order achromats** can be obtained by canceling D in the total map. Computer results indicate that this property holds for the fifth and sixth orders, and it is likely a general solution for arbitrary orders.

Utilizing the previous theory, a variety of third- fourth-, and fifth-order achromats have been designed. In the following sections, we provide three examples that show the basic features.

6.4.4 A Snake-Shaped Third-Order Achromat

The first example is **fully achromatic to third order**. It is based on a snake-shaped single-pass system (Fig. 6.10) consisting of four cells. As such, it could

ABERRATION CORRECTION VIA REPETITIVE SYMMETRY 241

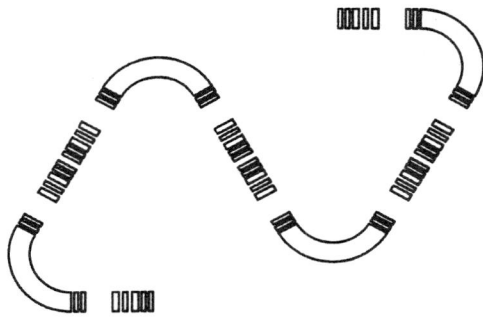

FIGURE 6.10. Third-order layout.

be used as either a beam transport line or a time-of-flight mass spectrometer when it is run on energy-isochronous mode. The layout shows that it is a rather compact system. The first-order forward cell contains a 120° inhomogeneous bending magnet and four quadrupoles (Fig. 6.11). The whole cell is symmetric around the midpoint, which entails that $(x|x) = (a|a)$ and $(y|y) = (b|b)$. Therefore, it is only necessary to fit five instead of seven conditions in order to obtain a map of the form shown in Table XI. Two extra conditions are used to fit the linear matrix to a purely rotational matrix. During the process of fitting, the drifts between each quadrupole and the dipole are fixed and the variables are the field index of the dipole, the two field strengths of the quadrupoles, and the drifts before and between the quadrupoles. The long drifts are places where higher order elements will be placed.

Instead of utilizing the minimum of four sextupoles, for purposes of convenience a total of 10 sextupoles were utilized to correct the second-order aberrations. They were inserted symmetrically with respect to the dipole as shown in Figure 6.12. The required values are weak, which indicates that the first-order

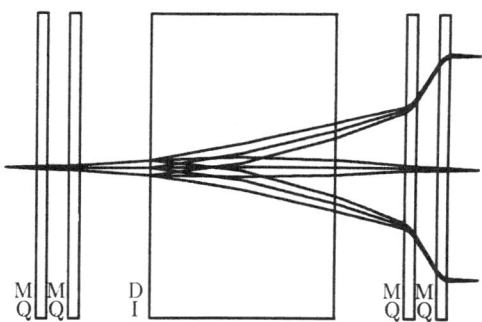

FIGURE 6.11. First Order Forward Cell

TABLE XI
First Order Map of the Forward Cell

```
-1.000000        0.0000000E+00  0.0000000E+00  0.0000000E+00  0.2463307E-15 100000
 0.0000000E+00  -1.000000       0.0000000E+00  0.0000000E+00 -7.366987       010000
 0.0000000E+00   0.0000000E+00  0.1387779E-15  1.000000       0.0000000E+00 001000
 0.0000000E+00   0.0000000E+00 -1.000000       0.1526557E-15  0.0000000E+00 000100
 0.0000000E+00   0.0000000E+00  0.0000000E+00  0.0000000E+00  1.000000      000010
 7.366987       -0.2775558E-15  0.0000000E+00  0.0000000E+00  1.733828      000001
```

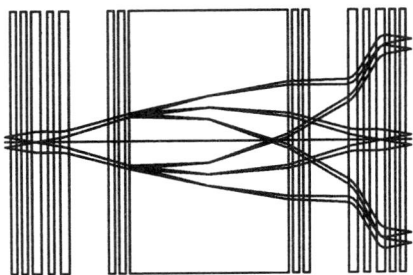

FIGURE 6.12. Second-order layout of the forward cell. The short elements are the sextupoles.

layout gives weak aberrations and the newly introduced sextupoles will also not produce strong third-order aberrations.

The last step is to correct the third order aberrations. Fourteen octupoles are superimposed in the existing quadrupoles and sextupoles because this tends to require weaker octupole fields. An octupole component is added to the inhomogeneous dipole field, which is the 15th and last variable needed for the corrections. Figure 6.13 shows the beam envelope of the whole system. Table XII presents the third-order map, which shows that $(t|d^n)$ ($n = 1, 2, 3$) are the nonzero terms left, which would make this system an effective time-of-flight spectrometer.

Using an emittance of 1 mm mrad and a momentum spread of 1%, Fig. 6.14 shows the motion of the beam to eighth order around the final focal point. The uncorrected aberrations of orders higher than three at the focal point add up to about 10 μm horizontally and 3 μm vertically, which is indeed quite small.

TABLE XII
COSY Output of Third-Order Map of the Four-Cell System

```
 1.000000       0.0000000E+00  0.0000000E+00  0.0000000E+00  0.0000000E+00 100000
 0.0000000E+00  1.000000       0.0000000E+00  0.0000000E+00  0.0000000E+00 010000
 0.0000000E+00  0.0000000E+00  1.000000       0.0000000E+00  0.0000000E+00 001000
 0.0000000E+00  0.0000000E+00  0.0000000E+00  1.000000       0.0000000E+00 000100
 0.0000000E+00  0.0000000E+00  0.0000000E+00  0.0000000E+00  1.000000      000010
 0.0000000E+00  0.0000000E+00  0.0000000E+00  0.0000000E+00  6.935312      000001
 0.0000000E+00  0.0000000E+00  0.0000000E+00  0.0000000E+00 -21.18904      000002
 0.0000000E+00  0.0000000E+00  0.0000000E+00  0.0000000E+00  59.36542      000003
```

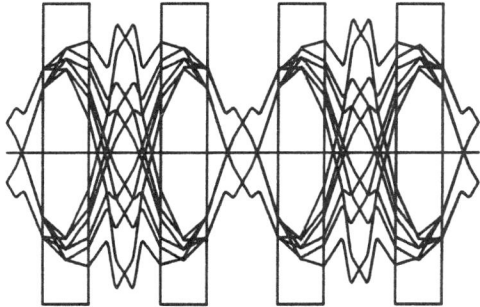

FIGURE 6.13. Third-order beam envelope with the bending magnets shown indeed quite small.

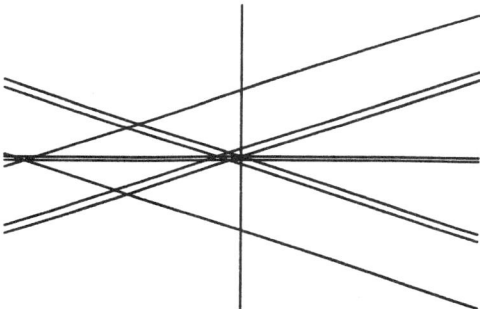

FIGURE 6.14. Remaining aberrations up to order eight (scale: $30\mu m \times 20\mu m$).

6.4.5 Repetitive Third- and Fifth Order Achromats

In the previous example, a compact design resulted due to the use of cells bending in two different directions. In the case of circular accelerators or storage rings, however, all bending usually occurs in only one direction. For this purpose, the only possible choice for the layout of a third-order achromat is the **FRFR configuration**.

For our particular case, we utilize the first-order layout of the ESR ring at GSI and convert it into a time-of-flight mass spectrometer. Because the design is limited to the FRFR configuration, the first-order map of the forward cell has to satisfy $(x|x) = (a|a) = (a|d) = 0$, as shown in Table X. The ESR ring contains six dipoles, 20 quadrupoles, and eight sextupoles as well as RF cavities, beam cooling devices, the injection–extraction system, and beam monitoring devices. Two long, straight sections divide it into two identical parts, each of which is symmetric about its center (Franzke, 1987), as shown in Figure 6.15. Half of the ring is chosen as the forward cell and the quadrupoles are excited symmetrically, which ensures that $(x|x) = (a|a)$ and $(y|y) = (b|b)$ and reduces the number of first-order knobs from five to three. In order to make a system a mass spectrometer,

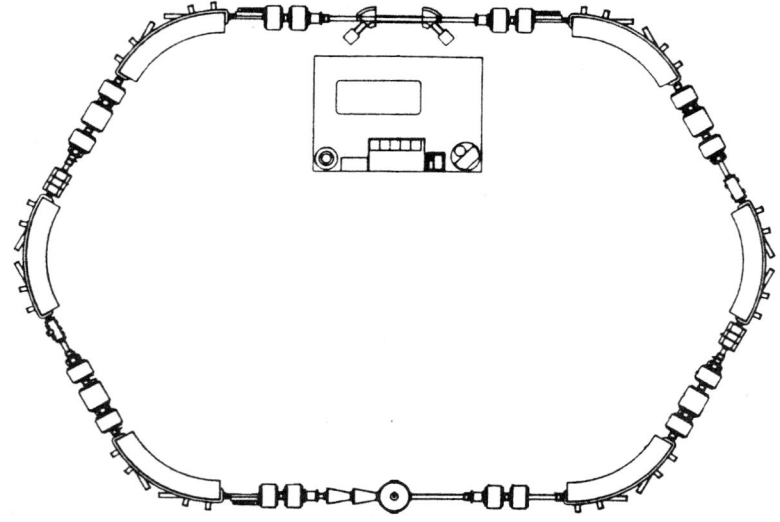

FIGURE 6.15. The original ESR.

$(t|d)$ must also be canceled, which increases the number to four. After the fine-tuning of the free knob, the third quadrupole, the best behaved first-order solution was found and the horizontal beam envelope is shown in Figure 6.16.

According to the previous theory, four independent sextupoles are required to obtain a second-order achromat. However, because of the fact that to first order, the cell R is identical to the cell F, a simplification is possible based on Brown's theory of second-order achromats (Brown, 1979b; Carey, 1981). In this theory it

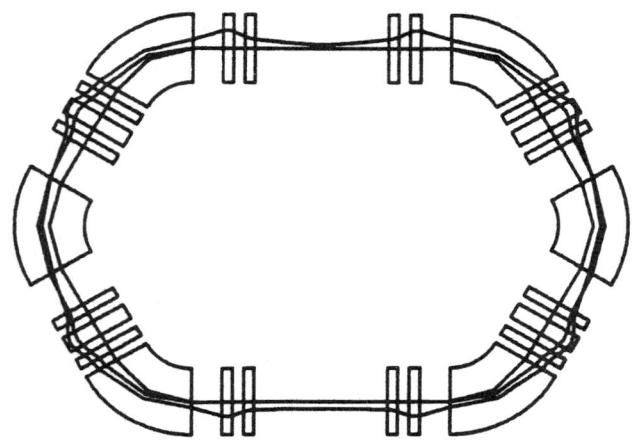

FIGURE 6.16. The first-order beam envelope.

TABLE XIII
FIRST ORDER MAP OF HALF THE RING.

```
0.4551914E-14  0.2306961       0.0000000E+00  0.0000000E+00  0.0000000E+00  1000000
-4.334708      0.4163336E-14   0.0000000E+00  0.0000000E+00  0.0000000E+00  0100000
0.0000000E+00  0.0000000E+00  -0.1443290E-14  0.5119279      0.0000000E+00  0010000
0.0000000E+00  0.0000000E+00  -1.953400      -0.1817990E-14  0.0000000E+00  0001000
0.0000000E+00  0.0000000E+00   0.0000000E+00  0.0000000E+00  1.000000           0000100
0.0000000E+00  0.0000000E+00   0.0000000E+00  0.0000000E+00  0.1679212E-14  0000010
0.0000000E+00  0.0000000E+00   0.0000000E+00  0.0000000E+00 -4.187160           0000001
```

is shown that a second-order achromat can be achieved by placing two pairs of sextupoles in dispersive regions and separating the two in each pair by a negative identity in both transverse planes. In our case, the first-order one-turn map is a negative identity, and thus the same sextupole can be used in two consecutive turns to satisfy the previous requirements. This is the second reason why the forward cell is made symmetric and $(y|y)$ is canceled instead of $(y|b)$ and $(b|y)$. Although in principle a second-order achromat can be achieved with two sextupoles per cell (half ring), the constraint that the second half be the reversion of the first requires that the sextupoles be split into symmetrically excited pairs. Again, one more pair is inserted to remove the term $(t|d^2)$ as shown in Table XIII. Sixteen octupoles were introduced to correct the remaining third-order aberrations including $(t|d^3)$. The positions of some of the multipoles were carefully chosen to minimize the required field strengths, which resulted in a realistic setting.

Since our goal is to make ESR a multipass time-of-flight mass spectrometer, its **dynamic aperture** has to be studied in detail. This was done by tracking the number of turns using its eighth-order one-turn map. For particles of momentum spread of ±0.25% to survive 100 turns, the horizontal and vertical apertures must be approximately 1π mm mrad.

The resolution was determined in a statistical manner by sending a large number of particles inside a certain phase space area through the eighth-order one-turn map n times, from which the n-turn time-of-flight of each particle was computed. Considering the random errors of the detector, which was assumed to be about 100 ps, the predicted mass deviations of all particles were calculated. Finally, the difference between the predicted and initial energy deviations was obtained and the resolution of the ring was determined by calculating the inverse of the average difference. The dependence of the resolution on the number of turns is presented in Fig. 6.17.

As a proof of principle, a circular **fifth-order achromat** is designed by **canceling the term** D in the total map described previously. The first-order layout should avoid large changes in the beam width in order to minimize nonlinear aberrations; furthermore, there should be enough room for the insertion of correction multipoles. Another consideration is that, if possible, the number of first-order conditions should be further reduced through symmetry arrangements inside a cell.

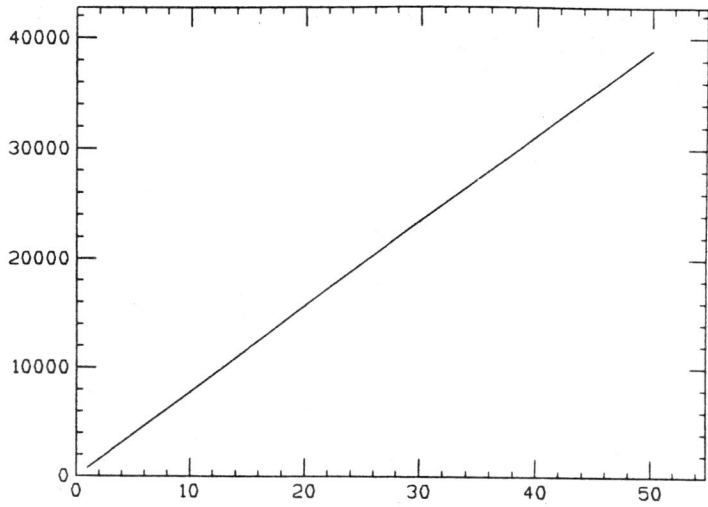

FIGURE 6.17. Resolution versus number of turns.

The result of these considerations is a ring shown in Fig. 6.18, which consists of 16 FODO cells plus two dispersion correction sections each including two quadrupoles. The left half is the forward cell and the right half is the reversed cell. Achromaticity is obtained after two turns. The forward cell consists of two parts, one of which is the reversion of the other. This guarantees that $(x|x) = (a|a)$ and $(y|y) = (b|b)$. All four FODO cells within one part of a cell are identical except that the last one has an extra quadrupole for dispersion correction. Hence, there are three knobs for the first-order design which can zero $(x|x)$, $(a|a)$, $(y|y)$, $(b|b)$, $(x|d)$ and $(a|d)$ at the same time. Figure 6.18 shows that the beam travels around the ring in a uniform manner, avoiding large ray excursions. As described previously, a second-order achromat is achieved by placing and exciting symmetrically two pairs of sextupoles in each half.

After investment in a careful first- and second-order layout, the third-, fourth-, and fifth-order corrections are conceptually straightforward, even though they are computationally more demanding. In the whole process of optimization, only two aspects are worth mentioning. First, the required multipole strengths strongly depend on the average distance between multipoles of the same order. In order to keep their strength limited, it is important to dimension the total size of the ring and the dispersive region sufficiently large and distribute multipoles of the same order approximately uniformly. Second, all decapoles have to be placed in regions with sufficient dispersion because all the fourth-order aberrations remaining after third-order corrections are chromatic aberrations. The combination of these considerations results in weak multipole strengths for third-, fourth-, and fifth-order corrections.

ABERRATION CORRECTION VIA REPETITIVE SYMMETRY 247

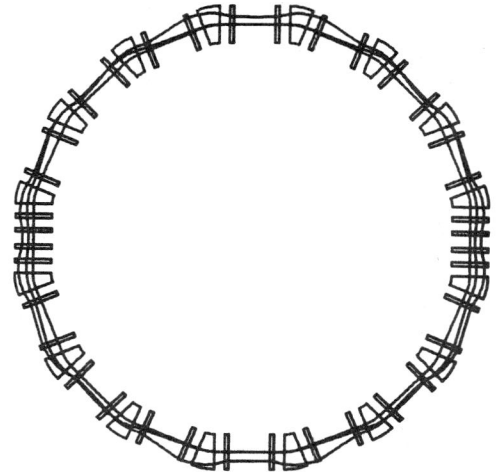

FIGURE 6.18. Layout, beam envelope and dispersive ray of the achromatic ring. Circumference: 266.64 m; Emittance: 30π mm mrad; Dispersion: 0.3%

The 1000-turn dynamic apertures for both horizontal and vertical motions are studied using 11th-order maps. For particles of momentum spread of $\pm 0.5\%$ to survive 1000 turns, the dynamical apertures are at least 100 π mm mrad both horizontally and vertically. They are much larger than the acceptance of the ring, which is approximately 30–40 π mm mrad. Figure 6.19 shows the horizontal motion of on-energy particles up to 1000 turns. The absence of linear effects as well as any nonlinearities up to order five leads to a very unusual behavior that is entirely determined by nonlinearities of order six and higher.

The time-of-flight energy resolution of this ring was determined in a statistical manner similar to that of the previous example except that the one-turn map is of ninth order. The dependence of resolution on the number of turns is presented in Fig. 6.20.

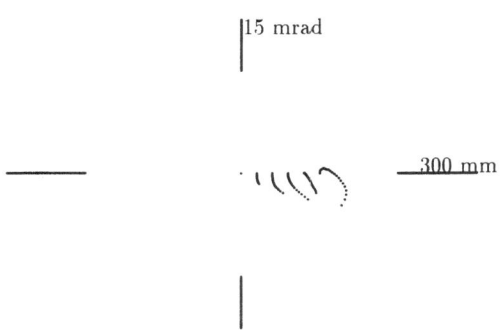

FIGURE 6.19. 1000-turn tracking of the x–a motion of on-energy particles.

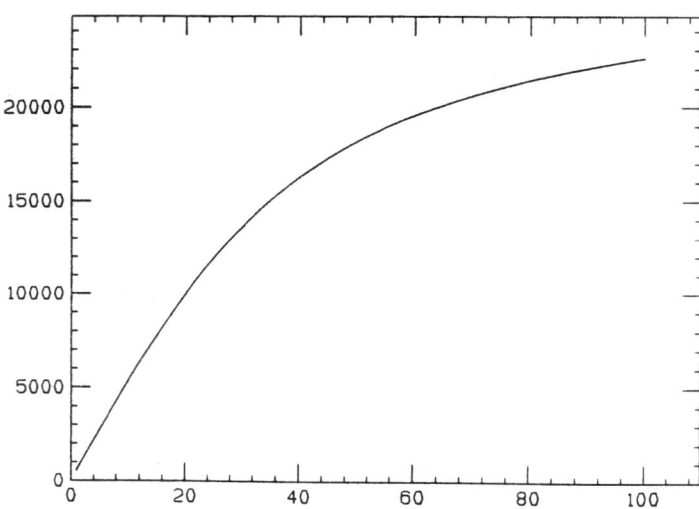

FIGURE 6.20. Resolution vs. numbers of turns at the acceptance.

Chapter 7

Repetitive Systems

For the past three decades, repetitive beam optical systems such as synchrotrons and storage rings have represented one of the major tools for high-energy physics studies. Recently, synchrotron radiation from electron storage rings has been increasingly used in fields such as condensed matter physics, chemistry, and biological sciences. All these repetitive devices require that the particles remain confined in the ring, which means that the motion is **stable** in that their coordinates do not exceed a certain bound d. Therefore, the study of the **dynamics of repetitive systems** has long been a key concern of beam physics.

The description of these repetitive systems in terms of maps relating initial phase space coordinates to final phase coordinates is much the same as for the case of single-pass systems. For the repetitive systems, it is most convenient to compute the map for one complete revolution of the device, the **one-turn map**. Instead of studying the long-term behavior of the differential equations governing the motion, we study the repeated application of a map. This idea has been used since the beginning of the study of dynamical systems, which originated with the extensive work of Henri Poincare (1893), and in acknowledging this is usually referred to as the **Poincare Map**.

The study of the Poincare map is indeed **sufficient** for the understanding of the stability of the entire device. Should the repetitive application of the Poincare map result in unstable motion, then surely the underlying differential equations are unstable because the repeated application of the Poincare map merely corresponds to the evaluation of the solutions of the differential equations at certain fixed times $n \cdot \Delta t$. On the other hand, should the repetitive application of the Poincare map show stability for a certain region G, then one merely has to verify that this region G transports through the system once without exceeding d in between, which can usually be achieved without high requirements for accuracy.

The questions associated with the stability of **repetitive systems are usually of a quite different nature** as in the case of single-pass systems. In case a system is transversed only once, the correction or control of aberrations is the dominating question, and the knowledge of aberrations to sufficiently high order allows a complete understanding of the system under consideration. In the case of repetitive systems the situation is more complicated and subtle since for sufficiently large numbers of iterations, even slight approximations that are made in representing the system may build up over many turns.

To understand this fundamentally important effect qualitatively, we consider a repetitive system that is well behaved in the sense that the particles are stable. Consider the motion of one particle through the Poincare map, and record its phase space position over many turns. Now assume the system is perturbed in such a way that after each turn, the particle's coordinates are moved by a small amount depending on the position of the particle. In some instances, the various displacements may **average out**, and the qualitative behavior of the particle is the same as before. On the other hand, it may also happen that at the locations the particle visits, the displacements have a tendency to move in the same direction and thus **repeatedly enhance** each other, which over longer terms may fundamentally change the character of the motion and even lead to its instability. While qualitatively it is very difficult to characterize this effect in more detail, it is to be expected that consequences are more likely if the particle's motion is restricted to only a few locations in phase space. In this case, the average of the effect has to be taken over a much smaller region, and hence the likelihood of it averaging out decreases. This effect is particularly pronounced if the particle in fact exactly repeats its motion after k turns, which we call a kth-order **resonance**. Many of these effects can be described very well in a quantitative manner with the so-called normal form methods discussed in detail later.

7.1 LINEAR THEORY

While the understanding of the detailed effects of repetitive motion is usually very complicated, as long as the motion is linear, it is completely straightforward. Since the work of Courant and Snyder (1958), the linear theory of repetitive systems has been fully developed and has become a basic tool in the design and understanding of circular machines. In this section, we discuss the stability of the linear Hamiltonian motion and the invariant ellipses and the behavior of the elliptical beam phase space. An overview of various types of lattices used in circular machines can be found in (Berz *et al.*, 1999).

7.1.1 The Stability of the Linear Motion

As discussed previously, the stability of the motion of charged particles is the most important problem for a repetitive system because particles usually have to stay in the system for millions of turns. The stability of the linear motion is determined by the **eigenvalues** of the transfer matrix; if any of them has an absolute value >1, the motion is unstable. This can be seen as follows. Suppose the initial state of a particle is an eigenvector $\vec{V}_j$ with eigenvalue λ_j. After n turns, the final state is

$$\vec{r}_f = \lambda_j^n \cdot \vec{V}_j; \tag{7.1}$$

LINEAR THEORY 251

therefore, by one of the properties of a linear vector-space norm,

$$||\vec{r}_f|| = |\lambda_j|^n \cdot ||\vec{V}_j||.$$

Hence, if $|\lambda_j| > 1$, the "length" of $\vec{r}_f$ goes to infinity as n becomes large. In the more general case that $\vec{r}_i$ is a linear combination of eigenvectors, $\vec{r}_i = \sum_k a_k \vec{V}_k$, its component in direction of $\vec{V}_j$ will still increase with a factor of $|\lambda_j|^n$; therefore, the motion appears to be "drawn" toward this dominating eigenvector and ultimately will behave very similarly to Eq. (7.1). Hence, if $|\lambda_{\max}| > 1$, some particles will be lost, and the device is **linearly unstable**.

As shown in Chapter 1, Section 1.4.4, the system of a particle moving in electromagnetic field can be described by a Hamiltonian, which according to Liouville's theorem (Eq. 1.184) entails that the phase space volume is conserved, and according to (Eq. 1.239) the transfer map is symplectic, which means that its Jacobian M satisfies

$$\hat{M}\hat{J}\hat{M}^t = \hat{J}, \qquad (7.2)$$

where $\hat{J}$ is the matrix from Eq. (1.86).

A direct result of symplecticity is that the eigenvalues always appear in **reciprocal pairs**. This can be shown from the facts that $\det(\hat{M}) = \det(\hat{J}) = 1$, $\hat{J}^2 = -\hat{I}$ and $\det(\hat{M}^t) = \det(\hat{M})$. From the previous symplectic condition, we obtain $\hat{M} = -\hat{J}(\hat{M}^t)^{-1}\hat{J}$. Hence,

$$\det(\hat{M} - \lambda \hat{I}) = \det(-\hat{J}(\hat{M}^t)^{-1}\hat{J} - \lambda \hat{I}) = \det(-(\hat{M}^t)^{-1} + \lambda \hat{I})$$
$$= \det(-\hat{I} + \lambda \hat{M}^t) = \lambda^{2v}\det\left(\hat{M} - \frac{1}{\lambda}\hat{I}\right)$$

where $2v$ is the dimensionality of phase space. Moreover, the fact that $\hat{M}$ is real requires that any complex eigenvalues also appear in conjugate pairs. Therefore, when $|\lambda| \neq 1$ and $\lambda \neq 0$, they are always in **quadruples** arranged symmetrically with respect to the real axis and the unit circle (Fig. 7.1). The discussion at the beginning of this section shows that the system is linearly stable when all eigenvalues of $\hat{M}$ satisfy $|\lambda| \leq 1$. This implies that the system is **stable when all eigenvalues lie on the unit circle**. In this case, reciprocity and conjugacy are the same, and thus the eigenvalues always occur in pairs.

Next, we study the stability of a system with respect to a small **perturbation**, and under what conditions its stability is sensitive to the perturbation. We first discuss the case in which $\hat{M}$ is nondegenerate, i.e., all $2v$ eigenvalues are distinct. Then it is possible to draw circles centered at each eigenvalue such that none of the circles intersect (Fig. 7.2). If the perturbation is small enough, all eigenvalues of the perturbed system still lie in their respective circles. Suppose that one eigenvalue λ moves off the unit circle, which means that $|\lambda| \neq 1$. Symplecticity

252 REPETITIVE SYSTEMS

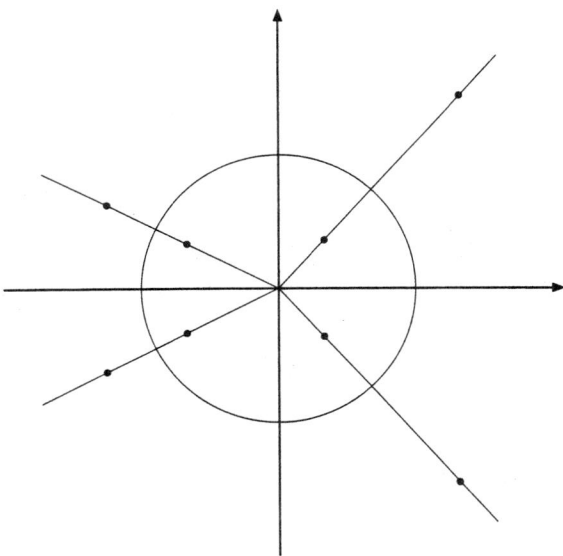

FIGURE 7.1. Eigenvalues of a symplectic matrix.

would then require that another eigenvalue $\lambda' = 1/\lambda$ exists in the same circle in which λ lies. As a result, the total number of eigenvalues is $>2v$, which is not possible. This shows that nondegenerate stable system remains **stable under small perturbations**.

The previous arguments prove that perturbing a nondegenerate stable system causes the eigenvalues to move along the unit circle. What happens when two **eigenvalues collide**? First, when two eigenvalues collide, their complex conjugates also collide. These four eigenvalues can form a quadruple, can leave the unit circle, and can make the system **unstable** (Fig. 7.3). Collisions also occur when two complex–conjugate eigenvalues meet on the real axis. These collisions may lead to instabilities because $\text{Im}(\lambda)$ vanishes at the collision point (Fig. 7.4).

Now, we discuss a method to **decouple** a general, not necessarily symplectic, linear matrix into 2×2 blocks, which will make it much easier to study the evolution of the system. In the following, we assume that the linear part of the phase space map has **distinct eigenvalues**, which is the case for most important systems. Then the linear part of the map can be diagonalized. We now group the eigenvalues such that complex–conjugate eigenvalues form pairs; we group any remaining real eigenvalues into pairs by demanding that the elements of a pair have the same sign. This is always possible since the determinant is positive and thus there is an even number of negative eigenvalues.

LINEAR THEORY 253

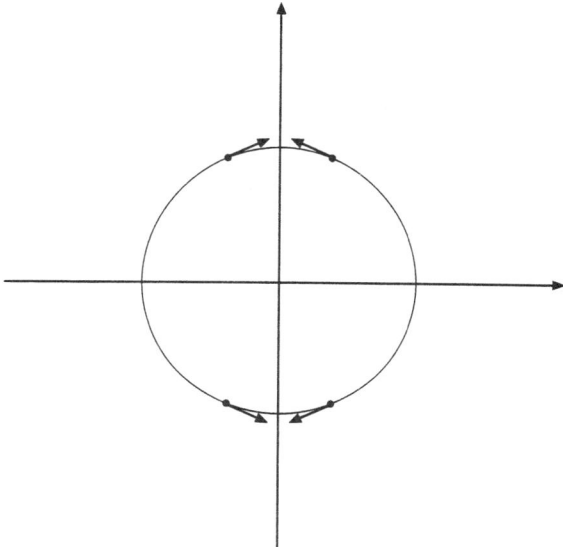

FIGURE 7.2. Behavior of nondegenerate eigenvalues under small perturbation.

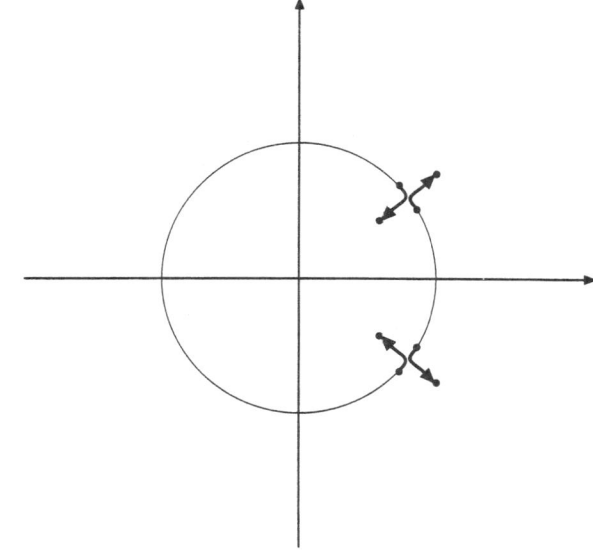

FIGURE 7.3. Possible behavior of degenerate eigenvalues under small perturbation.

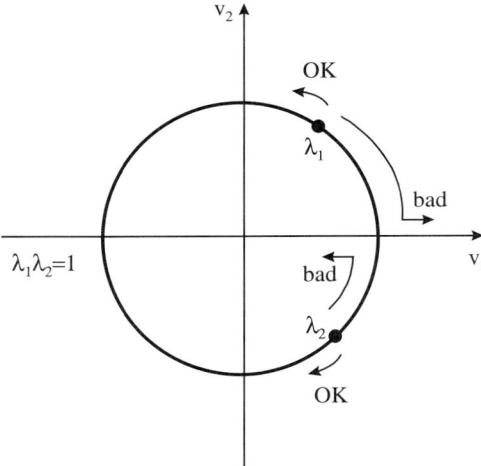

FIGURE 7.4. Collision of complex–conjugate eigenvalues on the real axis.

Each pair we write as $r_j \cdot e^{\pm i\mu_j}$. In the case of a complex pair, this is readily accomplished by choosing r_j and μ_j as the modulus and phase. In the case of a real pair R_{1j} and R_{2j}, we choose $r_j = \pm\sqrt{R_{1j}R_{2j}}$, where the sign is determined to be the same as that of R_{1j} and R_{2j}. We then choose $\mu_j = i \cdot \log(\sqrt{R_{1j}/R_{2j}})$. Since the determinant is nonzero and R_{1j} and R_{2j} are of the same sign, r_j and μ_j are always well-defined.

Denoting the eigenvectors corresponding to $r_j e^{\pm i\mu_j}$ by $s_j^{\pm}$, we obtain that in the eigenvectors basis, the linear part of the map has the form

$$\begin{pmatrix} r_1 e^{+i\mu_1} & & & & \\ & r_1 e^{-i\mu_1} & & & \\ & & \ddots & & 0 \\ & 0 & & r_v e^{+i\mu_v} & \\ & & & & r_v e^{-i\mu_v} \end{pmatrix}. \quad (7.3)$$

We note that if the jth eigenvalue pair consists of complex conjugate eigenvalues, then so are the associated eigenvectors, and if the jth eigenvalue pair is real, so are the eigenvectors.

We now perform another change of basis after which the matrix is real. For each complex conjugate pair of eigenvalues, we choose the real and imaginary parts of an eigenvector as two basis vectors. For each pair of real eigenvalues, we choose the two real eigenvectors themselves.

LINEAR THEORY

The result of this change of basis is a matrix having 2×2 subblocks along the diagonal. A subblock originating from a complex eigenvalue pair will have four nonzero entries, and a subblock originating from a real eigenvalue pair will be diagonal. Therefore, the matrix has the form

$$\begin{pmatrix} a_1 & b_1 & & & \\ c_1 & d_1 & & 0 & \\ & & \ddots & & \\ & 0 & & a_v & b_v \\ & & & c_v & d_v \end{pmatrix}. \quad (7.4)$$

We note that if the underlying matrix is symplectic, it is possible to scale the transformation matrix such that it is also symplectic via a Gram-Schmidt like procedure with the antisymmetric scalar product $<,>$. Since products of symplectic matrices are symplectic, so is the transformed matrix.

This decoupling provides a convenient stepping stone to the computation of the quantities relevant for a study of stability. With the linear matrix decoupled by the method described previously, or by the symmetry of the system (e.g., the midplane symmetry in Section 3.1.2), it is sufficient to study only one of the 2×2 blocks to understand the linear behavior of the system. The characteristic polynomial of a 2×2 matrix is

$$\lambda^2 - T \cdot \lambda + D = 0, \quad (7.5)$$

where T and D are its trace and determinant, respectively. An immediate result is that $\lambda_1 \lambda_2 = D$. When $D > 1$, the system is always unstable because there would be at least one eigenvalue whose absolute value is >1. However, in practice this case will never be chosen deliberately in single-particle systems and therefore will not be discussed further. When $D < 1$, the system is stable not only when the eigenvalues form a complex–conjugate pair but also when they are real numbers and both of them have absolute values smaller than 1; this case occurs when damping is present. When $D = 1$, the system is symplectic; it is stable only when the eigenvalues form a complex–conjugate pair.

The eigenvalues of Eq. (7.5) are given by

$$\lambda_{1,2} = \frac{T}{2} \pm \sqrt{\left(\frac{T}{2}\right)^2 - D}. \quad (7.6)$$

The nature of the eigenvalues depends on the relation between T and D.

As the **first case**, we consider $|T| > 2\sqrt{D}$. Then $\lambda_{1,2}$ are real numbers. If $D < 1$, the system can be stable or unstable depending on the value of T. If $D = 1$, the system is unstable and a phase space point from it moves along a hyperbola, which can be seen as follows (Fig. 7.5). Considering motion in eigenvector coordinates

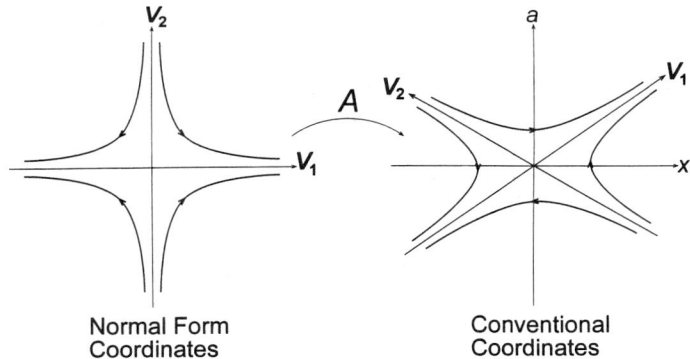

FIGURE 7.5. Trajectories of phase space points in the eigenvector coordinates in case of $|tr\hat{M}| > 2$.

$\vec{V}_1$ and $\vec{V}_2$, we have

$$\vec{x}_0 = a_0 \vec{V}_1 + b_0 \vec{V}_2,$$
$$\vec{x}_1 = \hat{M}\vec{x}_0$$
$$= \lambda_1 a_0 \vec{V}_1 + \lambda_2 b_0 \vec{V}_2$$
$$= a_1 \vec{V}_1 + b_1 \vec{V}_2.$$

Thus, the relations between the components are

$$a_1 = \lambda_1 a_0, \tag{7.7}$$
$$b_1 = \lambda_2 b_0. \tag{7.8}$$

Since $\lambda_1 \cdot \lambda_2 = D = 1$, we obtain that

$$a_1 b_1 = a_0 b_0 = \text{const.} \tag{7.9}$$

This curve shows that the particle moves further out and the system is **unstable**.

As the **second case**, we consider $|T| = 2\sqrt{D}$. The eigenvalues are degenerate, where

$$\lambda_{1,2} = \pm\sqrt{D}. \tag{7.10}$$

Therefore, the motion is stable when $D < 1$. If $D = 1$, it is **marginally stable**; that is, it can become unstable under perturbation as discussed previously.

As the **third case**, we study $|T| < 2\sqrt{D}$. The eigenvalues are

$$\lambda_{1,2} = \sqrt{D}\left(\frac{T}{2\sqrt{D}} \pm i\sqrt{1-\left(\frac{T}{2\sqrt{D}}\right)^2}\right). \tag{7.11}$$

We introduce

$$r = \sqrt{D}$$
$$\mu = \text{sign}(b) \cdot \text{acos}\left(\frac{T}{2\sqrt{D}}\right), \tag{7.12}$$

where $\text{sign}(b) = +1$ if $b \geq 0$, and $\text{sign}(b) = -1$ if $b < 0$. The solutions can be written as

$$\lambda_{1,2} = r \cdot e^{\pm i\mu}. \tag{7.13}$$

It is clear that $|\lambda_{1,2}| = r$, and hence the motion is stable when $D \leq 1$.

It is useful and customary (Courant and Snyder, 1958) to introduce a set of new quantities, the so-called **Twiss parameters**

$$\alpha = \frac{a-d}{2r \sin \mu}$$
$$\beta = \frac{b}{r \sin \mu}$$
$$\gamma = -\frac{c}{r \sin \mu}. \tag{7.14}$$

The Twiss parameters satisfy $\beta\gamma - \alpha^2 = 1$, so they are not independent. Two of them, together with r and μ, determine the matrix $\hat{M}$. We note that β is always positive.

Using the Twiss parameters, the matrix can be written as

$$\hat{M} = r \cdot \begin{pmatrix} \cos \mu + \alpha \sin \mu & \beta \sin \mu \\ -\gamma \sin \mu & \cos \mu - \alpha \sin \mu \end{pmatrix}. \tag{7.15}$$

For the sake of completeness we note that in the important case of $\beta \neq 0$, the eigenvectors assume the compact form

$$v_{1,2} = (i\beta, -i\alpha \mp 1). \tag{7.16}$$

The Twiss parameters define the similarity transformation in which the map is diagonal, and we obtain

$$\hat{A} \cdot \hat{M} \cdot \hat{A}^{-1} = \begin{pmatrix} r \cdot e^{+i\mu} & 0 \\ 0 & r \cdot e^{-i\mu} \end{pmatrix}, \tag{7.17}$$

where

$$\hat{A}^{-1} = \begin{pmatrix} i\beta & i\beta \\ -1-i\alpha & 1-i\alpha \end{pmatrix} \quad \text{and} \quad \hat{A} = \begin{pmatrix} (1-i\alpha)/2i\beta & -1/2 \\ (1+i\alpha)/2i\beta & +1/2 \end{pmatrix}. \tag{7.18}$$

Since the case $r > 1$, and thus $D > 1$, leads to unstable motion, it is not of interest to us. When $r < 1$, and thus $D < 1$, the phase space area shrinks with each turn. This is the case when damping is present. If the damping is strong and r is significantly smaller than unity, the nonlinear motion is simple, whereas if the damping is weak and thus r is very close to unity, the motion is similar to the one with $r = 1$. Thus, in the rest of this section only stable symplectic motion is studied.

First, the transfer matrix can be written as

$$\hat{M} = \begin{pmatrix} \cos\mu + \alpha\sin\mu & \beta\sin\mu \\ -\gamma\sin\mu & \cos\mu - \alpha\sin\mu \end{pmatrix}, \quad (7.19)$$

which has the eigenvalues $\lambda_{1,2} = e^{\pm i\mu}$. Here, the quantity μ is a very important characteristic parameter of a repetitive system, and is called the **tune**. The eigenvectors satisfy the same relation as the eigenvalues, $\vec{V}_1 = \vec{V}_2$. We now introduce a new basis that is defined by the unit vectors $\vec{V}_+ = Re(\vec{V}_1)$ and $\vec{V}_- = Im(\vec{V}_1)$. Expressed in these unit vectors, the motion is said to be in **linear normal form.** In this coordinate system, we have

$$\hat{M}\vec{z} = \begin{pmatrix} \cos\mu & \sin\mu \\ -\sin\mu & \cos\mu \end{pmatrix}\vec{z}, \quad (7.20)$$

which means that any particle moves in a circle; therefore, the motion is stable under iteration (Figure 7.6). The angular advance in the normal coordinates is constant for each iteration and equals the tune μ.

Note that in the original coordinates, the angular advance does not have to be constant from turn to turn; but it can quickly be determined that over many turns, the **average angular advance equals the tune**. For this purpose we observe that the angle advance in the original coordinates can be described by the angle advance due to the transformation to linear normal form coordinates plus the angle advance in linear normal form coordinates plus the angle advance of the reverse transformation. Since for many iterations, the transformation advances contribute only an insignificant amount, the average approaches the angle in the normal form coordinates.

It is very important in practice to ensure the stability of the accelerator when it is under small perturbation. As shown previously, this requires that the **traces** of any submatrix lie between -2 and 2 and also that the traces be different from one another. Furthermore, the choice of the traces, and thus the tunes, is crucial to the nonlinear behavior of a repetitive system and often a delicate one in the sense of trying to avoid the so-called resonances.

To conclude this discussion, we address another aspect of repetitive motion, namely, the behavior of **off-energy particles**. In case there are no accelerating structures, such as in a plain FODO lattice (*f*ocusing element, *d*efocusing element, and drifts denoted by O), the energies of the individual particles stay constant in time but affect the motion in that particles with higher energy are bent less in dipoles and in quadrupoles. Therefore, the existence of dipoles, which are always present in repetitive systems, causes the off-energy particles that move along the center to drift away from it. Likewise, the focusing power of quadrupoles is changed for off-energy particles. Consequently, closed orbit as well as the tunes of the off-energy particles are likely to differ from those of on-energy particles. The tune shifts caused by the energy deviation are customarily called **chromaticities**.

The presence of different off-energy closed orbits and tunes leads to potentially different stability properties of off-energy particles. Therefore, it is a standard practice in accelerator design to eliminate at least first-order chromaticities, i.e., the linear dependence of the tunes on energy, in order to increase the range of **energy acceptance**.

We now address the linear behavior of the off-energy particles, which can be described within a matrix theory. The full nonlinear behavior of these particles will be discussed in detail in Section 7.2.3. For a lattice satisfying midplane symmetry, $(y|d)$ and $(b|d)$ vanish; hence, a 3×3 matrix can describe the motion:

$$\hat{M}_x = \begin{pmatrix} (x|x) & (x|a) & (x|d) \\ (a|x) & (a|a) & (a|d) \\ 0 & 0 & 1 \end{pmatrix}. \tag{7.21}$$

When studying the eigenvalue spectrum of this matrix, different from the 2×2 case, there is now an additional eigenvalue, which can be read to be 1. The corresponding eigenvector is a **fixed point** of the matrix and is denoted as $(\eta, \eta', 1)$. It is easy to show that

$$\eta = \frac{[1 - (a|a)](x|d) + (x|a)(a|d)}{2(1 - \cos\mu)}, \tag{7.22}$$

$$\eta' = \frac{[1 - (x|x)](a|d) + (a|x)(x|d)}{2(1 - \cos\mu)}. \tag{7.23}$$

The values of η and η' describe position and momentum of the fixed point of an off-energy particle with $d = 1$. Similarly, the fixed point of an off-energy particle with another value of d is given by $(\eta d, \eta' d)$ because $(\eta d, \eta' d, d)$ is also a fixed point of $\hat{M}_x$. Apparently, the value of η for different positions around the accelerator is an important quantity for its description, and it is commonly referred to as the **eta-function**.

The knowledge of η and η' allows the introduction of new relative coordinates around the fixed point. In these coordinates, the matrix of the linear motion assumes the form

$$\hat{M}_x^* = \begin{pmatrix} (x|x)^* & (x|a)^* & 0 \\ (a|x)^* & (a|a)^* & 0 \\ 0 & 0 & 1 \end{pmatrix}, \tag{7.24}$$

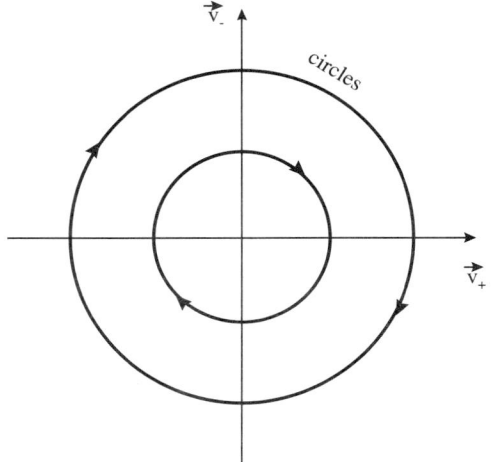

FIGURE 7.6. Trajectories in the case of $|tr\hat{M}| < 2$.

and in particular has the same 2×2 block form as before. This allows the computation of tunes of off-energy particles as $\arccos(((x|x)^* + (a|a)^*)/2)$ and hence their dependence on different values of d.

7.1.2 The Invariant Ellipse of Stable Symplectic Motion

In this section we study the dynamics in a linear stable symplectic system. As shown in Eq. (7.19), the matrix has the form

$$\hat{M} = \begin{pmatrix} \cos\mu + \alpha\sin\mu & \beta\sin\mu \\ -\gamma\sin\mu & \cos\mu - \alpha\sin\mu \end{pmatrix}, \quad (7.25)$$

and we have $\det(\hat{M}) = 1$, which implies that $\beta\gamma - \alpha^2 = 1$.

First, we prove that

$$\hat{M}^t \begin{pmatrix} \gamma & \alpha \\ \alpha & \beta \end{pmatrix} \hat{M} = \begin{pmatrix} \gamma & \alpha \\ \alpha & \beta \end{pmatrix}. \quad (7.26)$$

Let $\hat{T} = \begin{pmatrix} \gamma & \alpha \\ \alpha & \beta \end{pmatrix}$. We introduce the matrix $\hat{K} = \begin{pmatrix} \alpha & \beta \\ -\gamma & -\alpha \end{pmatrix} = \hat{J}\hat{T}$.
Then $\hat{M}$ can be expressed in terms of $\hat{K}$ and the 2×2 unity matrix $\hat{I}$:

$$\hat{M} = \hat{I}\cos\mu + \hat{K}\sin\mu. \quad (7.27)$$

LINEAR THEORY

We have

$$\hat{T}\hat{K} = \begin{pmatrix} \gamma_i & \alpha_i \\ \alpha_i & \beta_i \end{pmatrix} \begin{pmatrix} \alpha_i & \beta_i \\ -\gamma_i & -\alpha_i \end{pmatrix} = \hat{J}$$

$$\hat{K}^t\hat{T} = \hat{K}^t\hat{T}^t = \hat{J}^t = -\hat{J}$$

$$\hat{K}^t\hat{T}\hat{K} = -\hat{J}\hat{K} = \hat{T}$$

and so

$$\hat{M}^t\hat{T}\hat{M} = (\hat{I}\cos\mu + \hat{K}^t\sin\mu)\hat{T}(\hat{I}\cos\mu + \hat{K}\sin\mu)$$
$$= \hat{T}.$$

It follows that the quadratic form,

$$\vec{z}^t T \vec{z} = \varepsilon, \tag{7.28}$$

remains **invariant** under the transformation $\hat{M}$. This relation entails that if the envelope of a beam at the beginning fills the ellipse described by the quadratic form, it will remain the same after one turn and hence after any number of turns. This ellipse is an invariant of motion, which is called the invariant ellipse or the **Courant–Snyder invariant**. A beam that fills an invariant ellipse is said to be **matched** to the repetitive system. If the beam is not matched, the beam size at a certain point will vary from turn to turn, which is referred to as **beating** (Fig. 7.7).

The invariant ellipse of a stable repetitive system is a characteristic of the system and is as important as the tune. Also, it has universal importance, independent of the particular location of the Poincare section. Consider two Poincare sections at different locations around a repetitive system, and let $\hat{M}_1$ and $\hat{M}_2$ be the one-turn maps describing the motion from the respective Poincare sections. Let $\hat{M}_{12}$ and $\hat{M}_{21}$ be the maps from section 1 to 2 and from section 2 to 1, respectively. Then $\hat{M}_1$ and $\hat{M}_2$ can be expressed in terms of $\hat{M}_{12}$ and $\hat{M}_{21}$ via the equations

$$\hat{M}_1 = \hat{M}_{21} \cdot \hat{M}_{12} \tag{7.29}$$

$$\hat{M}_2 = \hat{M}_{12} \cdot \hat{M}_{21}. \tag{7.30}$$

By eliminating $\hat{M}_{21}$, we obtain the following relationship between $\hat{M}_1$ and $\hat{M}_2$:

$$\hat{M}_2 = \hat{M}_{12} \cdot \hat{M}_1 \cdot \hat{M}_{12}^{-1}. \tag{7.31}$$

This entails that $\hat{M}_1$ and $\hat{M}_2$ have the same trace and hence the same tune. Furthermore, if the system is matched at one Poincare section, it is matched at the other.

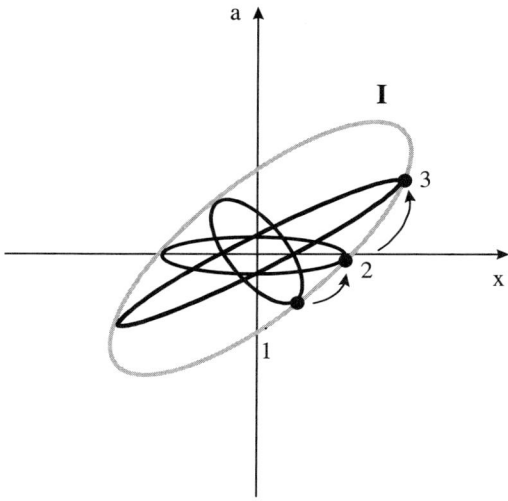

FIGURE 7.7. The relationship between the beam ellipse and the machine ellipse.

7.1.3 Transformations of Elliptical Phase Space

Here, we discuss how a beam traverses phase space. In reality, the shape of the phase space of a beam can be irregular, and there exist various ways to model it. For example, parallelograms and other **polygons** are sometimes used to represent beams in single-pass systems. The study of the transformation properties of such a polygon is straightforward; since the transformation preserves straight lines, it is sufficient to study the transformation of the corner points of the polygon.

Another important approach is to describe the beam by an **ellipse**. Similar to polygons, the transformation of ellipses occurs through linear transformations, and it is merely necessary to study the transformation properties of the parameters describing the ellipse. This is particularly advantageous for repetitive systems that are operated most efficiently if the beam fills the invariant ellipse of the system.

In the following we study the propagation of such beam ellipses through linear symplectic transformations. To begin, we study some basic geometric properties of ellipses. An arbitrary ellipse centered at the origin is described by

$$(x, a) \begin{pmatrix} \gamma & \alpha \\ \alpha & \beta \end{pmatrix} \begin{pmatrix} x \\ a \end{pmatrix} = \varepsilon. \tag{7.32}$$

The quantity ε serves as a scaling parameter, and without loss of generality we thus require the determinant of the ellipse to be unity, i.e., $\beta\gamma - \alpha^2 = 1$. After simple arithmetic, we have

$$\gamma x^2 + 2\alpha x a + \beta a^2 = \varepsilon, \tag{7.33}$$

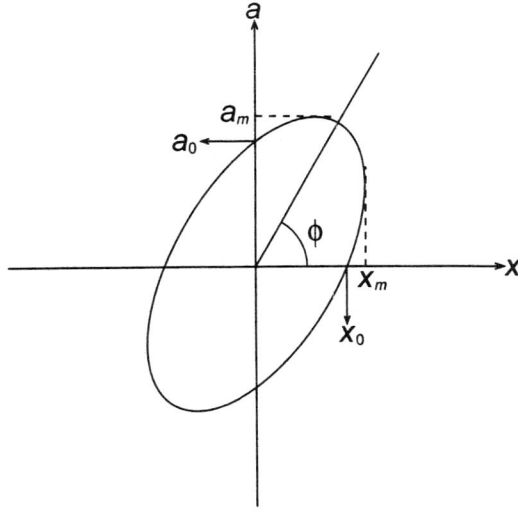

FIGURE 7.8. Elliptic phase area.

from which we determine that the **axis intersections** of the ellipse are

$$x_0 = \pm\sqrt{\frac{\varepsilon}{\gamma}} \quad \text{and} \quad a_0 = \pm\sqrt{\frac{\varepsilon}{\beta}}. \tag{7.34}$$

To determine the **maxima** in the x and a directions, let

$$f(x, a) = \gamma x^2 + 2\alpha x a + \beta a^2;$$

then we have

$$\vec{\nabla} f = (2\gamma x + 2\alpha a, 2\alpha x + 2\beta a). \tag{7.35}$$

At $x_m (\vec{\nabla} f)_a = 0$; thus, at this point, $a = -\alpha/\beta \cdot x$. We define the slope S of the ellipse as

$$S = -\frac{\alpha}{\beta}. \tag{7.36}$$

Inserting a back into Eq. (7.33), we obtain x_m; in a similar way, we can obtain a_m. Altogether, we have

$$x_m = \pm\sqrt{\beta \varepsilon} \quad \text{and} \quad a_m = \pm\sqrt{\gamma \varepsilon}. \tag{7.37}$$

As a function of the position s around the ring, the quantity β is a measure for the beam width and is thus important. The function β is often called the **betatron** or **beta function** and $x_{\max}$ the **beam envelope**. It is worth noting that the area of the beam ellipse described in Eq. (7.33) is $\pi\varepsilon$, which is called the **emittance** (Fig. 7.8).

Now we return to the question of beam ellipses transformations. Suppose at one point a beam fills the ellipse

$$(x_1, a_1) \begin{pmatrix} \gamma_1 & \alpha_1 \\ \alpha_1 & \beta_1 \end{pmatrix} \begin{pmatrix} x_1 \\ a_1 \end{pmatrix} = \varepsilon. \tag{7.38}$$

Under a map $\hat{M}$, the point (x_1, a_1) moves to

$$\begin{pmatrix} x_2 \\ a_2 \end{pmatrix} = \hat{M} \cdot \begin{pmatrix} x_1 \\ a_1 \end{pmatrix}. \tag{7.39}$$

Conversely, we have

$$\begin{pmatrix} x_1 \\ a_1 \end{pmatrix} = \hat{M}^{-1} \cdot \begin{pmatrix} x_2 \\ a_2 \end{pmatrix}. \tag{7.40}$$

Inserting this relation back into Eq. (7.38), we obtain the new beam ellipse as

$$(x_2, a_2) \left[(\hat{M}^{-1})^t \begin{pmatrix} \gamma_1 & \alpha_1 \\ \alpha_1 & \beta_1 \end{pmatrix} \hat{M}^{-1} \right] \begin{pmatrix} x_2 \\ a_2 \end{pmatrix} = \varepsilon. \tag{7.41}$$

Since

$$\left[(\hat{M}^{-1})^t \begin{pmatrix} \gamma_1 & \alpha_1 \\ \alpha_1 & \beta_1 \end{pmatrix} \hat{M}^{-1} \right]^t = \left[(\hat{M}^{-1})^t \begin{pmatrix} \gamma_1 & \alpha_1 \\ \alpha_1 & \beta_1 \end{pmatrix} \hat{M}^{-1} \right],$$

the beam still fills an ellipse at point (x_2, a_2). The relations between the initial and final beam parameters are explicitly given as

$$\begin{pmatrix} \beta_2 \\ \alpha_2 \\ \gamma_2 \end{pmatrix} = \begin{pmatrix} (x|x)^2 & -2(x|x)(x|a) & (x|a)^2 \\ -(x|x)(a|x) & (x|x)(a|a) + (x|a)(a|x) & -(x|a)(a|a) \\ (a|x)^2 & -2(a|x)(a|a) & (a|a)^2 \end{pmatrix}$$

$$\cdot \begin{pmatrix} \beta_1 \\ \alpha_1 \\ \gamma_1 \end{pmatrix}. \tag{7.42}$$

For the special case that the transformation describes one turn around a ring of a beam that is matched to the machine, we would have that $(\beta_2, \alpha_2, \gamma_2) = (\beta_1, \alpha_1, \gamma_1)$. Viewed in terms of Eq. (7.42), this represents an eigenvalue problem and offers another way to compute the invariant ellipse of a given system.

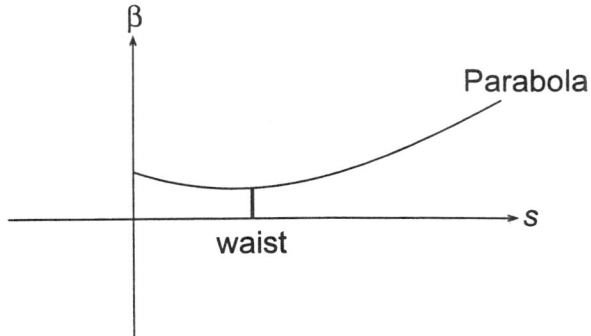

FIGURE 7.9. The behavior of β through a drift.

As an example for the transformation properties of a beam ellipse as a function of s, we consider the transform through a drift, which has the transfer matrix

$$\hat{M} = \begin{pmatrix} 1 & s \\ 0 & 1 \end{pmatrix}.$$

Plugging the matrix elements into Eq. (7.42), we have

$$\beta_2 = \beta_1 - 2s\alpha_1 + s^2\gamma_1. \tag{7.43}$$

Therefore, the beta function of the beam varies **quadratically** in a drift. The minimum of β_2 occurs at $s = \alpha_1/\gamma_1$, and this point is called the **waist** (Fig. 7.9). The betatron function around the waist is often written as

$$\beta(s) = \beta_* + \frac{s^2}{\beta_*}. \tag{7.44}$$

Note that the location of the waist does not have to coincide with the location of an image. Furthermore, we observe that a waist is characterized by $\alpha_2 = 0$.

7.2 PARAMETER-DEPENDENT LINEAR THEORY

So far we have discussed only the linear properties of ideal repetitive systems and their impacts on design considerations. When studying the nonlinear effects of the motion, there are two separate aspects that can be considered. The first is the influence of **parameters** including energy spread and also magnet misalignments and magnetic field errors on the **linear motion**, which will be discussed here. The second consideration addresses the full **nonlinear dynamics** of the particles, which requires more sophisticated tools including normal form methods and is addressed in Section 7.3.

Deviations of various kinds, such as energy spread, magnet misalignments, and magnet field errors, change the linear dynamics in possibly complicated ways. It is very important to study these changes in order to decide on error tolerances for a specific design. Among all the linear quantities discussed previously, changes in tunes usually affect the motion most severely. As discussed previously, it is important to "sample" phase space uniformly in order to not be susceptible to resonances. For tunes, this means that the tune should not be a rational number, and the ratio of tunes should not be rational. Therefore, any shifts in tune due to changes in parameters may potentially destroy the stability of a system, and so it is important to control them. Among all changes in tune, tune shifts caused by energy spread are the most important because the energy spread is usually large compared to other factors.

Differential algebra (DA) methods provide a powerful tool to study the influences of perturbations to linear quantities such as tunes and lattice parameters to very high accuracy. Here, we first study the method of finding parameter-dependent fixed points of a map, which is the foundation for the algorithm of the parameter-dependent tune shifts and normal form theory (see Section 7.3). The algorithm for computing tune shifts is outlined later, as is correction of chromaticities, which is a direct application.

7.2.1 The Closed Orbit

As discussed in Section 7.1, the energy dependence of fixed points identifying the closed orbit is very important to accelerator design. Similarly, if large errors are present, it is important to study their influence as well. In general, knowledge of the parameter dependency of fixed points is needed to compute parameter-dependent tune shifts to high orders. Let us assume we are given a set of parameters $\vec{\delta}$, which may include the energy of the particle but also important system parameters. As discussed in Chapter 5, the map $\mathcal{M}(\vec{z}_F, \vec{\delta})$ of the system depending on the variables $\vec{z}$ as well as the system parameters $\vec{\delta}$ can be obtained to any order using DA methods. Our goal is to establish a **nonlinear relationship** between the fixed point $\vec{z}_f$ and the values of the parameters $\vec{\delta}$. As always within the perturbative approach, we are interested only in the nth order expansion of this relationship.

In order to determine how the fixed point $\vec{z}_F$ depends on the parameters $\vec{\delta}$, we observe that the fixed point satisfies

$$(\vec{z}_F, \vec{\delta}) = \mathcal{M}(\vec{z}_F, \vec{\delta}). \tag{7.45}$$

This fixed-point equation can be rewritten as

$$(\mathcal{M} - I_H)(\vec{z}_F, \vec{\delta}) = (\vec{0}, \vec{\delta}) \tag{7.46}$$

where the map I_H contains a unity map in the upper block corresponding to the variables $\vec{z}$ and zeros everywhere else. This form of the fixed-point equation clearly shows how the parameter-dependent fixed point $\vec{z}_F$ can be obtained: It is necessary to **invert** the map $\mathcal{M} - \mathcal{I}_H$. Since we are interested only in the properties of the inverse up to order n, we work with the equivalence class $[\mathcal{M} - I]_n$ of the map and apply the technique given in Section 2.3.1 to obtain the equivalence class $[\vec{z}_F]_n$.

As previously determined, we can compute the nth order inverse if and only if the linear part of the map is invertible. While this is always the case for transfer maps, here the situation is more subtle; the map is invertible if and only if the phase space part of $\mathcal{M}$ does not have **unit eigenvalue**. However, since such an eigenvalue can lead to an unstable map under small errors as discussed previously, it is avoided in the design of repetitive systems. Altogether, up to order n, the fixed point is given as the upper part of

$$(\vec{z}_F, \vec{\delta})_n = [\mathcal{M} - \mathcal{I}_H]_n^{-1} (\vec{0}, \vec{\delta}). \tag{7.47}$$

7.2.2 Parameter Tune Shifts

As the first step in the computation of tune shifts, we change coordinates by performing a **transformation to the parameter-dependent fixed point**. In these coordinates, the map is origin preserving, i.e., $\mathcal{M}(\vec{0}, \vec{\delta}) = \vec{0}$. Furthermore, the linear motion around this new fixed point follows ellipses, as discussed previously. This also implies that all partial derivatives of the final coordinates with respect to parameters alone vanish.

The key consequence of this transformation is that we can now view the map in such a way that the partial derivatives of the final phase space variables with respect to the initial phase space variables and hence **the aberrations depend on the system parameters**, but the system parameters do not influence the map otherwise. Therefore, in this view, our map now relates initial phase space coordinates to final phase space coordinates, and the expansion coefficients depend on the parameters.

We now compute the linear matrix of motion in which all matrix elements depend on system parameters to order n. If the motion is in the form of 2×2 subblocks, we obtain a matrix $\hat{M}$ consisting of matrix elements that are now equivalence classes of order $m = n - 1$ depending on the parameters

$$\hat{M} = \begin{pmatrix} [a]_m & [b]_m \\ [c]_m & [d]_m \end{pmatrix}. \tag{7.48}$$

Note that one order is lost in the process: Since a was a first derivative, its (m)th derivatives with respect to the parameters are certain nth derivatives of the original map $\mathcal{M}$.

The computation of the parameter dependence of the tunes is now quite straightforward. Following common DA practice, we replace all real operations in the computation of the tunes and dampings by the corresponding DA operations.

268   REPETITIVE SYSTEMS

After the motion is decoupled, i.e., the map is already in 2×2 block form, this merely involves the computation of the class of μ from the determinant and trace. In particular, we obtain

$$[\mu]_m = \text{sign}([b]_m) \, \text{acos} \left(\frac{[a]_m + [b]_m}{2([a]_m \cdot [d]_m - [b]_m \cdot [c]_m)} \right), \qquad (7.49)$$

where the arccos and sign operations are acting in DA.

In the case of coupled motion, conceptually the strategy is the same. In addition, one simply replaces all operations in the whole eigenvalue package necessary for bringing the motion to block form by the corresponding ones in DA. Since for low dimensions, good eigenvalue and eigenvector algorithms are very efficient, the additional effort is minor. Altogether, a very compact algorithm results, and compared to the more general DA normal form algorithm discussed later which obtains the parameter-dependent tunes as a by-product, the tune shifts are obtained in an efficient and straightforward way.

7.2.3 Chromaticity Correction

As discussed in Section 7.1, the correction of chromaticities plays an important role in accelerator design. Here, we show an immediate and useful application of the algorithm outlined in the previous sections, namely, the correction of chromaticities using system parameters. To this end, we utilize the method of the previous section to write the v tunes in terms of the system parameters

$$\vec{\mu} =_m \mathcal{T}(\vec{\delta}). \qquad (7.50)$$

The map $\mathcal{T}$ contains a constant part (the linear tunes) as well as nonlinear parts, and the algorithm of the previous section allowed us to compute the class $[\mathcal{T}]_{m-1}$ of $\mathcal{T}$.

We now split the parameters into the energy deviation d_k and the true system parameters. For the subsequent discussion, we are interested only in the case of v true system parameters, i.e., one for each phase space pair. Furthermore, we choose the parameters such that they do not produce tune shifts by themselves but only in connection with energy deviations. For example, this can be achieved by using the strengths of v suitably chosen sextupoles as parameters. (Quadrupoles strengths are not useful because they produce tune shifts even without d_k, since they obviously affect the linear tune.) In this case, the tune equations reduce to

$$\vec{\mu} = \vec{\mu}_0 + d_k \cdot \vec{c} + d_k \cdot \mathcal{S}(\vec{\delta}), \qquad (7.51)$$

where $\mathcal{S}$ is a nonlinear map. To correct the chromaticities, i.e., make the tune independent of d_k, requires that the following be satisfied:

$$\vec{c} + \mathcal{S}(\vec{\delta}) = \vec{0}, \qquad (7.52)$$

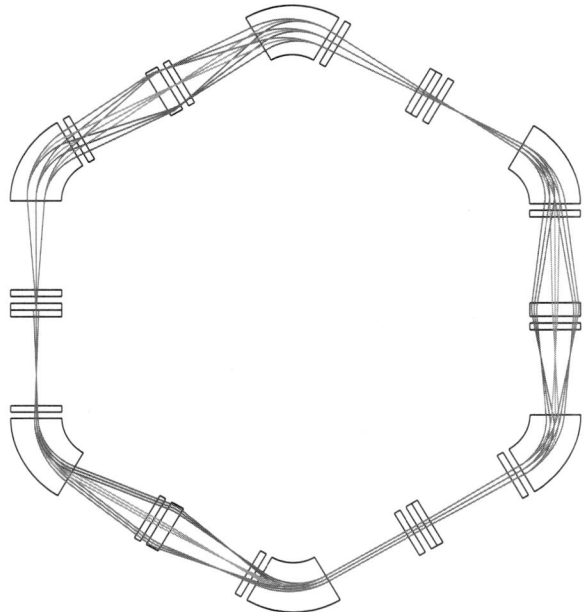

FIGURE 7.10. The layout of a simple superperiod 3 DBA lattice.

which can be obtained by choosing

$$\vec{\delta} = \mathcal{S}^{-1}(-\vec{c})$$

if the inverse of $\mathcal{S}$ exists. From $\mathcal{S}^{-1}$ we now pass to its equivalence classes and use the inversion algorithm (Eq. 2.83). This yields the classes $[\vec{\delta}]_m$ and hence the Taylor expansion of the strengths of v suitably chosen elements to correct the chromaticities. Using these Taylor expansions, an approximate value for $\vec{\delta}$ can be computed. Since the missing terms scale with the nth power of $\vec{\delta}$, iterating the procedure yields very fast convergence, requiring only very few steps in practice.

To **illustrate** the method, we use a simple storage ring designed for 1 GeV protons, with superperiodicity 3 and each cell being a double bend achromat (see (Berz et al., 1999)). The quadrupoles outside the dispersion region help to obtain the designed tunes (Fig. 7.10). From Table I, it is obvious that the chromaticities are large after the first-order design. Note that the so-called first-order chromaticities are in fact second-order quantities since the tunes are first-order quantities. The sextupoles inside the dispersive region are used to correct the first-order chromaticities. Using the methods discussed previously, the strengths of the two sextupoles are adjusted to make the dependencies vanish. After five steps the sequence converges; the results are shown in Table II.

TABLE I
CHROMATICITIES BEFORE CORRECTION

Energy tune shifts	Order	Exponents
0.6500000000412144	0	0 0 0 0 0
−2.487379243931609	1	0 0 0 0 1
0.3399999569645454	0	0 0 0 0 0
−1.491359645132462	1	0 0 0 0 1

TABLE II
CHROMATICITIES AFTER CORRECTION

Energy tune shifts	Order	Exponents
0.6500000000412144	0	0 0 0 0 0
$-0.9405857911266889E - 09$	1	0 0 0 0 1
0.3399999569645454	0	0 0 0 0 0
$-0.2365239133389891E - 08$	1	0 0 0 0 1

7.3 Normal Forms

In this section, we will discuss a method to obtain many quantities of interest for repetitive systems exactly. In particular, it will allow the computation of the **tune** of a given particle under **nonlinear motion** and thus the shifts from the linear tune due to the particle's distance from the origin. It will also provide a family of **approximate invariants** of the nonlinear motion as an extension of the linear ellipses. Finally, it will provide a mechanism for the detailed and quantitative study of **resonances** as well as a method for their correction.

Normal form ideas were introduced to the field by Dragt and Finn (1979) in the Lie algebraic framework (Dragt and Finn, 1976; Dragt, 1982). While the original paper (Dragt and Finn, 1979) contains all the core ideas, some simplifications were necessary before a first implementation for realistic systems was obtained by Neri and Dragt for orders of three and later five. Fundamental difficulties inherent in the Lie algebraic formulation limited the efforts to relatively low orders, and only the combined DA–Lie approach (Forest *et al.*, 1989) circumvented this problem, resulting in the first arbitrary order algorithm. The method we present in the following section is purely DA based and much more compact than the hybrid method, and it also allows the treatment of nonsymplectic systems.

7.3.1 The DA Normal Form Algorithm

The goal of the normal form algorithm is to provide a **nonlinear change of variables** such that the map in the new variables has a significantly simpler structure than before. In particular, it will attempt to find a transformation such that in the new variables, up to a certain order, the motion is **circular** with an **amplitude-dependent frequency**.

Therefore, we assume we are given the transfer map of a particle optical system

$$\vec{z}_f = \mathcal{M}(\vec{z}_i, \vec{\delta}), \tag{7.53}$$

where $\vec{z}$ are the $2v$ phase space coordinates and $\vec{\delta}$ are system parameters. Using the DA methods, we are able to compute the partial derivatives $[\mathcal{M}]_n$ of the map to any order n.

The normal form algorithm consists of a sequence of coordinate transformations $\mathcal{A}$ of the map

$$\mathcal{A} \circ \mathcal{M} \circ \mathcal{A}^{-1} \tag{7.54}$$

The **first** such coordinate transformation is the move to the **parameter-dependent fixed point** $\vec{z}_F$, which satisfies

$$\vec{z}_F = \mathcal{M}(\vec{z}_F, \vec{\delta}). \tag{7.55}$$

This transformation can be performed to arbitrary order using DA methods as shown in Eq. (7.47). After the fixed-point transformation, the map is origin preserving; this means that for any $\vec{\delta}$,

$$\mathcal{M}(\vec{0}, \vec{\delta}) = \vec{0} \tag{7.56}$$

As discussed previously, we note that the fixed-point transformation is possible if and only if 1 is not an eigenvalue of the linear map.

In the **next** step we perform a coordinate transformation that provides a **linear diagonalization** of the map. For this process, we have to assume that there are $2v$ distinct eigenvalues. This together with the fact that no eigenvalue should be unity and that their product is positive are the only requirements we have to demand for the map; under normal conditions, repetitive systems are always designed such that these conditions are met.

After diagonalization, the linear map assumes the form

$$M = \begin{pmatrix} r_1 e^{+i\mu_1} & & & & & \\ & r_1 e^{-i\mu_1} & & & & \\ & & \ddots & & 0 & \\ & & 0 & \ddots & & \\ & & & & r_v e^{+i\mu_v} & \\ & & & & & r_v e^{-i\mu_v} \end{pmatrix} \tag{7.57}$$

Here, the linear tunes μ_j are either purely real or purely imaginary. For stable systems, none of the $r_j e^{\pm i\mu_j}$ must exceed unity in modulus.

For **symplectic** systems, the determinant is unity, which entails that the product of the r_j must be unity. This implies that for symplectic systems, for any $r_j < 1$ there is another with $r_j > 1$. Thus, stable symplectic systems have $r_j = 1$ for all j because otherwise there would be one j for which r_j exceeds unity, and thus at least one $r_j e^{\pm i \mu_j}$ would have modulus larger than unity. This would also happen if a μ_j were imaginary. Therefore, all μ_j are real, and they are even nonzero because we demanded distinct eigenvalues.

To the eigenvector pair $s_j^\pm$ belonging to the eigenvalue $r_j e^{\pm i \mu_j}$, we associate another pair $t_j^\pm$ of variables as follows:

$$t_j^+ = (s_j^+ + s_j^-)/2$$
$$t_j^- = (s_j^+ - s_j^-)/2i. \tag{7.58}$$

In the case of complex $s_j^\pm$, which corresponds to the stable case, the $t_j^\pm$ are the real and imaginary parts and thus are real. In the unstable case, t_j^+ is real and t_j^- is imaginary. Obviously, the $s_j^\pm$ can be expressed in terms of the $t_j^\pm$ as

$$s_j^+ = t_j^+ + i\, t_j^-$$
$$s_j^- = t_j^+ - i\, t_j^- \tag{7.59}$$

Later, we will perform the manipulations in the $s_j^\pm$, whereas the results are most easily interpreted in the $t_j^\pm$.

We now proceed to a **sequence of order-by-order transformations** that will simplify the nonlinear terms. After the fixed-point transformation and the linear diagonalization, all further steps are purely nonlinear and do not affect the linear part anymore, but the mth step transforms only the mth order of the map and leaves the lower orders unaffected.

We begin the mth step by splitting the momentary map $\mathcal{M}$ into its linear and nonlinear parts $\mathcal{R}$ and $\mathcal{S}_m$, i.e., $\mathcal{M} = \mathcal{R} + \mathcal{S}_m$. The linear part $\mathcal{R}$ has the form of Eq. (7.57). Then we perform a transformation using a map that to mth order has the form

$$\mathcal{A}_m = \mathcal{I} + \mathcal{T}_m, \tag{7.60}$$

where $\mathcal{T}_m$ vanishes to order $m-1$. Because the linear part of $\mathcal{A}_m$ is the unity map, $\mathcal{A}_m$ is invertible. Moreover, inspection of the algorithm to invert transfer maps (Eq. 2.83) reveals that up to order m,

$$\mathcal{A}_m^{-1} =_m \mathcal{I} - \mathcal{T}_m. \tag{7.61}$$

Of course, the full inversion of $\mathcal{A}_m$ contains higher order terms, which is one of the reasons why iteration is needed.

NORMAL FORMS

It is also worth noting that in principle the higher order parts of $\mathcal{T}_m$ can be chosen freely. In most cases it is particularly useful to choose these terms in such a way that they represent the flow of a dynamical system by interpreting $\mathcal{T}$ as the first term in the **flow operator** (Eq. 2.120) and utilizing the arguments about flow factorizations from Section 4.3.1. This has the advantages that the computation of the inverse is trivial and also that the transformation map will automatically be **symplectic** as soon as the original map is symplectic.

To study the effect of the transformation, we infer up to order m:

$$\begin{aligned}
\mathcal{A} \circ \mathcal{M} \circ \mathcal{A}^{-1} &=_m (\mathcal{I} + \mathcal{T}_m) \circ (\mathcal{R} + \mathcal{S}_m) \circ (\mathcal{I} - \mathcal{T}_m) \\
&=_m (\mathcal{I} + \mathcal{T}_m) \circ (\mathcal{R} + \mathcal{S}_m - \mathcal{R} \circ \mathcal{T}_m) \\
&=_m \mathcal{R} + \mathcal{S}_m + (\mathcal{T}_m \circ \mathcal{R} - \mathcal{R} \circ \mathcal{T}_m)
\end{aligned} \quad (7.62)$$

For the first step, we used $\mathcal{S}_m \circ (\mathcal{I} - \mathcal{T}_m) =_m \mathcal{S}_m$, which holds because $\mathcal{S}_m$ is nonlinear and $\mathcal{T}_m$ is of order m. In the second step we used $\mathcal{T}_m \circ (\mathcal{R} + \mathcal{S}_m - \mathcal{R} \circ \mathcal{T}_m) =_m \mathcal{T}_m \circ \mathcal{R}$, which holds because $\mathcal{T}_m$ is of exact order m and all variables in the second term are nonlinear except $\mathcal{R}$.

A closer inspection of the last line reveals that $\mathcal{S}_m$ can be simplified by choosing the commutator $\mathcal{C}_m = \{\mathcal{T}_m, \mathcal{R}\} = (\mathcal{T}_m \circ \mathcal{R} - \mathcal{R} \circ \mathcal{T}_m)$ appropriately. Indeed, if the range of $\mathcal{C}_m$ is the full space, then $\mathcal{S}_m$ can be removed entirely. However, most of the time this is not the case.

Let $(\mathcal{T}_{mj}^\pm | k_1^+, k_1^-, \ldots, k_n^+, k_n^-)$ be the Taylor expansion coefficient of $\mathcal{T}_{mj}$ with respect to $(s_1^+)^{k_1^+} (s_1^-)^{k_1^-} \cdots (s_n^+)^{k_n^+} (s_n^-)^{k_n^-}$ in the jth component pair of $\mathcal{T}_m$. Therefore, $\mathcal{T}_{mj}^\pm$ is written as

$$\mathcal{T}_{mj}^\pm = \sum (\mathcal{T}_{mj}^\pm | k_1^+, k_1^-, \ldots, k_n^+, k_n^-) \cdot (s_1^+)^{k_1^+} (s_1^-)^{k_1^-} \cdots (s_n^+)^{k_n^+} (s_n^-)^{k_n^-}. \quad (7.63)$$

Similarly, we identify the coefficients of $\mathcal{C}_m$ by $(\mathcal{C}_{mj}^\pm | k_1^+, k_1^-, \ldots, k_n^+, k_n^-)$. Because $\mathcal{R}$ is diagonal, it is possible to express the coefficients of $\mathcal{C}_m$ in terms of the ones of $\mathcal{T}$. One obtains

$$\begin{aligned}
&(\mathcal{C}_{mj}^\pm | k_1^+, k_1^-, \ldots, k_n^+, k_n^-) \\
&= \left(\left(\prod_{l=1}^n r_l^{(k_l^+ + k_l^-)} \right) \cdot e^{i\vec{\mu} \cdot (\vec{k}^+ - \vec{k}^-)} - r_j \cdot e^{\pm i\mu_j} \right) \\
&\quad \cdot (\mathcal{T}_{mj}^\pm | k_1^+, k_1^-, \ldots, k_n^+, k_n^-) \\
&= C_{mj}^\pm (\vec{k}^+, \vec{k}^-) \cdot (\mathcal{T}_{mj}^\pm | k_1^+, k_1^-, \ldots, k_n^+, k_n^-). \quad (7.64)
\end{aligned}$$

Now it is apparent that a term in $\mathcal{S}_{mj}^{\pm}$ can be removed if and only if the factor $C_{mj}^{\pm}(\vec{k}^+, \vec{k}^-)$ is nonzero; if it is nonzero, then the required term in $\mathcal{T}_{mj}^{\pm}$ is the negative of the respective term in $\mathcal{S}_{mj}^{\pm}$ divided by $C_{mj}^{\pm}(\vec{k}^+, \vec{k}^-)$.

Therefore, the outcome of the whole normal form transformation depends on the conditions under which the term $C_{mj}^{\pm}(\vec{k}^+, \vec{k}^-)$ vanishes. This is obviously the case if and only if the moduli and the arguments of $r_j \cdot e^{\pm i\mu_j}$ and $(\prod_{l=1}^{n} r_l^{(k_l^+ + k_l^-)}) \cdot e^{i\vec{\mu}\cdot(\vec{k}^+ - \vec{k}^-)}$ are identical. Next, we discuss the conditions of this to happen for various special cases and draw conclusions.

7.3.2 Symplectic Systems

As discussed previously, in the stable symplectic case all the r_j are equal to 1, and the μ_j are purely real. Therefore, the moduli of the first and second terms in $C_{mj}^{\pm}(\vec{k}^+, \vec{k}^-)$ are equal if and only if their phases agree modulo 2π. This is obviously the case if

$$\vec{\mu} \cdot (\vec{k}^+ - \vec{k}^-) = \pm \mu_j \,(\text{mod}\, 2\pi), \tag{7.65}$$

where the different signs apply for $C_{mj}^+(\vec{k}^+, \vec{k}^-)$ and $C_{mj}^-(\vec{k}^+, \vec{k}^-)$, respectively. This can occur in two possible ways:

1. $k_l^+ = k_l^- \ \forall l \neq j$ and $k_j^+ = k_j^- \pm 1$.
2. $\vec{\mu} \cdot \vec{n} = 0 \,(\text{mod } 2\pi)$ has nontrivial solutions.

The first case is of mathematical nature and lies at the heart of the normal form algorithm. It yields terms that are responsible for **amplitude-dependent tune shifts**. We will discuss its consequences later. The second case is equivalent to the system lying on a higher order **resonance** and is of a more physical nature. In case the second condition is satisfied, there will be resonance-driven terms that cannot be removed and that prevent a direct computation of amplitude tune shifts.

Before proceeding, we note that the second condition entails complications even if it is almost, but not exactly, satisfied. In this case, the removal of the respective term produces a small denominator that generates terms that become increasingly larger, depending on the proximity to the resonance. For the removal process, in Eq. (7.64) this resonance proximity factor is multiplied by the respective expansion coefficient, and therefore this product is obviously an excellent **characteristic of resonance strength**.

With increasingly higher orders, i.e., larger k^+ and k^-, the number of relevant resonances increases. Since the resonances lie dense in tune space, eventually the growth of terms is almost inevitable and hence produces a map that is much more nonlinear than the underlying one. As we shall discuss in the next section, this problem is alleviated by damping.

We now discuss the form of the map if no resonances occur. In this case, the transformed map will have the form

$$\mathcal{M}_j^+ = s_j^+ \cdot f_j(s_1^+ s_1^-, \ldots, s_v^+ s_v^-)$$
$$\mathcal{M}_j^- = s_j^- \cdot \bar{f}_j(s_1^+ s_1^-, \ldots, s_v^+ s_v^-). \tag{7.66}$$

The variables $s_j^\pm$ are not particularly well suited for the discussion of the result, and we express the map in terms of the adjoined variables $t_j^\pm$ introduced in Eq. (7.58). Simple arithmetic shows that

$$s_j^+ \cdot s_j^- = (t_j^+)^2 + (t_j^-)^2. \tag{7.67}$$

It is now advantageous to write f_j in terms of amplitude and phase as $f_j = a_j \cdot e^{i\phi_j}$. Performing the transformation to the coordinates $t_j^\pm$, we thus obtain

$$\mathcal{M}_j^\pm = \begin{pmatrix} 1/2 & 1/2 \\ 1/2i & -1/2i \end{pmatrix}$$
$$\cdot \begin{pmatrix} (t_j^+ + it_j^-) \cdot f_j[(t_1^+)^2 + (t_1^-)^2, \ldots, (t_v^+)^2 + (t_v^-)^2] \\ (t_j^+ - it_j^-) \cdot \bar{f}_j[(t_1^+)^2 + (t_1^-)^2, \ldots, (t_v^+)^2 + (t_v^-)^2] \end{pmatrix}$$
$$= a_j \cdot \begin{pmatrix} \cos(\phi_j) & -\sin(\phi_j) \\ \sin(\phi_j) & \cos(\phi_j) \end{pmatrix} \cdot \begin{pmatrix} t_j^+ \\ t_j^- \end{pmatrix}. \tag{7.68}$$

Here, $\phi_j = \phi_j[(t_1^+)^2 + (t_1^-)^2, \ldots, (t_v^+)^2 + (t_v^-)^2]$ depends on a rotationally invariant quantity.

Therefore, in these coordinates, the motion is now given by a **rotation**, the frequency of which depends only on the amplitudes $(t_j^+)^2 + (t_j^-)^2$ and some system parameters and thus does not vary from turn to turn. Indeed, these frequencies are precisely the **tunes** of the nonlinear motion.

For any repetitive system, the tune of one particle is the total polar angle advance divided by the number of turns in the limit of turn number going to infinity, if this limit exists. If we now express the motion in the new coordinates, we pick up an initial polar angle for the transformation to the new coordinates; then, every turn produces an equal polar angle ϕ_j which depends on the amplitude and parameters of the particle. We produce a final polar angle for the transformation back to the old coordinates.

As the number of turns increases, the contribution of the initial and final polar angles due to the transformation becomes increasingly insignificant, and in the limit the tune results in only ϕ_j. Therefore, we showed that the limit exists and that it can be computed analytically as a by-product of the normal form transformation.

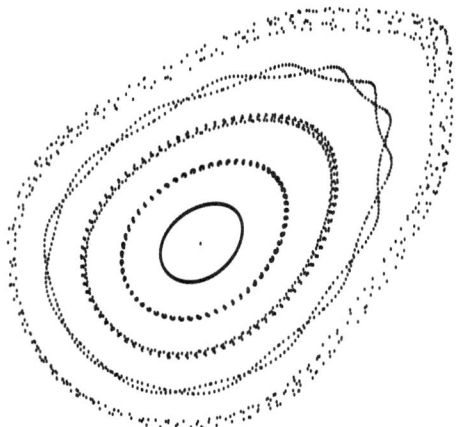

FIGURE 7.11. An x–a projection of a tracking picture in conventional coordinates.

To demonstrate the effects of nonlinear normal form transformations in the symplectic case, Fig. 7.11 shows a typical tracking picture for a stable symplectic system. The nonlinear effects disturb the elliptic structure expected from linear theory, and the cross-coupling between the two planes induces a substantial broadening.

On the other hand, Fig. 7.12 shows the same motion displayed in normal form coordinates. The motion is circular up to printer resolution, removing the cross-coupling as well as other nonlinear effects.

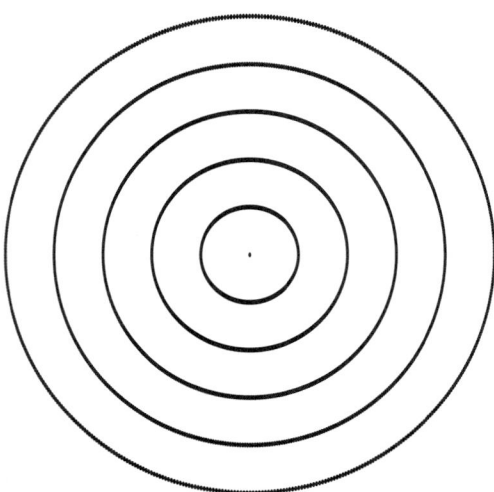

FIGURE 7.12. The tracking picture of the picture shown in Fig. 7.11 in normal form coordinates.

7.3.3 Nonsymplectic Systems

In the case of stable, nonsymplectic maps, all r_j must satisfy $r_j \leq 1$ because otherwise at least one of the $r_j e^{\pm i \mu_j}$ is larger than unity in modulus. Since in the normal form transformation, terms can be removed if and only if the phases or amplitudes for the two contributions in $C_{mj}^{\pm}(k^+, k^-)$ are different and the amplitudes contribute, more terms can be removed.

Of particular practical interest is the totally damped case in which $r_j < 1$ for all j and all μ_j are real, which describes damped electron rings. In this case an inspection of Eq. (7.64) reveals that now every nonlinear term can be removed. Then a similar argument as presented in the previous section shows that the motion assumes the form

$$\mathcal{M}_j^{\pm} = r_j \cdot \begin{pmatrix} \cos(\phi_j) & -\sin(\phi_j) \\ \sin(\phi_j) & \cos(\phi_j) \end{pmatrix} \cdot \begin{pmatrix} t_j^+ \\ t_j^- \end{pmatrix}, \qquad (7.69)$$

where the angle ϕ_j does not depend on the phase space variables anymore but only on the parameters. This means that the normal form transformation of a **totally damped system** leads to exponential spirals with constant frequency ϕ_j. In particular, this entails that a totally damped system has **no amplitude-dependent tune shifts**, and the motion eventually collapses into the origin. Since in practice the damping is of course usually very small, these effects are usually covered by the short-term sensitivity to resonances.

It is quite illuminating to consider the small denominator problem in the case of totally damped systems. Clearly, the denominator can never fall below $1-\max(r_j)$ in magnitude. This puts a limit on the influence of any low-order resonance on the dynamics; in fact, even being exactly on a low-order resonance does not have any serious consequences if the damping is strong enough. In general, the influence of a resonance now depends on two quantities: the distance in tune space and the contraction strength r_j. High-order resonances are suppressed particularly strongly because of the contribution of additional powers of r_j.

Because all systems exhibit a residual amount of damping, the arguments presented here are generally relevant. It is especially noteworthy that residual damping suppresses high order resonances by the previous mechanism even for proton machines, which entails that from a theoretical view, ultimately high-order resonances become insignificant.

Clearly, the normal form algorithm also works for unstable maps. The number of terms that can be removed will be at least the same as that in the symplectic case, and sometimes it is possible to remove all terms. Among the many possible combinations of r_j and μ_j, the most common case in which the μ_j are real is worth studying in more detail. In this case, all terms can be removed unless the

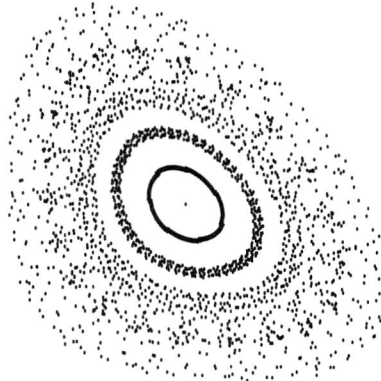

FIGURE 7.13. A tracking picture of the first two coordinates of a damped system.

logarithms of the r_j and the tunes satisfy the same resonance condition, i.e.,

$$\vec{n} \cdot (\log(r_1), \ldots, \log(r_v)) = 0$$
$$\vec{n} \cdot \vec{\mu} = 0 \,(\mathrm{mod}\, 2\pi) \qquad (7.70)$$

have simultaneous nontrivial solutions. This situation characterizes a new type of resonance, the coupled **phase-amplitude resonance**.

Phase-amplitude resonances can never occur if all r_j are greater than unity in magnitude. This case corresponds to a totally unbound motion, and the motion in normal form coordinates moves along growing exponential spirals.

Symplectic systems, on the other hand, satisfy

$$\prod_{l=1}^{n} r_l = 1.$$

Therefore, if there are r_j with both signs of the logarithm, the possibility for amplitude resonances exists. In fact, any symplectic system lies on the fundamental amplitude resonance characterized by $\vec{n} = (1, 1, \ldots, 1)$. In this light, the stable symplectic case is a degeneracy in which all logarithms vanish and so the system lies on every amplitude resonances. Thus, it is susceptible to any phase resonance, and it suffices to study only these.

To study the nonlinear normal form transformations in the case of damped systems, Fig. 7.13 shows a typical tracking picture of a very weakly damped system. A difference to the symplectic case is not visible, and there is familiar cross-coupling as well as nonlinear distortions.

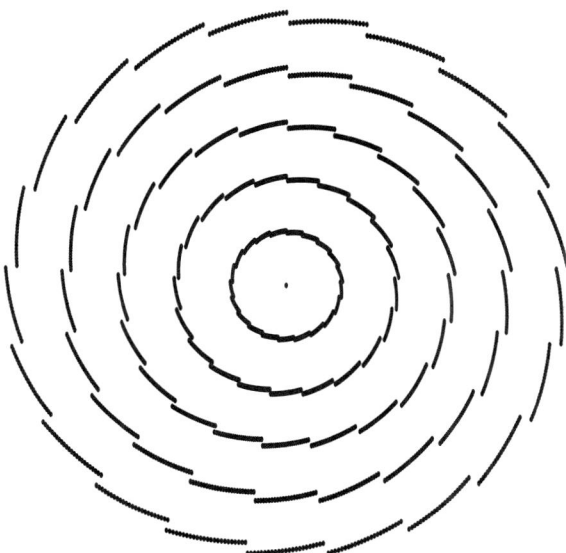

FIGURE 7.14. The tracking picture of that shown in Fig. 7.13 in normal form coordinates. The weak damping that is not directly visible becomes apparent, and the motion follows elliptical spirals.

On the other hand, Fig. 7.14 shows the same motion displayed in normal form coordinates. The cross-coupling and nonlinear distortions are removed, and the motion follows elliptic spirals.

7.3.4 Amplitude Tune Shifts and Resonances

It was shown in Eq. (7.68) that in the normal form coordinates, the motion of the particles has the form

$$\mathcal{M}_j^\pm = a_j \cdot \begin{pmatrix} \cos(\phi_j) & -\sin(\phi_j) \\ \sin(\phi_j) & \cos(\phi_j) \end{pmatrix} \cdot \begin{pmatrix} t^+ \\ t^- \end{pmatrix}, \qquad (7.71)$$

where the angles depend only on the rotationally invariant quantities $(t_1^+)^2 + (t_1^-)^2$ and possible system parameters, and thus the angle represent the tunes of the motion. This allows for a direct and accurate computation of the dependence of the tune on amplitude. As such, it offers a **fast and powerful** alternative to the conventional numerical computation of tune dependence on amplitude, which is based on tracking a seed of particles for a sufficient number of turns and then computing the tune for each as the average angle advance.

We want to study the practical aspects of the computation of the tunes using the DA normal form algorithm implemented in COSY INFINITY. As a first example, we calculate the tune shifts of the Saskatoon EROS storage ring with energy used as a parameter using both the DA normal form algorithm and the method to calculate high-order resonances. Table III shows the results. The chromaticity column shows only terms depending on energy, the exponents of which appear as the fifth entry in the exponent column. On the other hand, the normal form column shows terms depending on both energy and x^2, a^2 and y^2, b^2, which describe amplitude-dependent tune shifts. As can be seen, the energy-dependent terms that are calculated by both methods agree to nearly machine precision. However, since the two algorithms for their computation are different, this can be used to assess the overall accuracy of the methods.

An independent test was performed using the code DIMAD which calculates tunes numerically by applying the fast Fourier transform method to tracking data. The various orders of chromaticities were compared to those found by COSY, and the results are listed in Table IV. Agreement within the accuracy of the numerical methods used in DIMAD was obtained.

Next, we want to address some applications for the computation of **resonances**. By virtue of Eq. (7.64), the coefficient of $\mathcal{T}$ necessary to obtain removal of terms in $\mathcal{S}$ in the normal form process is given by

$$(\mathcal{T}_{mj}^{\pm}|\vec{k}^+, \vec{k}^-) = \frac{(\mathcal{S}_{mj}^{\pm}|\vec{k}^+, \vec{k}^-)}{\left(\left(\prod_{l=1}^{n} r_l^{(k_l^+ + k_l^-)}\right) \cdot e^{i\vec{\mu}\cdot(\vec{k}^+ - \vec{k}^-)} - r_j \cdot e^{\pm i\mu_j}\right)}. \quad (7.72)$$

Even if the system is not on a resonance and thus the denominator does not vanish, these terms can become large if the system is near a resonance and the term $(\mathcal{S}_{mj}^{\pm}|\vec{k}^+, \vec{k}^-)$ is large. If this is the case, then the normal form transformation map $\mathcal{A}_m$ will have large coefficients, which leads to large coefficients in the partially transformed map $\mathcal{A}_m \circ \mathcal{M} \circ \mathcal{A}_m^{-1}$ and entails that subsequent transformation attempts have to confront larger coefficients. The ultimate consequence is that the map in normal form coordinates, which is circular to order n, will have large coefficients beyond order n that induce deviations from circularity.

Therefore, the coefficients $(\mathcal{T}_{mj}^{\pm}|\vec{k}^+, \vec{k}^-)$ that appear in the normal form transformation process have very practical significance in that they quantify the total effect of the proximity to the resonance [via $(\prod_{l=1}^{n} r_l^{(k_l^+ + k_l^-)}) \cdot e^{i\vec{\mu}\cdot(\vec{k}^+ - \vec{k}^-)} - r_j \cdot e^{\pm i\mu_j}$] and the system's sensitivity to it [via $(\mathcal{S}_{mj}^{\pm}|\vec{k}^+, \vec{k}^-)$]. As such, the study and analysis of the resonance strengths $(\mathcal{T}_{mj}^{\pm}|\vec{k}^+, \vec{k}^-)$ is an important characteristic of the system. Figure 7.15 shows a graphical representation of the resonance strengths for the Large Hadron Collider, which allows for a fast identification of the dominating resonances that would benefit from correction.

Having assessed the strengths of the various resonances, a system can be optimized to suppress them. Since in the normal form picture, the resonance described

NORMAL FORMS

TABLE III
The Vertical Tunes of the EROS Ring, Computed with DA Normal Form Theory and the Chromaticity Tools[a]

Total Tune Shift (normal form)	Energy tune shift (chromaticity)	Order	Exponent
0.8800260865565617	0.8800260865565616	0	0 0 0 0 0
0.4044489434279511	0.4044489434279519	1	0 0 0 0 1
20.27300549111636		2	2 0 0 0 0
-12.70523391367074		2	0 0 2 0 0
19.93950369004059	19.93950369004062	2	0 0 0 0 2
511.1991794912686		3	2 0 0 0 1
-272.9888390984160		3	0 0 2 0 1
353.1162640647963	353.1162640647955	3	0 0 0 0 3
13502.07274250955		4	4 0 0 0 0
-4936.135382965938		4	2 0 2 0 0
1072.710823056354		4	0 0 4 0 0
10058.89904831771		4	2 0 0 0 2
-5699.740274508591		4	0 0 2 0 2
12684.92656366016	12684.92656366014	4	0 0 0 0 4
-5111660.969538658		5	4 0 0 0 1
-111233.3887058197		5	2 0 2 0 1
32336.40175202939		5	0 0 4 0 1
321102.2916307596		5	2 0 0 0 3
-313130.5638738061		5	0 0 2 0 3
487095.4522760212	487095.4522760201	5	0 0 0 0 5
10165111.83496411		6	6 0 0 0 0
7921291.158179961		6	4 0 2 0 0
753492.5011067157		6	2 0 4 0 0
-117199.2118569492		6	0 0 6 0 0
2005266230.252126		6	4 0 0 0 2
-497920.0057811320		6	2 0 2 0 2
1629044.258487686		6	0 0 4 0 2
4141105.646457314		6	2 0 0 0 4
-21315439.11959259		6	0 0 2 0 4
19692776.49371354	19692776.49371350	6	0 0 0 0 6

[a] While the chromaticity column contains only the energy-dependent tune shifts, the normal form column also shows amplitude-dependent terms.

TABLE IV
Comparison between Numerical and DA Computation of Tune Dependence on Energy for the Saskatoon EROS Ring

Order	Difference
0	1×10^{-5}
1	1×10^{-4}
2	$5 \times 5 \times 10^{-3}$
3	1×10^{-2}
4	$4 \times 4 \times 10^{-2}$

Resonances of Order 7

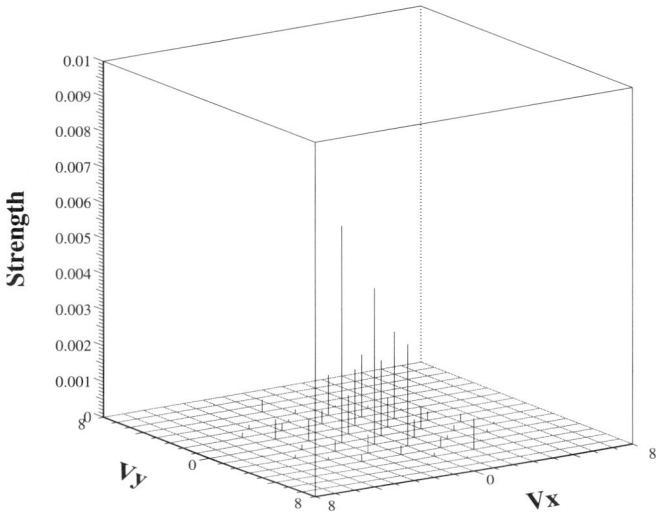

FIGURE 7.15. The resonance strengths up to order 7 for the LHC.

by the term $(\mathcal{T}_{mj}^{\pm}|\vec{k}^+,\vec{k}^-)$ is driven by terms of order $\sum k_i^+ + k_i^-$, by suitably placing correction elements of that order in the lattice, it is possible to reduce the magnitude of $(\mathcal{S}_{mj}^{\pm}|\vec{k}^+,\vec{k}^-)$ and hence $(\mathcal{T}_{mj}^{\pm}|\vec{k}^+,\vec{k}^-)$. In the context of such numerical optimization, both the speed of the normal form approach and the high computational accuracy are of crucial significance.

Figure 7.16 shows the effect of such a resonance correction procedure for the case of the Superconducting Supercollider low-energy booster. Tracking pictures as well as numerically obtained resonance strengths are shown before and after the resonance correction with COSY. As expected, the tracking picture shows a more regular behavior, and the main offending resonance decreased in magnitude and is now in line with others (note the change in scale).

7.3.5 Invariants and Stability Estimates

Because of the clean appearance of motion in normal form coordinates, a detailed study of the dynamics is possible. For example, if the normal form motion would produce perfect circles, we know we have found an **invariant** of the system for each degree of freedom and hence the system is **integrable**. Considering that the transformation of normal form circles to conventional variables is continuous, the

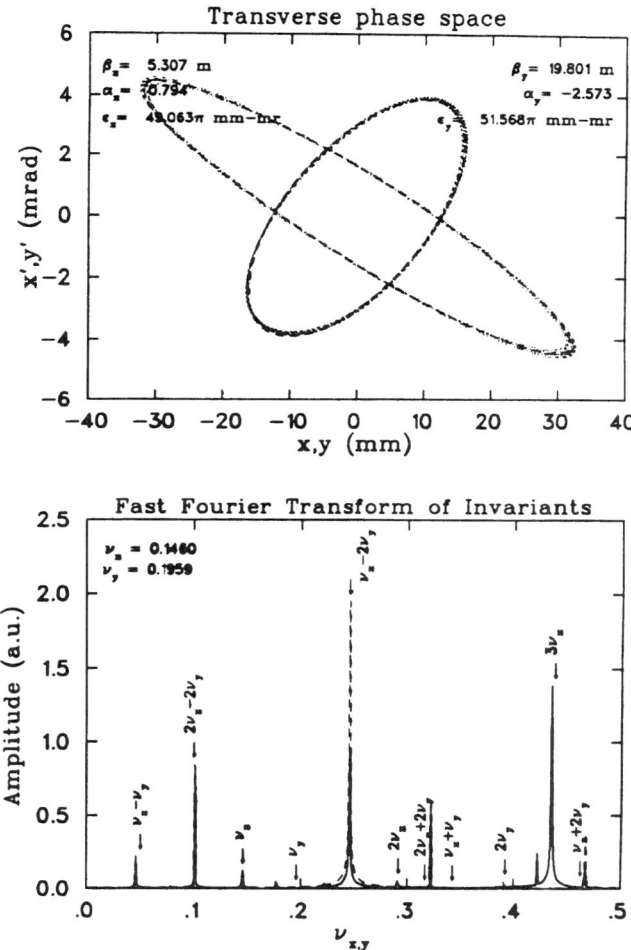

FIGURE 7.16. Normal form resonance suppression for the SSC low-energy booster. Shown are both tracking pictures and numerically obtained resonance strengths before and after correction.

resulting set is bounded and we can conclude that the motion is stable forever and particles can never get lost.

However, because most systems are not integrable, the order-by-order normal form transformations do not actually converge but rather approach a certain saturation beyond which a new order does not bring any improvement, or sometimes even worsens the situation. The mathematical reasons for this phenomenon lie in the resonance-driving terms and the associated resonance denominators

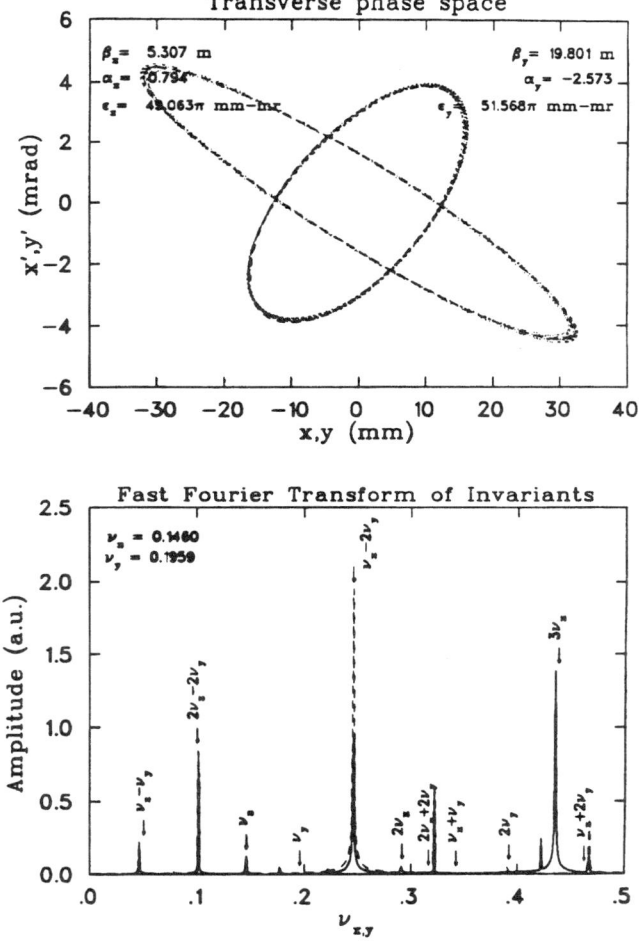

FIGURE 7.16. (*Continued*)

$\exp(i\vec{n} \cdot \vec{\mu}) - 1$ in Eq. (7.64), which, as order and hence the number of admissible resonances increase, become increasingly smaller. For certain systems it is possible to estimate the asymptotic behavior of such resonance-driving terms, and sometimes analytic statements of stability can be made.

In most practical cases, the normal form transformation leads to **near-circular motion** with varying amounts of deviation from circularity. To demonstrate the effect in more detail, we show the invariant defects, i.e., the residual deviation from perfect invariance, for several cases.

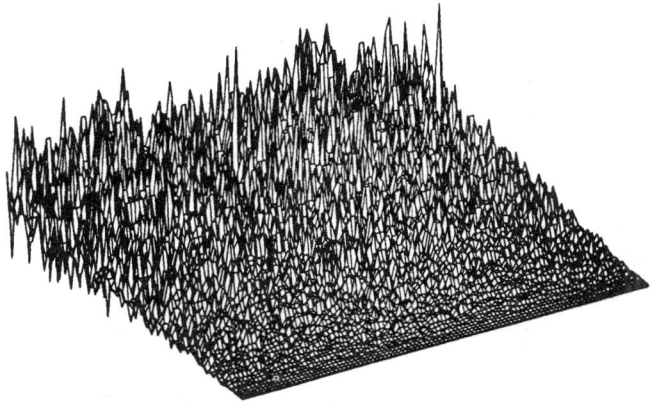

FIGURE 7.17. Normal form defect for a pendulum.

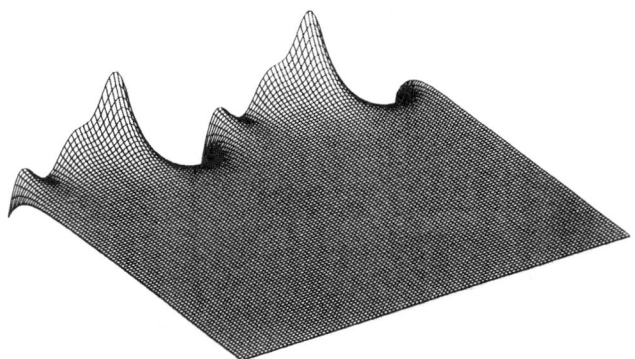

FIGURE 7.18. Normal form defect for a pendulum at a larger amplitude than that shown in Fig. 7.17.

We begin with the invariant defects as a function of normal form radius and angle for a simple one-dimensional pendulum. In the case of this nonlinear motion, there is an invariant (the energy). Thus, in principle, the normal form algorithm could converge. Figure 7.17 shows the invariant defect for amplitudes of 1/10 rad using a normal form map of order 16. In this case the scale is approximately 10^{-17}, and all the errors are of a computational nature. In Fig. 7.18 the amplitude of the pendulum is increased to 1/2 rad into a nonlinear regime. Again, a normal form transformation map of order 16 was used. Now the scale of the invariant defects is 10^{-13}, and some systematic effects due to the limited order become apparent.

In the next example, we study the **PSR II**, a ring accelerator proposed by Los Alamos National Laboratory. In this case, an invariant does not have to exist a

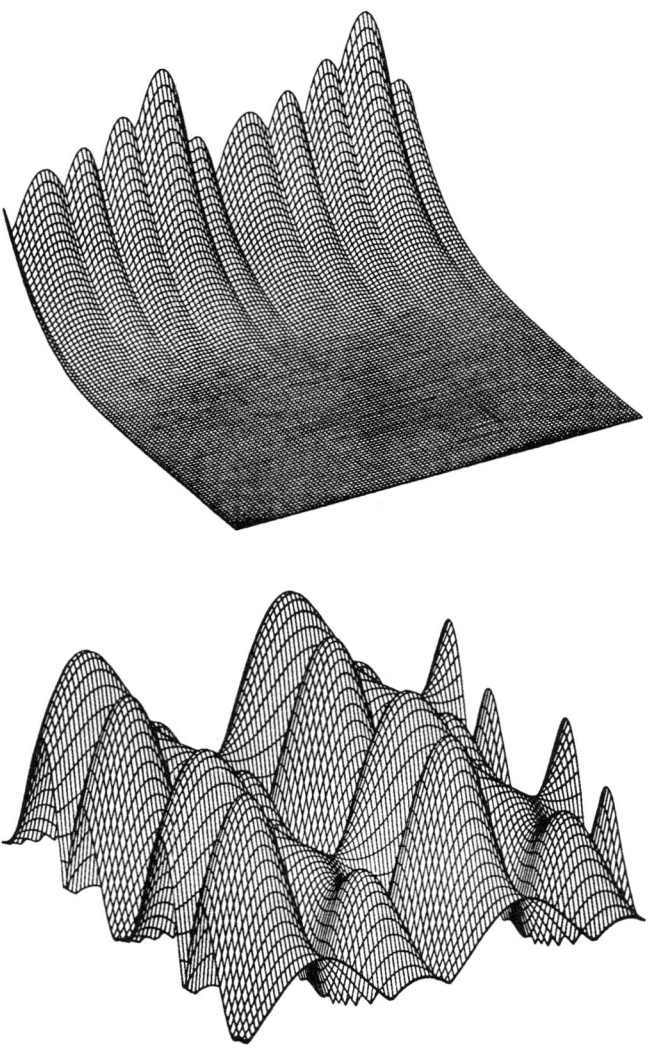

FIGURE 7.19. Normal form defect for the Los Alamos PSR.

priori. It is likely that the motion is indeed nonintegrable, preventing convergence of the normal form map. Figure 7.19 shows the invariant defect, i.e., the deviation from perfectly circular structure, in this case with a normal form transformation map of order 6. The defect is shown as a function of normal form radius and angle, and the increase with radius due to the limited convergence is clearly visible. The scale is approximately 10^{-6}.

While the existence of invariant defects makes statements about general stability impossible, it allows an estimate of stability for finite but long times (Berz and Hoffstätter, 1994; Hoffstätter, 1994; Hoffstätter and Berz, 1996). This approach has been used repeatedly in theoretical studies including Nekhoroshev (1977) and was later studied for several cases (Warnock et al., 1988; Warnock and Ruth, 1991; Warnock and Ruth, 1992; Turchetti, 1990; Bazzani et al., 1992) by trying to fit models for approximate invariants to numerical data. Since in practice the invariant defects are very small, we can make estimates of how long a particle takes to traverse a certain region of normal form coordinate space for all the n subspaces in which the motion follows approximately circular shape. This method is illustrated in Fig. 7.20 for one of these subspaces. Let us assume that the whole region of normal form coordinates up to the maximum radius r_{max} corresponds to coordinates within the area which the accelerator can accept in its beam pipes. Let us further assume that nowhere in the $r - \phi$ diagram is the invariant defect larger than Δr. If we launch particles within the normal form region below r_{min}, then all these particles require at least

$$N = \frac{r_{max} - r_{min}}{\Delta r} \tag{7.73}$$

turns before they reach r_{max}. Considering the small size of Δr in practical cases, this can often ensure stability for a large number of turns.

In most cases, the invariant defects grow quickly with increasing values of r, as shown in Fig. 7.17 and 7.18. Therefore, this estimate can be refined in the following obvious way. Suppose the region of r values between r_{min} and r_{max} is subdivided in the following manner:

$$r_{min} = r_1 < r_2 < \cdots < r_l = r_{max}. \tag{7.74}$$

Let us assume that in each of these regions the maximum invariant defect is bounded by Δr_i. Then we can predict stability for

$$N = \sum_{i=1}^{l-1} \frac{r_{i+1} - r_i}{\Delta r_i} \tag{7.75}$$

turns. Since in practice, at least for the first values of i, Δr_i can be substantially less than Δr, this lower bound can be much greater than N in Eq. (7.73).

To determine the survival time of a particle, one can determine the corresponding numbers of N for all the n normal form subspaces and then take the smallest N as a lower bound.

In practice, the appeal of the mathematically rigorous method outlined here hinges critically on the ability to determine rigorous bounds for the Δr_i, and its practical usefulness is directly connected to the sharpness of these bounds. However, in practice these functions have a large number of local maxima, and a computation of their bounds requires much care. For all the $l - 1$ regions in phase

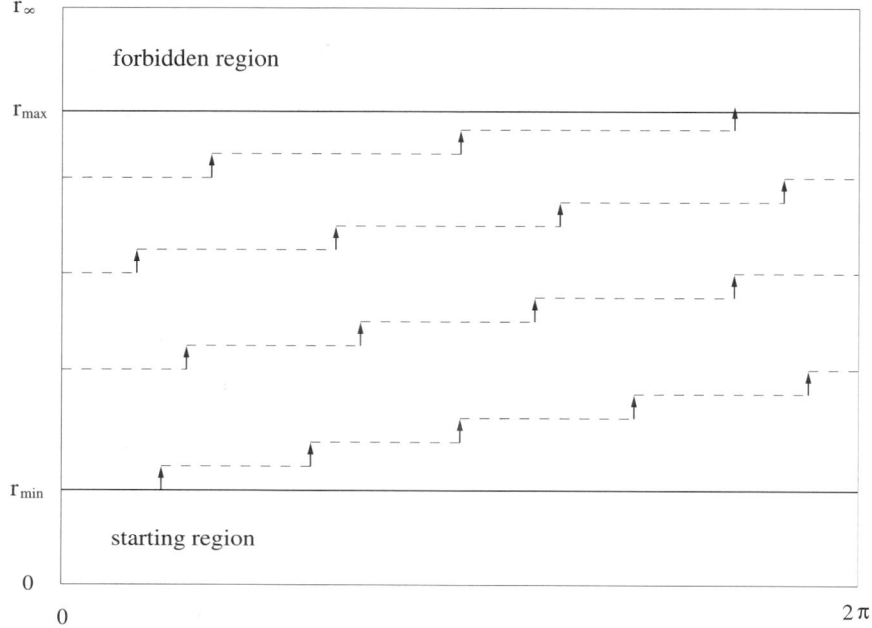

FIGURE 7.20. The principle of the pseudoinvariant stability estimate.

space, we are faced with the task of finding n bounds for the maxima $\Delta r^{(j)}$ of deviation functions:

$$\Delta r^{(j)} \geq \max[r^{(j)}(\vec{M}(\vec{x})) - r^{(j)}(\vec{x})], \qquad (7.76)$$

where $r^{(j)}(\vec{x})$ is the normal form radius in the jth normal form subspace of a particle at position $\vec{x}$. The regions in which the bounds for the maxima have to be found are the regions where $r^{(j)}(\vec{x}) \in [r_i^{(j)}, r_{i+1}^{(j)}]$. As shown previously, these functions exhibit several local maxima in a six-dimensional space. To be useful, the bounds for the maxima have to be sharp to about 10^{-6}, and for some applications to 10^{-12}. Methods are under development that allow both the rigorous bounding of the deviation function (Berz and Hoffstätter, 1994; Hoffstätter, 1994; Makino, 1998; Makino and Berz, 1996; Berz and G. Hoffstätter, 1998; Makino and Berz, 1998) and the computation of rigorous bounds for the remainders of maps (Makino, 1998; Berz and Makino, 1998; Berz, 1997b).

7.3.6 Spin Normal Forms

Similar to the case of the orbit motion, for spin dynamics many quantities of interest can be obtained directly within a normal form framework. In particular, these include the so-called invariant polarization axis as well as the spin tunes. Recall that according to Eq. (5.60), the motion assumes the special form

$$\begin{cases} \vec{z}_f = \vec{\mathcal{M}}(\vec{z}_i, s), \\ \vec{S}_f = \hat{A}(\vec{z}_i, s) \cdot \vec{S}_i. \end{cases}$$

In particular, the orbit motion is unaffected by the spin, and the spin motion is linear with an orbit-dependent orthogonal matrix. This entails that compared to the situation in the previous sections, there are certain differences.

As a **first step** of the normal form analysis, we want to perform an orbit-dependent transformation of the spin variables to coordinates that **preserves one component** of the spin vector at every value of $\vec{z}$. To this end, we will determine an orbit-dependent axis $\bar{n}(\vec{z})$, the **invariant polarization axis**, such that if the spin is described in a coordinate system in which one axis is $\bar{n}(\vec{z})$, the spin's component in this direction is constant from turn to turn. In the first step we do not concern ourselves with the specific orientation of the coordinate system around the axis $\bar{n}(\vec{z})$. This question will be addressed in the second step in connection with the spin tune.

As in the case of orbit motion, the first step is similar to the transformation to the **fixed point**. In the orbit case, particles on or near that fixed point stay on or near it, and for practical purposes it is thus advantageous to inject near the fixed point to limit the regions visited by the beam particles during further revolutions. In the spin case, spin vectors on or near the invariant polarization axis will stay on or near it, and for practical purposes it is advantageous to inject near the invariant polarization axis to **preserve overall polarization** of the beam for further revolutions.

In addition to the significance of the invariant polarization axis for the preservation of polarization of the beam, it is also important in connection with electron or positron machines subject to synchrotron radiation damping. In this case, the damping leads to the buildup of polarization due to the non-random spin of emitted photons, the amount of which can be estimated via semiempirical arguments derived by Derbenev and Kondratenko (1973) from the knowledge of the motion of the axis $\bar{n}(\vec{z})$ around the ring. Methods for the computation of $\bar{n}$ were developed to first order in spin and orbit by Chao (1981), to third order by Eidelmann and Yakimenko (1991), to arbitrary order by Mane (1987) and Balandin and Golubeva (1992), and to arbitrary order including s-dependent elements by Berz (1995b). There are also various approaches to determining $\bar{n}$ and the equilibrium polarization from numerical tracking tools (Yokoya, 1992; Limberg et al., 1989).

If the spin matrix $\hat{A}$ is independent of $\vec{z}$, the axis $\bar{n}$ is apparently merely the rotation axis of $\hat{A}$. Also, since the matrix $\hat{A}$ has a constant part $\hat{A}_0$ that is independent of $\vec{z}$, the **constant part** $\bar{n}_0$ of $\bar{n}$ is just the **rotation axis of** $\hat{A}_0$. However, for a $\vec{z}$-dependent $\hat{A}$, the axis $\bar{n}$ depends on $\vec{z}$, and the requirement is different. Indeed, to preserve the projection of the spin on $\bar{n}(\vec{z})$, we have to satisfy

$$\vec{S}_i^t(\vec{z}_i) \cdot \bar{n}(\vec{z}_i) = \vec{S}_f^t(\vec{z}_f) \cdot \bar{n}(\vec{z}_f)$$

$$= \left(\hat{A}(\vec{z}_i) \cdot \vec{S}_i(\vec{z}_i)\right)^t \cdot \bar{n}(\mathcal{M}(\vec{z}_i))$$

$$= \vec{S}_i^t(\vec{z}_i) \cdot \hat{A}^t(\vec{z}_i) \cdot \bar{n}(\mathcal{M}(\vec{z}_i)).$$

Since this must hold for every vector $\vec{S}_i^t(\vec{z}_i)$, we infer that $\bar{n}$ must satisfy $\bar{n}(\vec{z}_i) = \hat{A}^t(\vec{z}_i) \cdot \bar{n}(\mathcal{M}(\vec{z}_i))$, or

$$\hat{A}(\vec{z}) \cdot \bar{n}(\vec{z}) = \bar{n}(\mathcal{M}(\vec{z})). \quad (7.77)$$

In the following, we will discuss an algorithm for the determination of the equilibrium polarization $\bar{n}(\vec{z})$ similar to the DA normal form approach. We assume the orbital map is already in normal form and the linear spin map is diagonalized. We proceed in an iterative way.

For 0th order, observe that $\mathcal{M}(\vec{z}) =_0 0$, and thus the equation reads $\hat{A}_0 \cdot \bar{n}_0 = \bar{n}_0$. Since $\hat{A}(\vec{z}_i, s)$ and thus $\hat{A}_0$ are in $SO(3)$, the eigenvalue spectrum of $\hat{A}_0$ contains the value 1 as well as a complex conjugate pair of unity modulus. The eigenvector corresponding to the eigenvalue 1 represents the **axis** $\bar{n}_0$, and the phase of the complex conjugate pair represents the **linear spin tune** μ_s.

For higher orders, assume we already know $\bar{n}$ to order $m-1$ and want to determine it to order m. Assume $\mathcal{M}(\vec{z})$ is in normal form, i.e., $\mathcal{M}(\vec{z}) = \mathcal{R} + \mathcal{N}$. Write $\hat{A} = \hat{A}_0 + \hat{A}_{\geq 1}$, $\bar{n} = \bar{n}_{<m} + \bar{n}_m$, and obtain

$$(\hat{A}_0 + \hat{A}_{\geq 1}) \cdot (\bar{n}_{<m} + \bar{n}_m) = (\bar{n}_{<m} + \bar{n}_m) \circ (\mathcal{R} + \mathcal{N})$$

To order m, this can be rewritten as

$$\hat{A} \cdot \bar{n}_{<m} + \hat{A}_0 \cdot \bar{n}_m =_m \bar{n}_{<m} \circ (\mathcal{R} + \mathcal{N}) + \bar{n}_m \circ R \quad (7.78)$$

or

$$\hat{A}_0 \cdot \bar{n}_m - \bar{n}_m \circ \mathcal{R} =_m \bar{n}_{<m} \circ (\mathcal{R} + \mathcal{N}) - \hat{A} \cdot \bar{n}_{<m}.$$

The right hand side must be balanced by choosing $\bar{n}_m$ appropriately. However, as in Eq. (7.64) for the orbit case, the coefficients of $\hat{A}_0 \bar{n}_m - \bar{n}_m \circ \mathcal{R}$ differ from those of $\bar{n}_m$ only by resonance denominators, and so the task can be achieved as soon as these resonance denominators do not vanish. This requires the absence of spin-amplitude resonances in which, because of the purely linear structure of $\hat{A}_0 \cdot \bar{n}$, the spin tune ν_s appears only linearly. Therefore, spin-amplitude terms are characterized by the condition

$$\mu_s + \vec{n} \cdot \vec{\mu} = 0 \bmod 2\pi. \quad (7.79)$$

NORMAL FORMS

As a result of the iterative scheme, we obtain the value of the invariant polarization axis $\bar{n}(\vec{z})$ for all values of $\vec{z}$, probably the single most important nonlinear characteristic of spin motion.

In order to proceed further and determine a nonlinear normal form as well as the associated tune shifts for symplectic orbit motion, we perform a coordinate **rotation** of the spin variables that is linear in spin and nonlinear in orbit, such that at each point $\vec{e}_z$ is **parallel to** $\bar{n}(\vec{z})$ and $\vec{e}_x$ is in the horizontal plane. Expressed in these coordinates, the spin matrix then has the form

$$\hat{A}(\vec{z}) = \begin{pmatrix} a_{11}(\vec{z}) & a_{12}(\vec{z}) & 0 \\ a_{21}(\vec{z}) & a_{22}(\vec{z}) & 0 \\ 0 & 0 & 1 \end{pmatrix}. \quad (7.80)$$

Furthermore, $a_{11}(\vec{z}) \cdot a_{22}(\vec{z}) - a_{12}(\vec{z}) \cdot a_{21}(\vec{z}) = 1$.

We now drop the invariant third component of the spin coordinates and introduce

$$\hat{A}_r(\vec{z}) = \begin{pmatrix} a_{11}(\vec{z}) & a_{12}(\vec{z}) \\ a_{21}(\vec{z}) & a_{22}(\vec{z}) \end{pmatrix}, \quad (7.81)$$

the reduced matrix for the remaining spin coordinates. The resulting **spin–orbit motion** in these eight coordinates $(\vec{S}_r, \vec{z})$ has the form

$$\vec{S}_{r,f} = \hat{A}_r(\vec{z}) \cdot \vec{S}_{r,i} \quad (7.82)$$

$$\vec{z}_f = \mathcal{M}(\vec{z}_i). \quad (7.83)$$

where for every $\vec{z}$, the matrix $\hat{A}_r(\vec{z})$ has a complex conjugate pair of eigenvalues with product 1. For the particular case of $\vec{z} = \vec{0}$, these two eigenvalues have the form $\exp(\pm i \cdot \mu_s)$.

The resulting linearization of the system is apparently

$$\begin{pmatrix} \vec{S}_{r,f} \\ \vec{z}_f \end{pmatrix} = \begin{pmatrix} \hat{A}_{r0} & \hat{0} \\ \hat{0} & \hat{M} \end{pmatrix} \cdot \begin{pmatrix} \vec{S}_{r,i} \\ \vec{z}_i \end{pmatrix}. \quad (7.84)$$

In the absence of spin-orbit resonances, this eight-dimensional linearization is of the form in Eq. (7.57) required for the conventional DA normal form algorithm. In the absence of spin–orbit resonances, the linear eigenvalues are all disjoint and it is thus possible to execute the **conventional DA normal form algorithm** discussed in Section 7.3.1. In this scenario, it is most useful that this algorithm is not restricted to merely symplectic systems but readily adapts to handle this case in which one part of the map satisfies the symplectic symmetry and one part the orthogonal symmetry.

Because of the fact that in each step of the transformation, the spin coordinates occur only linearly, the resonance-driving terms for all **spin resonances** of higher order in spin vanish, and the only requirement is that

$$\mu_s + \vec{n} \cdot \vec{\mu} = 0 \mod 2\pi \tag{7.85}$$

cannot be satisfied for any choice of $\vec{n}$. As a result of the overall transformation, we obtain that the motion in spin–orbit normal form variables has the form

$$\mathcal{M}_j^\pm = \begin{pmatrix} \cos(\phi_j) & -\sin(\phi_j) \\ \sin(\phi_j) & \cos(\phi_j) \end{pmatrix} \cdot \begin{pmatrix} t_j^+ \\ t_j^- \end{pmatrix} \tag{7.86}$$

$$\hat{\mathcal{A}}_r^\pm = \begin{pmatrix} \cos(\phi_s) & -\sin(\phi_s) \\ \sin(\phi_s) & \cos(\phi_s) \end{pmatrix} \cdot \begin{pmatrix} t_s^+ \\ t_s^- \end{pmatrix}. \tag{7.87}$$

As a result, the motion in both the orbit and spin blocks now follows **plain rotations**. Since the entire orbit block was independent of the spin block, the ϕ_j are again only dependent on orbit amplitudes. On the other hand, since the spin motion is linear and dependent on the orbit motion, the ϕ_s are only dependent on orbital amplitudes. Therefore, the newly obtained values ϕ_s describe **amplitude spin tune shifts**.

7.4 SYMPLECTIC TRACKING

The potential buildup of inaccuracies in the dynamical system characterizing an accelerator is a fundamental and unavoidable problem that prevents exact predictions of the motion. While from a purist's point of view this fact represents the point of retreat, in many practical situations it is nevertheless often possible to obtain a qualitative understanding of the phenomena that can occur by **enforcing symmetries** that the system is known to have. An example that is particularly clearly understandable is demonstrated by a system that conserves **energy** (or any other function on phase space such as the Hamiltonian). If during the simulation one enforces that despite the presence of errors the energy of the system is indeed preserved, then this at least prevents the system from drifting off into regions with the wrong energy. Depending on the shape of the energy function, this sometimes ensures that the system stays in the right territory; of course, if the energy surfaces have a complicated topology, even this does not have to be the case.

Another symmetry of systems that are Hamiltonian that is often favored for artificial enforcement is the **symplectic symmetry** discussed in Section 4.3.2, which entails that if $\hat{M}$ is the Jacobian of the map, then

$$\hat{M} \cdot \hat{J} \cdot \hat{M}^t = \hat{J}. \tag{7.88}$$

The geometric ramifications of enforcing this symmetry are more difficult to understand than those of energy conservation; in fact, one of the very unfortunate theoretical results is that for most systems, it is indeed **impossible to simultaneously preserve symplecticity and energy**, unless the simulation by accident tracks the completely correct system.

7.4.1 Generating Functions

Enforcing the symplectic symmetry (Eq. 7.88) directly is not straightforward since the symplectic condition represents a complicated **nonlinear implicit partial differential equation** for the map. The question is whether for a map that is nearly symplectic, be it by truncation or because of errors, there is a symplectic function nearby that can be used instead. A particularly robust method for this purpose is to fabricate such a function through the use of **generating functions** in mixed variables. The four most commonly used generators are

$$F_1(\vec{q}_i, \vec{q}_f), F_2(\vec{q}_i, \vec{p}_f),$$
$$F_3(\vec{p}_i, \vec{q}_f), \text{ and } F_4(\vec{p}_i, \vec{p}_f); \qquad (7.89)$$

which are known to represent symplectic maps via the implicit equations

$$(\vec{p}_i, \vec{p}_f) = (\vec{\nabla}_{q_i} F_1, -\vec{\nabla}_{q_f} F_1)$$
$$(\vec{p}_i, \vec{q}_f) = (\vec{\nabla}_{q_i} F_2, \vec{\nabla}_{p_f} F_2)$$
$$(\vec{q}_i, \vec{p}_f) = (-\vec{\nabla}_{p_i} F_3, -\vec{\nabla}_{q_f} F_3)$$
$$(\vec{q}_i, \vec{q}_f) = (-\vec{\nabla}_{p_i} F_4, \vec{\nabla}_{p_f} F_4). \qquad (7.90)$$

However, to assert universal existence of a generator, as shown in Chapter 1, Section 1.4.6, a wider class of generators is necessary.

Furthermore, the map represented by any generating function, be it the right one or not, is always symplectic. An approximative generating function can be used to perform symplectic tracking in the following way: Use the underlying approximative map to compute first values of the final coordinates $(\vec{q}_f, \vec{p}_f)$. Depending on the accuracy of the map, the quadruple $(\vec{q}_i, \vec{p}_i, \vec{q}_f, \vec{p}_f)$ is already very close to a solution of the implicit Eq. (7.90). It is used as a starting point of a numerical solution of the implicit equations, and $\vec{q}_f$ and/or $\vec{p}_f$ are varied to determine an exact solution. This can be done by Newton's method, and usually one iterative step is enough to obtain machine accuracy.

To illustrate the practical use of the method, we apply it to a nonlinear dynamics problem in which the motion is well understood. We utilize the map of a homogeneous sector magnet with a bending magnet of deflection angle $2\pi/n$, where n is an integer. The resulting map is nonlinear, but since the motion inside a homogeneous bending magnet is purely circular, after 2π and hence n iterations,

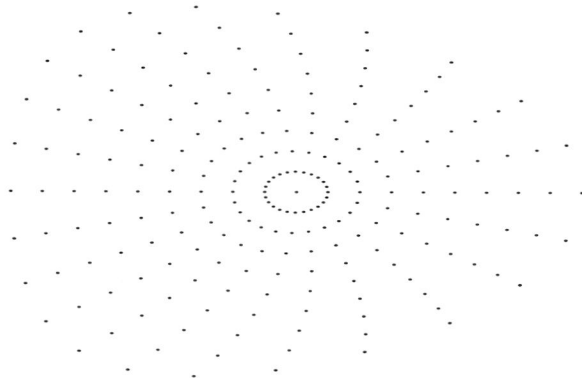

FIGURE 7.21. Motion through a nonlinear system free of any amplitude-dependent tune shifts, described by a map of order 20.

the overall map will be unity. In particular, while nonlinear, the system does not exhibit any amplitude dependent tune shifts.

Figure 7.21 shows map tracking for large amplitudes obtained with a nonsymplectified Taylor map of order 20. After many turns, individual separate points are reproduced faithfully, as should be the case because of the complete lack of amplitude-dependent tune shifts. Figure 7.22 shows the same tracking with a fifth-order map.

Because of the inaccuracy of the fifth-order approximation, errors in both tune and amplitude result. Figure 7.23 shows the motion using two different generating function symplectifications of the Taylor map. In both cases, after a suitable number of points, the overall deviation from the correct position is similar to that for

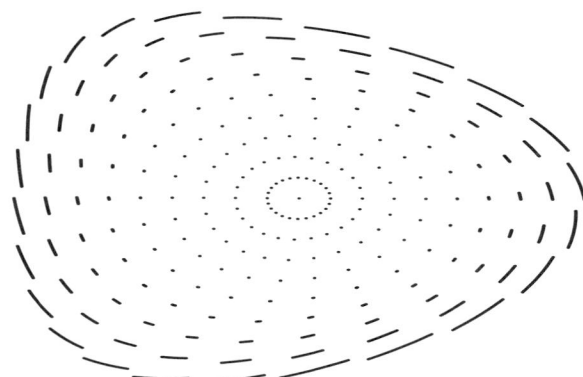

FIGURE 7.22. Motion through a nonlinear system free of any amplitude-dependent tune shifts, described by a map of order 5. Errors due to truncation are visible.

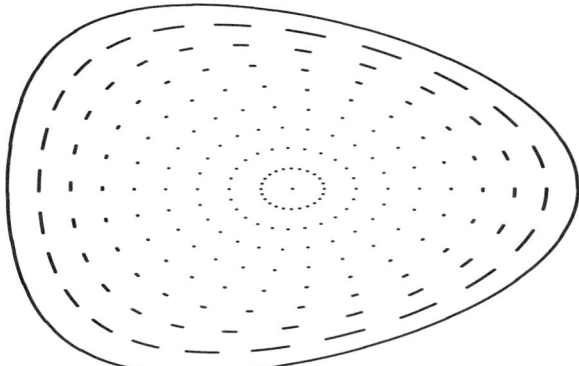

FIGURE 7.23. Symplectifed fifth-order tracking. While radial accuracy increases, angular and overall accuracy are comparable to those for the nonsymplectifiied case.

the case of the nonsymplectified tracking; on the left, the accuracy is higher because the generating function is less nonlinear as it is on the right. It is noteworthy that the resulting corrections due to symplectification lead mainly to deviations in angular direction, whereas the radial direction is apparently well preserved. This phenomenon is mostly due to the fact that symplectification preserves phase space volume, and so in this two-dimensional case no shrinking in phase space is possible.

The observations in this particular example are found in a similar way as those in other cases; symplectification has a tendency to **preserve radial information** but does not help in the reproduction of **angular information**; in fact, it sometimes leads to larger errors in angular direction. While tracking the preservation of radial information perhaps leads to a more realistic looking picture, for analytical causes such as the normal form methods, not much is necessarily gained. In particular, the errors in angular direction are usually approximately as strong in **amplitude tune shifts** as they are without symplectification. Similarly, since the **resonance strengths** in Eq. (7.64) are a measure of the proximity to resonances for various amplitudes, the errors in resonance strengths are usually not affected beneficially by symplectification. For the previous example, the structure of tune shifts and resonance strengths after symplectification is not reproduced faithfully, and although symplectification indeed preserves the symplectic symmetry of the system, it is not able to preserve the underlying resonance symmetry properly.

7.4.2 Prefactorization and Symplectic Extension

For all practical applications, the behavior of the nonlinear generating function is very important; in particular, it is important that its nonlinearities are not too large. While the nonlinearities of the transfer map are a more or less direct measure of the nonlinearity of the system, this is not the case for generating functions. By inspecting the algorithm for the computation of the generating function, it becomes apparent that the new results are concatenated over and over with the **inverse** of the linear relationship in mixed variables. Although, as shown in Chapter 1, Section 1.4.6, a set of mixed variables can be found such that the linearization is nonsingular, it may be the case that the determinant is close enough to zero that the inverse has large terms.

Therefore, while symplectic matrices are always restricted to unity determinant, this is not the case for the matrices that have to be inverted here, and so the resulting generating functions sometimes have stronger nonlinearities than the underlying maps.

This problem can be alleviated largely by **factoring out the linear part** of the map

$$\mathcal{M}_n = \mathcal{M}_1 \circ \mathcal{M}_n^*$$

and then treating $\mathcal{M}_1$ and $\mathcal{M}_n^*$ separately. For the map $\mathcal{M}_n^*$, the linear part is now unity, which implies that two of the four generating functions can be computed without the risk of small determinants. The **symplectification of the linear map** $\mathcal{M}_1$ represented by the matrix $\hat{M}_1$, which is also its Jacobian, is a linear algebra problem. It suffices to satisfy $\hat{M}_1 \cdot \hat{J} \cdot \hat{M}_1^t = \hat{J}$ which using the scalar product (Eq. 1.173) with metric $\hat{J}$ defined by $< x, y > = x \cdot \hat{J} \cdot y^t$ can be enforced by procedures such as the Gram–Schmidt.

Using the Taylor series expansion terms of the generating function as described previously, it is straightforward to compute a transfer map that agrees with the given transfer map to order n yet is symplectic to higher orders compared to the old transfer map. Even if the old transfer map violates symplecticity noticeably because of truncation errors, it is possible to have the extended map satisfy the symplectic condition to increasingly higher orders. Depending on the case, it is often possible to obtain machine accuracy symplecticity for the phase space regions of interest by the method of **symplectic extension**.

To this end, one first computes a suitable generating function to the same order as that of the original map, following the algorithm discussed in the last section. While the result is not the proper generating function for the true map, it has the same Taylor expansion as the proper one and agrees with it increasingly better the higher the order. One now approximates the real generating function by its Taylor series and computes the map that is generated from the approximate generating function using the previous algorithm. Up to order n, the old map is reproduced; continuing beyond order n produces increasingly higher orders extending the original transfer map, and the map can be made symplectic to any order $k > n$.

If the missing higher order terms are sufficiently small, what we have produced is an explicit symplectic integrator. This algorithm is particularly useful for systems in which the computation of the map is expensive to high orders, but

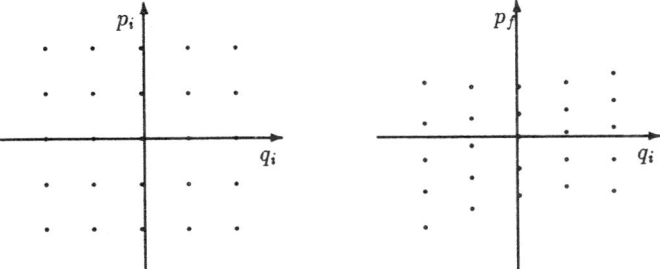

FIGURE 7.24. Local nodes around which Taylor expansion is performed.

whose inherent nonlinearity is not too high. In particular, this could be the case for machines consisting of many different elements or compact machines with very complicated fields requiring detailed integrations.

7.4.3 Superposition of Local Generators

Here, we discuss the symplectification of a map that is not given by one Taylor map but rather by a group of Taylor maps around different expansion points. The goal is to find a global generating function that locally reduces to the respective Taylor maps but that is globally symplectic.

When using generating functions for symplectic tracking, it is of course not mandatory that they actually have the same Taylor expansion as that of the true generating function. Indeed, in the case of nonlinear problems in which the function $\mathcal{N}_n$ is not well behaved, it may be advantageous to produce generating functions that are smoother overall. This can be achieved by a superposition of local generating functions.

To this end, a representative **ensemble of nodes** in phase space is chosen, preferably in a regular way as in Fig. 7.24. For each of these nodes, a transfer map is computed to a certain order. Then, for each of the nodes the respective generating function is computed. Each of these generating functions is uniquely determined except for its constant value c_i.

A total generating function can now be determined by a smooth interpolation of the local polynomial-type generating functions in a nonequidistant mesh. This has the form

$$F(\vec{q}_i, \vec{p}_f) = \sum_{j=1}^{n} F_j(\vec{q}_i, \vec{p}_f) \cdot w_j(\vec{q}_i, \vec{p}_f), \qquad (7.91)$$

where the w_i are smooth weighting factor functions that ensure that the influence of F_i only extends near the respective next nodes and not far beyond. For example, they can be Gaussians centered at $(\vec{q}_i, \vec{p}_f)$, with widths determined by the distances to the next nodes and a height chosen accordingly.

While in the case of a single generating function, the unknown constant term was irrelevant, here it is significant since it is multiplied by the position dependent weighting function and thus appears in the implicit solution $(\vec{q}_f, \vec{p}_f)$. Therefore, it is necessary to choose the c_i **in a self-consistent way**.

One solution to this problem is to demand that at each node, the predictions of all the neighboring nodes are as close as possible (which is similar to the finite element potential-solving problem). This yields a least squares problem for the c_i which can be solved using conventional techniques. Naturally, the higher the orders of the individual F_i, the better will be their prediction at the neighboring nodes and the smaller will be the resulting sum of squares.

Therefore, one obtains a generating function that is not Taylor series like, and one can cover large and rather nonlinear areas of phase space.

7.4.4 Factorizations in Integrable Symplectic Maps

In this section, we discuss symplectification techniques of other methods that allow the representation of a map by a symplectic map. The idea is to first generate a large **pool of symplectic maps** S_i that depend on certain parameters and **that can be evaluated explicitly**; such maps are sometimes called Cremona maps. Important examples of such a pool are drifts, rotations in phase space, and **kick maps** that affect only the momenta of particles as proposed by Irwin and Dragt. Other important classes of such explicitly integrable functions were proposed, including those by Shi (Shi and Yan, 1993; Shi, 1994).

Then this pool is used to model the original function as a **composition** of such maps as

$$\mathcal{M}_n =_n \Pi_{i=1}^k \mathcal{S}_i,$$

where the parameters are chosen such that the overall map $\mathcal{M}_n$ is reproduced. In all cases, it is important and not easily achieved that the map pool is rather **exhaustive** and allows the representation of the desired map through maps from the pool with **well-behaved parameter settings**. Furthermore, since the number of parameters that have to be adjusted is very large and usually equal to the number of free coefficients in a symplectic map, it is very advantageous if the system of equations that has to be solved is **linear**.

In the case of kick factorizations, a frequently used approach is to attempt to represent the map by kicks that affect only one particular order each and that are separated by rotations of fixed angles, all of which are fractions of 2π. The kick strengths are chosen as the free parameters $\vec{p}$, and they are determined in an order-by-order way. Specifically, for order n, all kicks of orders $<n$ are already known and only parameters $\vec{p}_n$ of order n are adjusted. Denoting by the vector $\vec{m}_n$ the nth order map properties that have to be adjusted, this leads to a linear system of

equations of the form

$$\vec{m}_n = \hat{A}_n \cdot \vec{p}_n. \tag{7.92}$$

Here, the information of the specific map is entirely contained in $\vec{m}_n$, whereas the coupling matrix $\hat{A}$ depends only on the placement of the kicks within the fixed rotations. In order to limit spurious higher order nonlinearities introduced by the approach, it is highly desirable that the matrix $\hat{A}$ that couples the coefficients to the nonlinearities of the map has a well-behaved inverse. A measure of this is the spectrum of the k eigenvalues λ_i of $\hat{A}$. The spacing and angles of rotations are adjusted in such a way as to maximize the **minimum** eigenvalue modulus

$$\lambda_{\min} = \min(|\lambda_i|); \tag{7.93}$$

obviously, a vanishing eigenvalue prevents inversion of the matrix, and it is expected that the coefficients of the inverse are more well behaved if $\lambda_{\min}$ is as large as possible. Likewise, it is also desirable that the **average**

$$\lambda_a = \frac{\sum_{i=1}^k \lambda_i}{k}. \tag{7.94}$$

be as large as possible. To satisfy both of these demands is rather nontrivial, and optimal and robust solutions for these questions have not been found for the general case.

7.4.5 Spin Tracking

We address another type of tracking that, although not preserving the symplectic symmetry, conceptually can be treated in a very similar way: the tracking of spin. As discussed in Chapter 5, Section 5.2, the map of a system consisting of orbit motion and spin motion is described by

$$\begin{cases} \vec{z}_f = \mathcal{M}(\vec{z}_i), \\ \vec{S}_f = \hat{A}(\vec{z}_i) \cdot \vec{S}_i \end{cases}, \tag{7.95}$$

where $\mathcal{M}$ is the symplectic orbit map, and $\hat{A}(\vec{z}_i) \in SO(3)$ is the spin matrix. For the purpose of tracking of spin, it is necessary to successively obtain $(\vec{z}_n, \vec{S}_n)$ from $(\vec{z}_{n-1}, \vec{S}_{n-1})$. This requires the determination of $\vec{z}_n$ via

$$\vec{z}_n = \mathcal{M}(\vec{z}_{n-1}), \tag{7.96}$$

for which symplectic integration techniques can be used. However, to compute $\vec{S}_n$ requires that first the orbit-dependent spin map $\hat{A}(\vec{z}_{n-1})$ be evaluated. Since the full dependence of $\hat{A}$ on $\vec{z}$ is not known, but only its polynomial representation to a given order, the evaluation will in general result in a matrix that is almost, but not exactly, orthogonal. For the same reason, symplectification may be desirable; one may thus want to make $\hat{A}$ orthogonal before computing the new spin.

While the preservation of the symplectic symmetry required the determination of a sufficiently close-by symplectic map, here it is necessary to find a sufficiently close-by orthogonal matrix. This, however, is straightforward, because it can be achieved with **orthonormalization** techniques such as Gram-Schmidt. Furthermore, if the starting map is nearly orthogonal, the procedure will not result in large deviations and it will produce a nearby matrix. After the matrix $\hat{A}(\vec{z}_n)$ is orthogonalized to become the matrix $\hat{A}^*$, the new spin coordinates are determined via

$$\vec{S}_n = \hat{A}^* \cdot \vec{S}_{n-1}. \tag{7.97}$$

REFERENCES

Abate, J., Bischof, C., Roh, L., and Carle, A. (1997). Algorithms and design for a second-order automatic differentiation module. In *Proceedings, ISSAC 97, Maui*. ACM, New York. [Also available at ftp://info.mcs.anl.gov/pub/tech_reports/reports/P636.ps.Z]

Adams, F., Gijbels R., and Van Grieken, R. (Eds.). (1988). *Inorganic Mass Spectrometry*. Wiley, New York.

Alefeld G. and Herzberger, J. (1983). *Introduction to Interval Computations*. Academic Press.

Ascher, M., Uri, M., and Petzold, L. R. (1998). *Computer Methods for Ordinary Differential Equations and Differential-Algebraic Equations*. SIAM, Philadelphia.

Balandin, V. and Golubeva, N., *International Journal of Modern Physics A*, 2B:998, 1992.

Bazzani A., Marmi, S., and Turchetti, G. (1998). *Nekhoroshev Estimate for Isochronous Nonresonant Symplectic Maps*. University of Bologna, Bologna, Spain.

Berz, M. (1986). The new method of TPSA algebra for the description of beam dynamics to high orders. Technical Report No. AT-6:ATN-86-16. Los Alamos National Laboratory, Los Alamos, NM.

Berz, M. (1987a). The differential algebra FORTRAN precompiler DAFOR, Technical Report No. AT-3:TN-87-32, Los Alamos National Laboratory, Los Alamos, NM.

Berz, M. (1987b). The method of power series tracking for the mathematical description of beam dynamics. *Nuclear Instr. Methods* **A258**;431.

Berz, M. (1988a). Differential algebraic description and analysis of trajectories in vacuum electronic devices including spacecharge effects. *IEEE Trans. Electron Devices*; **35**(11);2002.

Berz, M. (1988b). Differential algebraic description of beam dynamics to very high orders, Technical Report No. SSC-152. SSC Central Design Group, Berkeley, CA.

Berz, M. (1989). Differential algebraic description of beam dynamics to very high orders. *Particle Accelerators* **24**;109.

Berz, M. (1990a). Analysis auf einer nichtarchimedischen Erweiterung der reellen Zahlen, Report No. MSUCL-753. Department of Physics, Michigan State University, East Lansing. [In German]

Berz, M. (1990b). Arbitrary order description of arbitrary particle optical systems. *Nuclear Instr. Methods*, **A298**;426.

Berz, M. (1990e). Differential algebra precompiler version 3 reference manual, Technical Report No. MSUCL-755, Michigan State University, East Lansing.

Berz, M. (1991a). Forward algorithms for high orders and many variables. In *Automatic Differentiation of Algorithms: Theory, Implementation and Application*. SIAM, Philadelphia.

Berz, M. (1991b). High-order computation and normal form analysis of repetitive systems. In *Physics of Particle Accelerators* (M. Month, Ed.), Vol. AIP 249, p. 456. American Institute of Physics, New York.

Berz, M. (1992a). Automatic differentiation as nonarchimedean analysis. In *Computer Arithmetic and Enclosure Methods*, p. 439, Elsevier, Amsterdam.

Berz, M. (1992b). COSY INFINITY Version 6. In *Proceedings of the Nonlinear Effects in Accelerators* (M. Berz, S. Martin and K. Ziegler, Eds.), p. 125. IOP.

Berz, M. (1993a). COSY INFINITY Version 6 reference manual, Technical Report No. MSUCL-869, National Superconducting Cyclotron Laboratory, Michigan State University, East Lansing.

Berz, M. (1993b). New features in COSY INFINITY. In *Third Computational Accelerator Physics Conference*. AIP Conference Proceedings No. 297, p. 267. American Institute of Physics, New York.

Berz, M. (1994). Analysis on a nonarchimedean extension of the real numbers. Lecture Notes, 1992 and 1995 Mathematics Summer Graduate Schools of the German National Merit Foundation, MSUCL-933. Department of Physics, Michigan State University, East Lansing.

Berz, M. (1995a). COSY INFINITY Version 7 reference manual, Technical Report No. MSUCL-977. National Superconducting Cyclotron Laboratory, Michigan State University, East Lansing. [See also http://www.beamtheory.nscl.msu.edu/cosy]

Berz, M. (1995b). Differential algebraic description and analysis of spin dynamics. In *Proceedings, SPIN94*.

Berz, M. (1996). Calculus and numerics on Levi–Civita fields. In *Computational Differentiation: Techniques, Applications, and Tools* (M. Berz, C. Bischof, G. Corliss, and A. Griewank, Eds.), pp. 19–35. SIAM, Philadelphia.

Berz, M. (1997a). COSY INFINITY Version 8 reference manual, Technical Report No. MSUCL-1088, National Superconducting Cyclotron Laboratory, Michigan State University, East Lansing. [See also http://www.beamtheory.nscl.msu.edu/cosy]

Berz, M. (1997b). From Taylor series to Taylor models. *AIP* **405**;1–25.

Berz, M. (1998). Differential Algebraic Techniques. In *Handbook of Accelerator Physics and Engineering* (M. Tigner and A. Chao, Eds.). World Scientific, New York.

Berz, M. (1999). Computational Differentiation. In *Encyclopedia of Computer Science and Technology*. Marcel Dekker, New York.

Berz, M. et al. (1998). The COSY INFINITY web page: http://www.beamtheory.nscl.msu.edu/cosy.

Berz, M., Fawley, B., and Hahn, K. (1991). High order calculation of the multipole content of three dimensional electrostatic geometries. *Nuclear Instr. Methods* **A307**;1.

Berz, M. and Hoffstätter, G. (1994). Exact bounds of the long term stability of weakly nonlinear systems applied to the design of large storage rings. *Interval Computations* 2;68–89.

Berz, M. and Hoffstätter, G. (1998). Computation and Application of Taylor Polynomials with Interval Remainder Bounds. *Reliable Computing* **4**;83–97.

REFERENCES

Berz, M., Bischof, C., Griewank, A., Corliss, G. (Eds.). (1996). *Computational Differentiation: Techniques, Applications, and Tools*. SIAM, Philadelphia.

Berz, M., Hoffstätter, G., Wan, W., Shamseddine, K., and Makino, K. (1996). COSY INFINITY and its applications to nonlinear dynamics. In *Computational Differentiation: Techniques, Applications, and Tools* (M. Berz, C. Bischof, G. Corliss, and A. Griewank, Eds.), pp. 363–365. SIAM, Philadelphia.

Berz, M., Hofmann, H. C., and Wollnik, H. (1987). COSY 5.0, the fifth order code for corpuscular optical systems. *Nuclear Instr. Methods*, **A258**;402.

Berz, M., Joh, K., Nolen, J. A., Sherrill, B. M., and Zeller, A. F. (1993). Reconstructive correction of aberrations in nuclear particle spectrographs. *Phys. Rev. C*, **47**(2);537.

Berz, M. and Makino, K. (1998). Verified integration of ODEs and flows with differential algebraic methods on Taylor models. *Reliable Computing* **4**;361–369.

Berz, M., Makino, K., and Wan, W. (1999). *An Introduction to the Physics of Beams*. IOP, London.

Berz, M. and Wollnik, H. (1987). The program HAMILTON for the analytic solution of the equations of motion in particle optical systems through fifth order. *Nuclear Instr. Methods*, **A258**;364.

Bischof, C. and Dilley, F. A compilation of automatic differentiation tools. [See http://www.mcs.anl.gov/autodiff/AD_Tools/index.html]

Blosser, H. G., Crawley, G. M., DeForest, R., Kashy, E., and Wildenthal, B. H. (1971). *Nuclear Instr. Methods*, **91**;197.

Brenan, K. E., Campbell, S. L., and Petzold, L. R. (1989). *Numerical Solution of Initial-Value Problems in Differential-Algebraic Equations*. North-Holland.

Brown, K. L. (1979a). The ion optical program TRANSPORT, Technical Report No. 91, SLAC.

Brown, K. L. (1979b). A second order magnetic optical achromat. *IEEE Trans. Nucl. Sci.* **NS-26**(3);3490.

Brown, K. L. (1982). A first- and second-order matrix theory for the design of beam transport systems and charged particle spectrometers, Technical Report No. 75, SLAC.

Brown, K. L., Belbeoch, R., and Bounin, P. (1964). First- and second- order magnetic optics matrix equations for the midplane of uniform-field wedge magnets. *Rev. Sci. Instr.*, **35**;481.

Browne, C. P., and Buechner, W. W. (1956). Broad-range magnetic spectrograph. *Rev. Sci. Instr.* **27**;899.

Carey, D. C. (1987, 1992). *The Optics of Charged Particle Beams*. Harwood Academic, New York.

Carey, David C. (1981). Why a second-order magnetic optical achromat works. *Nuclear Instr. Methods* **189**;365.

Chao, A. W. (1981). *Nuclear Instr. Methods* **29**;180.

Conte, S. D. and de Boor, C. (1980). *Elementary Numerical Analysis*. McGraw Hill, New York.

Cotter, R. J. (Ed.). (1994). Time-of-Flight Mass Spectrometry. In *ACS Symposium Series No. 549*. American Chemical Society, Washington, DC.

Courant, E. D. and Snyder, H.S. (1958). Theory of the alternating gradient synchrotron. *Annals of Physics* **3**;1.

REFERENCES

Daubechies, I. (1992). *Ten Lectures on Wavelets*. SIAM, Philadelphia.

Davis, W. G., Douglas, S. R., Pusch, G. D., and Lee-Whiting, G. E. (1993). The Chalk River differential algebra code DACYC and the role of differential and Lie algebras in understanding the orbit dynamics in cyclotrons. In *Proceedings Workshop on Nonlinear Effects in Accelerators* (M. Berz, S. Martin, and K. Ziegler Eds.). IOP.

Derbenev, Ya. S. and Kondratenko, A. M. (1973). *Sov. Phys. JETP* **35**;968.

Dragt, A. J. (1982a). Lectures on nonlinear orbit dynamics. In *1981 Fermilab Summer School*. AIP Conference Proceedings, Vol. 87, American Institute of Physics, New York.

Dragt, A. J. (1987). Elementary and advanced Lie algebraic methods with applications to accelerator design, electron microscopes, and light optics. *Nuclear Instr. Methods* **A258**;339.

Dragt, A. J. and Finn, J. M. (1976). Lie series and invariant functions for analytic symplectic maps. *J. Math. Phys.* **17**;2215.

Dragt, A. J. and Finn, J. M. (1979). Normal form for mirror machine Hamiltonians. *J. Math. Phys.* **20**(12);2649.

Dragt, A. J., Healy, L. M., Neri, F., and Ryne, R. (1985). MARYLIE 3.0—A program for nonlinear analysis of accelerators and beamlines. *IEEE Trans. Nuclear Sci.* **NS-3**;5:2311.

Eidelmann, Y. and Yakimenko, V. (1991). The spin motion calculation using Lie method in collider magnetic field. In *Proceedings of the 1991 IEEE Particle Accelerator Conference*. IEEE, San Francisco.

Fenselau, C. (Ed.) (1994). Mass Spectrometry for the Characterization of Microorganisms. In *ACS Symposium Series No. 549*. American Chemical Society, Washington, DC.

Fischer, G. E., Brown, K. L., and Bulos, F. (1987) *Proceedings of the 1987 Particle Accelerator Conference*, p. 139. IEEE, San Francisco.

Flanz, J. B. (1989) *Proceedings of the 1989 Particle Accelerator Conference*, p. 1349. IEEE, San Francisco.

Forest, E., Berz, M., and Irwin, J. (1989). Normal form methods for complicated periodic systems: A complete solution using differential algebra and Lie operators. *Particle Accelerators* **24**;91.

Franzke, B. (1987). The heavy ion storage and cooler ring project ESR at GSI. *Nuclear Instr. Methods*, **B24/25**;18.

Gijbels, R., Adams, F., and Van Grieken, R. (Eds.) (1988). *Inorganic Mass Spectrometry*. Wiley, New York.

Glish, G. L., Busch, K., and McLuckey, S.A. (1988). *Mass Spectrometry/Mass Spectrometry: Applications of Tandem Mass Spectrometry*. VCH, New York.

Gordon, M. M. and Taivassalo, T. (1986). The z^4 orbit code and the focusing bar fields used in beam extraction calculations for superconducting cyclotrons. *Nuclear Instr. Methods* **247**;423.

Griepentrog, E. and Roswitha, M. (1986). *Differential-Algebraic Equations and their Numerical Treatment*. Teubner.

Griewank, A. and Corliss, G. F. (Eds.) (1991). *Automatic Differentiation of Algorithms*. SIAM, Philadelphia.

Hansen, E. (1969). *Topics in Interval Analysis*. Oxford University Press, London.

Hartmann, B., Berz, M., and Wollnik, H. (1990). The computation of fringing fields using Differential Algebra. *Nuclear Instr. Methods* **A297**;343.

Helm, R. H. (1963). First and second order beam optics of a curved inclined magnetic field boundary in the impulse approximation. Technical Report No. 24, SLAC.

Hoffstätter, G. and Berz, M. (1996a). Rigorous lower bounds on the survival time in particle accelerators. *Particle Accelerators* **54**;193–202.

Hoffstätter, G. and Berz, M. (1996b). Symplectic scaling of transfer maps including fringe fields. *Phys. Rev. E* **54**;4.

Hoffstätter, G. H. (1994). Rigorous bounds on survival times in circular accelerators and efficient computation of fringe–field transfer maps. Ph.D. thesis, Michigan State University, East Lansing.

Iselin, F. C. (1996). The CLASSIC project. In *Fourth Computational Accelerator Physics Conference*, AIP Conference Proceedings. American Institute of Physics, New York.

Jackson, J. D. (1975). *Classical Electrodynamics*. Wiley, New York.

Kaucher, E. W. and Miranker, W. L. (1984). *Self-Validating Numerics for Function Space Problems: Computation with guarantees for differential and integral equations*. Academic Press, New York.

Kearfott, R. B. and Kreinovich, V. (Eds.) (1996). *Applications of Interval Computations*, Vol. 3. Kluwer.

Kolchin, E. R. (1973). *Differential Algebra and Algebraic Groups*. Academic Press, New York.

Limberg, T., Rossmanith, R., and Kewisch, J. (1989). *Phys. Rev. Lett.*, **62**;62.

Makino, K. (1998). *Rigorous analysis of nonlinear motion in particle accelerators*. Ph.D. thesis, Michigan State University, East Lansing. [Also MSUCL-1093]

Makino, K. and Berz, M. (1996a). COSY INFINITY Version 7. In *Fourth Computational Accelerator Physics Conference*, AIP Conference Proceedings, Vol. 391, p. 253.

Makino, K. and Berz, M. (1996b). Remainder differential algebras and their applications. In *Computational Differentiation: Techniques, Applications, and Tools* (M. Berz, C. Bischof, G. Corliss, and A. Griewank, Eds.), pp. 63–74. SIAM, Philadelphia.

Makino, K. and Berz, M. (1998). Efficient control of the dependency problem based on taylor model methods. *Reliable Computing*, **5**;3–12.

Makino, K. and Berz, M. (1999). COSY INFINITY version 8. *Nuclear Instr. Methods* Vol. A427, pp. 338–343.

Mane, S. R. (1987). *Phys. Rev.* **A36**;120.

Matsuda, M. (1980). *First Order Algebraic Differential Equations: A Differential Algebraic Approach*. Springer-Verlag, Berlin.

Matsuo, T. and Matsuda, H. (1976). Computer program TRIO for third order calculations of ion trajectories. *Mass Spectrometry* **24**.

Michel, L., Bargmann, V., and Telegdi, V. L. (1959). Precession of the polarization of particles moving in a homogeneous electromagnetic field. *Phys. Rev. Lett.*, **2**;435.

Michelotti, L. (1990). MXYZTPLK: A practical, user friendly C++ implementation of differential algebra. Technical Report, Fermilab, Naperville, IL.

Moore, R. E. (1979). *Methods and Applications of Interval Analysis*. SIAM, Philadelphia.

Moore, R. E. (Ed.) (1988). *Reliability in computing: the role of interval methods in scientific computing*. Academic Press.

Morris, C. (1989). High resolution, position-sensitive detectors. In *Proceedings of the International Conference on Heavy Ion Research with Magnetic Spectrographs*. Technical Report No. MSUCL-685. National Superconducting Cyclotron Laboratory, Michigan State University, East Lansing.

Murray, J. J., Brown, K. L., and Fieguth, T. (1987) *Proceedings of the 1987 Particle Accelerator Conference*, p. 1331. IEEE, San Francisco.

Nekhoroshev, N. N. (1977) An exponential estimate of the time of stability of nearly integrable Hamiltonian systems, *Uspekhi Mat. Nauk.* **32**; 6, English translation Russ. Math. Surv., 32:6,5, 1977, p. 1.

Neri, F. Third Order Achnoinals (1991). In *Proceedings of the 1990 Workshop on High Order Effects in Accelerators and Beam Optics* (M. Berz, and J. McIntyre, Eds.). Technical Report No. MSUCL-767. National Superconducting Cyclotron Laboratory, Michigan State University, East Lansing.

Nolen, J., Zeller, A. F., Sherrill, B., DeKamp, J. C., and Yurkon, J. (1989). A proposal for construction of the S800 spectrograph. Technical Report No. MSUCL-694. National Superconducting Cyclotron Laboratory, Michigan State University, East Lansing.

Parrott, S. (1987). *Relativistic Electrodynamics and Differential Geometry*. Springer-Verlag, New York.

Poincare, H. (1893). *New Methods of Celestial Mechanics*, Vols. 1–3. American Institute of Physics, New York.

Pusch, G. D. (1990). Differential Algebraic Methods for Obtaining Approximate Numerical Solutions to the Hamilton-Jacobi Equation. Ph.D. thesis, Virginia Polytechnic Institute and State University, Blacksburg, VA.

Risch, R. H. (1969). The problem of integration in finite terms. *Trans. Am. Math. Soc.* **139**;167–189.

Risch, R. H. (1970). The solution of the problem of integration in finite terms. *Bull. Am. Math. Soci.*, **76**;605–608.

Risch, R. H. (1979). Algebraic properties of elementary functions of analysis. *Am. J. Math.* **101**(4);743–759.

Ritt, J. F. (1932). *Differential Equations from the Algebraic Viewpoint*. American Mathematical Society, Washington, DC.

Ritt, J. F. (1948). *Integration in Finite Terms—Liouville's Theory of Elementary Methods*. Columbia Univ. Press, New York.

Rohrlich, F. (1965, 1990). *Classical Charged Particles*. Addison-Wesley, Reading, MA.

Russenschuck, S. (1995). A computer program for the design of superconducting accelerator magnets. Technical Report No. CERN AT/95-39. CERN.

Schwarzschild, B. (1994) *Physics Today*, **47.7**; 22.

Shamseddine, K. and Berz, M. (1996). Exception handling in derivative computation with nonarchimedean calculus. In *Computational Differentiation: Techniques, Applications, and Tools* (M. Berz, C. Bischof, G. Corliss, and A. Griewank, Eds.), pp. 37–51. SIAM, Philadelphia.

Shamseddine, K. and Berz, M. (1997). Nonarchimedean structures as differentiation tools. In *Proceedings, Second LAAS International Conference on Computer Simulations*, pp. 471–480.

Shi, J. (1994). Integrable polynomial factorization for symplectic maps. *Physical Review E*, **50**(1).

Shi, J. and Yan, Y. (1993). Explicitly integrable polynomial Hamiltonians and evaluation of Lie transformations. *Physical Review E*, **48**(5).

Thomas, L. H. (1927). The kinematics of an electron with an axis. *Philos. Magazine*, **3**(13);1–22.

Turchetti, G. (1990). Nekhoroshev stability estimates for symplectic maps and physical applications. *Springer Proc. Phys.*, **47**;223.

van Zeijts, J. (1993). New features in the design code TLIE. In *Third Computational Accelerator Physics Conference*, AIP Conference Proceedings No. 297, p. 285. American Institute of Physics, New York.

van Zeijts, J. and Neri, F. (1993). The arbitrary order design code TLIE 1.0. In *Proceedings Workshop on Nonlinear Effects in Accelerators* (M. Berz, S. Martin, and K. Ziegler, Eds.). IOP.

Wan, W. (1995). Theory and applications of arbitrary-order achromats. Ph.D. thesis, Michigan State University, East Lansing. [Also No. MSUCL-976]

Wan, W. and Berz, M. (1996). An analytical theory of arbitrary order achromats. *Physical Review E*, **54**(3);2870.

Warnock, R. L. and Ruth, R. D. (1991). Stability of orbits in nonlinear mechanics for finite but very long times. In *Nonlinear Problems in Future Accelerators* (W. Scandale and G. Turchetti, Eds.), pp. 67–76, World Scientific, New York.

Warnock, R. L. and Ruth, R. D. (1992). Long-term bounds on nonlinear Hamiltonian motion, *Physica D*. **56**(14), 1992, p. 188–215.

Warnock, R. L., Ruth, R. D., Gabella, W. and Ecklund, K. (1988). Methods of stability analysis in nonlinear mechanics. In *1987 Accelerator Physics Summer School* (SLAC Publ. No. 4846, 1989), AIP Conference Proceedings, American Institute of Physics, New York.

Watson, J. T. (1985). *Introduction to Mass Spectrometry*. Raven Press, New York.

White, F. A. and Wood, G. M. (1986). *Mass Spectrometry*. Wiley, New York.

Wollnik, H. (1965). Image aberrations of second order for magnetic and electrostatic sector fields including all fringing field effects. *Nuclear Instr. Methods*, **38**;56.

Wollnik, H. (1987a). *Nuclear Instr. Methods*, **B26**; 267.

Wollnik, H. (1987b) *Nuclear Instr. Methods*, **A258**; 289.

Wollnik, H. and Berz, M. (1985). Relations between the elements of transfer matrices due to the condition of symplecticity. *Nuclear Instr. Methods*, **238**; 127.

Wollnik, H., Brezina, J., and Berz, M. (1984). GIOS-BEAMTRACE, a program for the design of high resolution mass spectrometers. In *Proceedings AMCO-7*, p. 679, Darmstadt.

Wollnik, H., Brezina, J., and Berz, M. (1987). GIOS-BEAMTRACE, a computer code for the design of ion optical systems including linear or nonlinear space charge. *Nuclear Instr. Methods*, **A258**;408.

Wollnik, H., Hartmann, B., and Berz, M. (1988). Principles behind GIOS and COSY. *AIP Conf. Proc.*, **177**;74.

Wouters, J. M., Wollnik, H., and Vieira, D. J. (1987). Tofi: An isochronous time-of-flight mass spectrometer. *Nuclear Instr. Methods* **A258**;331.

Wynne, C.G. (1973). Data for some four-lens paraboloid field correctors. *Not. Royal Astronomical Society*, **165**;1–8.

Yan, Y. (1993). ZLIB and related programs for beam dynamics studies. In *Third Computational Accelerator Physics Conference*. AIP Conference Proceedings No. 297, p. 279. American Institute of Physics, New York.

Yan, Y. and Yan, C.-Y. (1990). ZLIB, a numerical library for differential algebra. Technical Report 300, SSCL.

Yokoya, K. (1992). Technical Report No. 92-6, KEK, Tsunba, Japan.

INDEX

Aberration, 212
Aberrations
   Correction by Symmetry, 228
   in Spectrometers, 215
   Order-by-Order Correction, 124
   reconstructive correction, 217
   under Symplecticity, 155
   with Midplane Symmetry, 149
Absolute Value, 95, 114
Accelerator Mass Spectrometer, 211
Accelerator Physics, 1
Acceptance, 217
Achromat
   and Symplecticity, 158
   Definition, 228
   Fifth Order, 245
   Second Order, 244
   Third Order, 240, 243
Action Integral, 16
Addition, 86, 91, 95, 96, 113
   of Sets, 113
Addition Theorems on DA, 103
Algebra, 86, 110, 113
   Classification, 86
   Lie, 44
   of Functions, 81, 84
Algebraically Closed, 115
Alpha
   (Twiss Parameter), 257
Ampere-Maxwell Law, 63, 71
Amplitude Tune Shift, 274
   for Spin Motion, 292
Analyticity
   and Complex Variables, 151
   of Map, 167
Angular Acceptance, 217

Anti-Symplecticity, 148, 234
Antiderivation, 85
   Fixed Point of, 98
   for PDE Solver, 111
   in ODE Solver, 109
Antisymmetric Bilinear Form, 44
Antisymmetric Flow
   Orthogonality of, 192
Antisymmetric Matrix J, 21
Antisymmetric Product, 39
   as Poisson Bracket, 42
Archimedicity, 114, 116
Arclength, 168, 191
   Independent Variable for Hamiltonian, 185
Area Conservation
   by Symplectic Maps, *see* Volume Conservation 41
Atlas
   Manifolds, 9
Automatic Differentiation, 84
Autonomous
   Flow For ODE, 6
   Hamiltonian System, 23
   Making Hamiltonian, 25
Axiom
   Governing Electrodynamics, 62
   Governing Mechanics, 16
   of Choice, 112
   of Determinism, 5

Backwards Integration, 219
Baker-Campbell-Hausdorff, 164, 234
Banach, 104
   Fixed Point Theorem, 99, 115
   Space, 104, 115

Barber's Rule, 212
Barber's rule, 198
Basis, 93, 96
BCH Formula, *see*
    Baker-Campbell-Hausdorff 164, 234
Beam
    Definition, 1
    Ellipse, 262
    Envelope, 264
Beating, 261
Beta
    (Twiss Parameter), 257
    Function, 264
Big Bang, 2
Biot-Savart, 139
Blessing of Beam Physics, 124
BMT Equation, 190
Bound, 112, 117
Boundary Value Problem, 76
    Dirichlet and Neumann, 78
    and Surface Charges, 130
Browne–Buechner Spectrograph, 198, 212
    Aberrations, 214
    Reconstructive Correction, 223

Camera
    Lenses, 2
Campbell-Baker-Hausdorff, *see*
    Baker-Campbell-Hausdorff 164, 234
Canonical
    Conjugate of Time, 25
    Momenta, 33
    Positions, 22
    System, 21
    Transformation
        Definition, 35
        Elementary, 35
        Exchanging p and q, 36
        in Normal Form, 273
        to Normal Form, 270
    Variables, 22
Capacitance Matrix, 130
Cauchy
    Completeness, 104, 114, 115
        of Norm on DA, 104
    Point Formula, 117
    Sequence, 104, 115

CBH Formula, *see*
    Baker-Campbell-Hausdorff 164, *see*
    Campbell-Baker-Hausdorff 234
Celestial Mechanics, 1
Change
    of Independent Variable, 23, 189
    of Variables in DA, 146
Charge Density, 63
Chart
    Manifolds, 9
Chromaticity, 268
    Linear Treatment, 259
Class of Equivalence, 70
Classical Mechanics
    Determinism, 4
    Hamiltonian, 32
    Hamiltonian for EM Fields, 33
    Lagrangian, 14, 16
    Lagrangian for EM Fields, 17
Closed Orbit, 271
    Nonlinear Off-Energy, 266
    Parameter Dependence, 266
Completeness (Cauchy), 114
Complex Coordinates
    for Normal Forms, 272
    for Rotational Symmetry, 151
Complex Numbers, 86
Composition, 100, 146
    in DA, 101
    of Spin Maps, 195
Computation
    Composition, 101
    Elementary Funtions in DA, 103
    Inverse Maps, 102
    Inverses in DA, 100
    ODE Solutions with DA, 108, 109
    PDE Solutions with DA, 111
    Roots, 100
Computational Differentiation, 84, 85
Concatenation, 146
Condition of Symplecticity, 37, 159
Conductivity, 64
Conjugate
    for Time, 23
    Momenta, 22
    Position, 22
Conservation
    of Energy and Symplecticity, 292

INDEX 311

of Hamiltonian, 23
of Poisson Bracket, 42
of Symplecticity, 292
of Volume, 41
Symplecticity and Energy, 293
Constitutive Relations, 63
　Free Space, 75
　Homogeneity and Isotropicity, 63
Constraints, 8
Continuity Equation, 64
Continuous Rotational Symmetry, 150
Contraction
　Factor, 115
　of DA Operators, 98
Convergence, 105, 115
　in DA, 104
　of Derivatives, 106
　of Map Taylor Expansion, 167
　of Power Series in DA, 105
　of Propagator Series in DA, 110
　Radius of, 105, 116
　Strong, 116
　Weak, 116
Coordinate Transformation
　for Hamiltonians, 8
　for Lagrangians, 8
Coordinates
　Particle Optical, 168
cos, 103
Coulomb Gauge, 74
Coulomb's Law, 63, 71
Courant Snyder Invariant, 261
Courant Snyder Theory, 250
Cremona Map, 298
Current Density, 63
Curvilinear Coordinates, 168, 169
　Generalized Momenta for, 179
　Hamiltonian expressed in, 182, 185
　Lagrangian in, 175
Cylindrical Coordinates, 120

Damping, 277
Decapole, 122
Depth of DA vector, 96
Derivation, 83, 85, 88, 91, 92, 113
Derivative
　as Differential Quotient, 116
　Calculus of, 117

Computation by DA, 88
Convergence Theorem, 106
DA Computation Example, 89
of Power Series with DA, 117
Determinant
　and Existence of Generator, 47
　of Reconstruction Map, 220
　of Symplectic Matrices, 40
Determinism, 4
Dielectric Constant, 63, 64
Diffeomorphism, 9
Difference Quotient, 116
Differentiability, 91, 116
Differentiable, 117
Differential, 84, 87
　Part, 90
Differential Algebra, 83, 85, 88, 92, 110, 113
Differential Algebraic Equation, 9
Differential Basis, 93
Differential Equation, 3
Differential Part, 87
Differential Quotient, 116
Differentiation, 83, 116
　of Measured Data, 127
　of Product, 116
　of Sum, 116
Dimension, 92, 94
　Reduction of, 8
Dipole, 122
Directional Derivative, see Vector Field 4
Dirichlet Boundary Condition, 78
Discrete Rotational Symmetry, 150
Dispersion Matching, 221
Displacement, Electric, 63
Double Midplane Symmetry, 150
Dreibein, 168, 169
Dual Numbers, 86
Dynamical Coordinate, 23

Edge Angle, 209
Electric Conductivity, 64
Electric Displacement, 63
Electric Field, 20, 63
Electric Flux Density, 63
Electric Permittivity, 64
Electric Polarization, 63
Electromagnetic Field, 19, 20

312 INDEX

Hamiltonian for Motion In, 33
Motion in, 17
Electrostatic Elemements
  Multipole Decomposition, 130
Elementary Canonical Transformation, 35
Elementary Function, 116
  in DA, 102
Elementary Functions
  Computation in DA, 103
Ellipse
  Invariant, 261
  of Beam, 262
Embedding of Reals, 86
Emittance, 264
Energy Acceptance, 259
Energy Conservation
  Impossibility under Symplecticity, 293
  Tracking, 292
Energy Loss Mode, 221
Energy Mass Spectrometer, 211
Enge Function, 225
Envelope of Beam, 264
Equation of Motion
  Hamiltonian, 22
Equations of Motion
  for Spin, 190
  Lagrangian, 7
  Non-Deterministic, 4
Equilibrium Polarization, 289
Equivalence
  Class, 70, 83, 91
  Relation, 70, 91
Eta Function, 259
Euler's Equations, 12
Existence and Uniqueness
  of Scalar Potential, 65
  of Solutions for ODEs, 3
    Violation, 4
  of Vector Potential, 68
exp, 103
Extended Phase Space, 25
Extension of Reals, 86

Factorization
  in Integrable Maps, 298
  of Map in Flows, 159
  of Symplectic Map, 162
  Superconvergent, of Map, 162
Factorizations
  in Kicks, 298
Faraday's Law, 63
Fermat's Principle, 195
Field, 111, 113
  As Real Vector Space, 94
  Electric, 63
  Measurement, 127
  Order, 87
Fifth Order Achromat, 245
Fixed Point, 168
  Linear Off-Energy, 259
  Nonlinear Off-Energy, 266
  of Antiderivation Operator, 98
  Parameter Dependence, 266, 271
  Problem, 115
  Similarity to Invariant Polarization Axis, 289
  Theorem, 113
Fixed Point Theorem, 115
  by Banach, 99
  for Operators on DA, 98
  Use for Inversion, 102
Floating Point Numbers, 82
Flow, 6
Flow Factorization, 159
  of Symplectic Maps, 162
  Superconvergent, 162
Flux Density, 63
Formula Manipulation, 84
Forward Cell, 229
Fourier Transform Ion Cyclotron Resonance Spectrometer, 211
Free Space Maxwell's Equations, 64
Fringe Field, 124, 204
Function, 102
  Delta, 115
  Elementary in DA, 102
  Intrinsic, 116, 117
  n Times Differentiable, 91
  On Levi-Civita Field, 115
  Rational, 115
  Root, 115
  Spaces, 81
Functional Relationships in DA, 103

Gamma
  (Twiss Parameter), 257
Gauge Transformation, 72
  as Equivalence Relation, 70
  Coulomb, 74
  Lorentz, 73
  of Potentials, 70
Gauss Theorem, 77
Gaussian
  Image Charge, 132
  Units, 62
  Wavelet, 127
Generalize Momenta
  in Cuvilinear Coordinates, 179
Generalized Coordinates, 9
Generating Function, 58
  Avoiding Divergence, 296
  Computation in DA, 165
  Representation of Map, 164
  Superposition of, 297
  Time Dependence, 59
  Use for Symplectic Tracking, 293
Generator of Algebra, 93
Geometric Coordinates, 206
Geometric Interpretation
  of Hamiltonian Dynamics, 25
Glass Optics, 195
Gram-Schmidt Method, 296, 299
Green's First Identity, 76
Green's Function, 130
Group
  Canonical Transformations, 35
  Symplectic Diffeomorphisms, 39
  Symplectic Matrices, 39

Hamilton's Equations, 42
Hamilton's Principle, 15
Hamilton-Jacobi Equation, 61
Hamiltonian, 22, 33
  Change of Variables, 31
  Equipotential Lines, 26
  Existence, 27
  in Curvilinear Coordinates, 182, 185
  Independent Variable Change, 23
  Making Autonomous, 25
  Uniqueness of, 26
Hamiltonian Flow
  Geometric Interpretation, 25

Symplecticity, 53, 192
Hamiltonian Function, *see* Hamiltonian 22
Hamiltonian System, 21
Hexapole, *see* Sextupole 122
High Energy Physics, 2

Ideal, 97
Idempotency
  of Legendre Transformation, 28
Image, 265
Image Charge Method, 129
  in Magnetic Case, 132
Imaginary, 115
Implicit Function Theorem, 101
Incoherent Spot Size, 221
Independent Variable
  Change of
    in Hamiltonian System, 23
Index of Refraction, 195
Induction, 108
  Magnetic, 63
Infinite Dimensional, 117
Infinitely
  Large, 114, 115
  Small, 87, 95, 115
Infinitely Small, 91, 116
Infinitesimal, 87, 94, 95, 114
  Nilpotency, 97
Instanteneous Coulomb Potential, 74
Integrability, 282
Integrable Symplectic Maps, 298
Integration, 83
Interaction Forces, 19
Intermediate Value Theorem, 117
Interpolation by Wavelets, 127
Interval, 82
Intrinsic Function
  in DA, 102
Intrinsic Functions, 116
  Computation in DA, 103
Invariant, 282
  as Variable, 9
  Ellipse, 261
  from Normal Form Theory, 270
  Polarization Axis, 289
  Subspace for Spin, 194
Inverse, 87, 113, 115

Inversion, 101, 147
Ion Trap Mass Spectrometer, 211
Isotope Production, 2

Jacobi Identity, 44
Jacobian
  of Transfer Map, 146

Kick Approximation, 201
Kick Factorization, 298

Lagrange's Equations, 14, 19, 20
  and Variational Principle, 12
Lagrangian, 7, 14, 19, 20
  Coordinate Transformation, 8
  Depending on Momentum, 31
  for Lightrays, 195
  in Curvilinear Coordinates, 175
Laplace Equation, 76, 119
Laplacian
  in Curvilinear Coordinates with
    Midplane Symmetry, 125
  in Cylindrical Coordinates, 120
Leading Coefficient, 94
Leading Term, 94
Left-Finite, 112, 113, 115
Legendre Transformation, 27, 180
  Idempotency, 28
  of Hamiltonian, 30
  of Lagrangian, 28
Lens, 195
Levi-Civita, 111
Lexicographic Order, 95, 112
Lie Algebra, 44
Lie Derivative, *see* Vector Field 4
Lie Factorization, 159, 229
  Computation with DA, 163
Lindeloff-Picard Theorem, 5
Linear Stability, 251
Liouville's Theorem, 41, 54, 251
log, 103
Logic
  Formal, 112
Lorentz Force, 17–20
Lorentz Gauge, 73

Magnet Field Errors, 266
Magnet Misalignments, 266

Magnetic Induction, 63
Magnetic Moment Density, 63
Magnetic Monopoles, 63
Magnetic Permeability, 63, 64
Magnetic Polarization, 63
Magnetization, 63
Manifold, 9
Map, 81
  Analyticity, 167
  Change of Variables, 146
  Factorization in Flows, 159
  Factorization in Integrable Symplectic
    Maps, 298
  One-Turn, 249
  Parameter Dependence, 266
  Poincare, 249
  Representation by Generating
    Function, 164
  Superconvergent Factorization, 162
  Symplectic Factorization, 162
  Transfer, 6
Mass Spectrometer, 211
Matching, 261
Matrix Elements
  Symplectic Relations, 155
Maxwell's Equations, 62, 119
  in Free Space, 64
Measured Fields, 127
  Application for S800, 225
  Midplane Data, 128
  Multipole Data, 127
  Threedimensional Data, 132
Microscope, 211
Midplane
  Field Expansion, 126
  Measurement, 128
  Symmetry, 148, 238
  and Symplecticity, 156
  Double, 150
Mirror Symmetry, 147
Misalignments, 266
Mismatching, 261
MKSA Units, 62
Momentum Acceptance, 217, 259
Momentum Bite, *see* Momentum
  Acceptance 217
Momentum Spectrometer, 211
  Browne-Buechner, 212

Q Value, 214
  Resolution, 212
Momentum-Position Interchange, 36
Monopoles, Magnetic, 63
Monotonicity
  of Indpendent Variable, 23
Multiplication, 95, 113
  Computational Cost, 96
  Scalar, 86, 91, 96
  Vector, 86, 91, 96
Multipole
  Direct Use of Data, 127
  in Cartesian Coordinates, 122
Multipole Order, 122
Multipole Strength, 122

N Bar, 289
N Zero, 290
Neumann Boundary Condition, 78
Newton's Equations, 16
Nilpotent, 97
Non Standard Analysis, 111, 117
Non-Archimedean, 114
Non-Autonomous
  ODE
    Expressed as Autonomous ODE, 3
Non-Autonomous Hamiltonian
  Making Autonomous, 25
Nonlinear Dynamics, 1
Nonrelativistic Motion, 14, 19
  in Electromagnetic Field, 17
  of Interacting Particles, 14, 32
Norm on DA, 97
  Cauchy Completeness, 104
Normal Form, 250
  Invariant, 282
Nuclear Physics, 2
Nuclear Spectrometers, 217
Nucleosynthesis, 2
Nuisance of Beam Physics, 123
Numbers, 82

Observable, 4, 42
Octupole, 122
ODE, *see* Ordinary Differential Equation 3
Off-Energy Particles
  Linear Treatment, 258

Ohm's Law, 64
One Turn Map, 249
Optics, 1
  Glass, 195
Order, 87, 94, 114
  Lexicographic, 95, 112
  of Multipoles, 122
  Total, 114
Ordianary Differential Equation
  Solution with DA Method, 108
Ordinary Differential Equation, 3
Origin Preservation, 168
Orthogonality
  of Antisymmetric Flow, 192
Orthogonalization of Spin Matrix, 299

Pairs of Eigenvalues, 251
Parameter Dependence
  of Fixed Point, 271
  of Map, 266
Parametrization, 23
Partial Differential Equation, 76
  Hamilton-Jacobi, 61
  Solution with DA Method, 111
Particle Optical Coordinates, 168
Particle Physics, 2
PDE
  Solution with DA Methods, 111
Permeability, 63, 64
Permittivity, 63, 64
Phase Space, 22, 26
  Other Independent Variable, 23
Phase, Multipole, 122
Picard-Lindeloff Theorem, 5
Plane, Expansion out of, 126
Poincare, 249
  Map, 168, 249
  Section, 7
Poisson Bracket, 40
  and Antisymmetric Product, 42
  Conservation, 42
  Elementary, 43, 57
  Waiting to Happen, 162
Poisson equation, 76
Polarization, 63
  Axis of Invariance, 289
Polynomial, 115
Position-Momentum Interchange, 36

316 INDEX

Positive, 114
Potential, 14, 19, 65
   as Equivalence Class, 70
   Scalar, 18, 20, 65
   Scalar and Vector, 71
   Scalar for Field, 119
   Vector, 20, 65
   Velocity-independent, 14, 20
Power Series, 116
   Convergence in DA, 105
Pre-Factorization
   for Generating Function, 296
Preservation, *see* Conservation 292
Principle of Least Action, 15
Propagator, 6, 110, 273
   for Spin Motion, 193

Q Value, 214
Quadruple of Eigenvalues, 251
Quadrupole, 122
   Fringe Field, 124
   Mass Spectrometer, 211
Quaternion, 94
Quotient Rule
   DA derivation of, 99

Rational Numbers, 112
Real Numbers, 104
   Embedding, 113
Real Part, 87, 90
Reconstructive Correction, 217
   Browne-Buechner, 223
   S800 Spectrograp, 225
Reconstructive correction
   Examples, 223
Reduction of Dimensionality, 8
Reference Curve, 168, 169
Refraction, 195
Relative Coordinates, 168
Relativistic Motion, 19, 20, 33
   in Electromagnetic Field; Hamiltonian, 33
Repetitive Systems, 249
Resolution, 212
   Linear, 212
   Nonlinear, 214
Resolving Power, 212
Resonance, 250, 274, 280

   and Symplectification, 295
   Denominator, 284
   for Spin Motion, 292
   Strength, 280
Resonances, 270
Reversed Cell, 229, 235
Reversed-Switched Cell, 229
Ring, 86
Rolle's Theorem, 117
Root, 87, 115
Root Rule
   DA Derivation of, 100
Rotation, 275
Rotational Symmetry, 150
   of Fields, 120
   Preserved by Fields, 125
Routhian, 30

S800
   Reconstructive Correction, 225
Scalar Multiplication, 86
Scalar Potential, 65
   Existence and Uniqueness, 65
   for Electrodynamics, 71
Scalar Product
   Antisymmetric, 39
Scaling
   Symplectic, 205
Schwartz Inequality
   of DA Norm, 98
Second Order Achromat, 244
Sector Field Mass Spectrometer, 211
Separable PDE, 62
Sequence of Derivatives
   Convergence, 106
Sextupole, 122
Shooting Method, 219
SI Units, 62
sin, 103
Single Exponent Lie Factorization, 163
Smoothing, 128, 129, 134, 136
Snell's Law, 195
Space-Time Interchange, 23
Spaces of Functions, 81
Spectrograph, *see* Spectrometer 211
Spectrometer, 211
   Mass, 211

INDEX 317

Momentum, 211
Resolution, 212
Spectrometers
    for Nuclear Physics, 217
    reconstructive correction, 217
Spectroscopy, 2
Spin
    Equations of Motion, 190
    Tracking, 299
Spin Amplitude Tune Shift, 292
Spin Invariance Axis, 289
Spin Resonance, 292
Spin Tune, 290
Square Root, 103
Stability of Motion, 249
Standard Form, 233
State Vector, 1
Stokes Theorem, 68
Strength
    Multipole, 122
Sub-Standard Form, 233
Superconvergent Factorization, 162
Support, 112, 115, 117
Switched Cell, 229, 235
Switched-Reversed Cell, 229, 236
Symmetry
    for Aberration Correction, 228
    Midplane, 148
    of Fields, 120
    Rotational, 150
    Symplectic, 155
Symplectic
    Extension, 296
    Matrix, 39, 40
    Tracking, 292
    Transformation
        and Poisson Bracket, 42
Symplectic Condition, 37
Symplectic Scaling, *see* SYSCA 205
Symplecticity, 229, 234, 272, 292
    and Achromaticity, 158
    and Matrix Elements, 155
    and Trajectory Reconstruction, 220
    Impossibility of Preserving Energy, 293
    in Flow Factorization, 162
    in Reconstruction, 220
    of Hamiltonian Flow, 53
    of Normal Form, 273
    Preservation under Scaling, 208
Symplectification
    via Generating Function, 165
SYSCA, 205
Systeme International d'Unites, 62

Taylor Map, 81
Taylor's Theorem, 117
Theorem
    Fixed Point, 115
    Intermediate Value, 117
    Liouville, 54
    Mean Value, 107
    Rolle's, 117
    Taylor's, 117
Third Order Achromat, 240, 243
Thomas Equation, 190
Time
    Canonical Conjugate of, 25
    Interchange with Space, 23
Time Dependent
    Canonical Transformations, 59
    Generating Function, 59
    ODE
        Made Time-Independent, 3
Time Independent
    Maxwell's Equations, 64
    ODE, Flow For, 6
Time-of-Flight Mass Spectrometer, 211
Time-Of-Flight Spectrometer, 241
TOFI Spectrometer, 230
Total Order, 87, 95
Tower of Ideals, 97
TPSA, 84
Tracking
    of Spin Motion, 299
    Pre-Factorization, 296
    via Integrable Maps, 298
    via Superimposed Local Generators, 297
Trajectory Reconstruction, 218
    Browne-Buechner, 223
    in Energy Loss Mode, 221
    S800, 225
Transfer Map, 1, 6, 145, 167
Transfer Matrix
    Multiplication, 145

TRANSPORT Coordinates, 206
Triangle Inequality, 97
Tune, 258, 275
   and Poincare Section, 261
   of Spin Motion, 290
Tune Shift
   and Symplectification, 295
   Depending on Amplitude, 274
   for Spin Motion, 292
   Nonlinear Motion, 270
   Nonlinear Treatment, 268
Twiss Parameters, 257
Two-Dimensional Algebras
   Classification, 86

Union, 113
Uniqueness
   of Fields, *see* 7
   of Hamiltonian, 26
   of Lagrangian, 7

Valuation, 97
Variable Change, *see* Transformation 31

Variational Principle, 12
Vector
   Multiplication, 86
Vector Field, 4, 109
   of Hamiltonian System, 42
Vector Potential, 65
   Existence and Uniqueness, 68
   for Electrodynamics, 71
Vector Space, 86, 94
   Finite Dimensional, 94
   Infinite Dimensional, 113
   of Functions, 84
Volume Conservation
   by Symplectic Maps, 41
von Neumann Condition, 78

Waist, 265
Wavelet, 127
Weak Convergence, 116

Zermelo, 94

ISBN 0-12-014750-5